FINITE-DIMENSIONAL LINEAR ANALYSIS:
A SYSTEMATIC PRESENTATION IN PROBLEM FORM

FINITE-DIMENSIONAL LINEAR ANALYSIS:
A SYSTEMATIC PRESENTATION IN PROBLEM FORM

I. M. Glazman

Ju. I. Ljubič

Translated and edited by

G. P. Barker

G. Kuerti

The MIT Press

Cambridge, Massachusetts, and London, England

Library of Congress catalog card number: 74-1603.

ISBN 0 262 07058 8

Printed in the United States of America

PREFACE

The composer Ludwig Spohr, who lived at the turn of the 18th century, is little known among concert-goers of today. It is true that Spohr's violin concertos can still be heard within the walls of our conservatories, but his symphonies and operas are hardly ever performed in public. Of course, his name as one of the founders of the German romantic opera, who also was one of the most prominent musical performers of his day, has not vanished from the handbooks of musical history, and there is yet another rather unessential circumstance which perpetuates his name: it is his incidental invention of the conductor's baton, which has remained unchanged since. Spohr waved the baton for the first time to conduct an orchestra, when he was court conductor in Kassel.

Unfortunately, in this manuscript, which is outsize anyway, we must not stop for a detailed analysis of Spohr's work, nor for many other questions that have little relation to our subject. Nevertheless, we have to mention one other work of Spohr's. It is a textbook for the student of music, called a "Violinschule" -- a *school of violin-playing*. His was the first such work, which aimed at the same time at the development of the pupil's technical ability and at his artistic education. After Spohr's example, similar schools written by other performers of the 19th and 20th centuries followed.

Such a school usually consists of several parts, each of which is devoted to a well-defined aspect of musical preparation; all these parts together form the basis for the performing activity of the future musician. The material of the school consists of comparatively short musical "texts"; they have the character of exercises, etudes, and pieces of varying musical form. All of them are simply called *numbers* and are arranged in such order that within the bounds of one thematic cycle each of them creates the basis for the performance of subsequent numbers; many of them would appear difficult by themselves if taken out of context.

The musical material of the school is accompanied by brief explanatory remarks, containing instructions or comments on the meaning of the single numbers. One should notice here that, with few exceptions, the numbers are not to be taken as original creations of the musician who wrote the *school*. Fragments from the works of others presented in their original or in changed form are the musical material. Often it is so well known that it would be difficult to attribute it to a particular composer. But some numbers do carry the name of the composer in spite of the evident essential simplification of the original version, in which polyphonic orchestration has been transformed into a single melodic line. The preservation of the original theme of a composer is considered reason enough for assigning his name to the altered version.

On the whole, the artistic-technical range of such schools and of related collections of etudes can be very broad; it may include practice numbers on the open strings, but also virtuoso pieces with flautato double-stops.§

Our book presents itself as an attempt to construct a course analogous in character and purpose to those schools of violin playing. We would have called it a School of Functional Analysis if we had sufficient reason to believe that our attempt has been successful. In fact, we can only claim somewhat less today: our attempt has finally come to an end.

There are 2405 problems in this book. Thematically they encompass linear algebra and a course of functional analysis "transcribed" for finite-dimensional space. As regards the degree of difficulty, we can say that the infimum mentioned before (playing on open strings) has actually been attained (see, e.g., problem I,(1)), but it is not possible to assert the same about the supremum.

In partitioning the material into individual problems a certain unevenness was intentionally admitted. Material that contains fundamental concepts, ideas, and theories is fractured into smaller pieces than material of special and more technical character.

Instructions which may make the use of the book easier can be collected in the following points:

1. No preliminary knowledge of linear algebra or functional analysis is required of the reader. The necessary algebraic and topological concepts that are a part of the general mathematical education are explained in a special section at the end of the book (Dictionary of General Terms),

---

§See "Encyclopedic Musical Dictionary", G.V. Keldysh, ed., Moscow, 1959. About the musical scale cf. G.E. Shilov, "The Simple Gamut", Fizmatgiz, 1963. One can show that the difficulty in the production of harmonics on string instruments arises from their instability in the sense of problem (293), Ch. I.

where only the language of elementary set theory is assumed known.

2. The book consists of problems and inserted accompanying text. The accompanying text contains all the necessary definitions, hints for some problems, and other comment. Because all problems require answers in form of proofs, answers have not been included.

3. (This paragraph concerns the printer's arrangement of the problems and accompanying text. Please, look here at the Foreword to the English edition.)

4. Sometimes auxiliary lemmas and directions for the solution of a given problem are found not before but after it. Thus, in reading the formulation of some single problem in a sequence of problems, it might be a good idea to look also at the beginning of the accompanying text that follows, and it is in general useful to glance through the whole context in which a problem is presented before attempting the proof.

5. The problems are followed by a bibliography. Its first part contains a list of general textbooks in algebra analysis, topology, linear algebra, and functional analysis. In the second part, the literature we have used and recommend is listed by chapters. To solve the problems it should not be necessary to consult the literature in the bibliography or in other sources. The main purpose of the bibliography is to familiarize the reader with the original literature that contains the development of the theory broached in the book.

6. At the end of the book you will find the Bibliography and the Dictionary of General Terms already mentioned. Then follows a List of Notations and a List of Theorems quoted by names. (The latter has been omitted in the English edition. About this and the Index, see the Foreword to the English edition.)

7. We have not formulated any unsolved problems, although at some places of the book such problems arise naturally in direct connection with those listed by us. However, search for and answers to such questions belong to the creative work of the reader, for which we wish him much success.

We wish to express our deep appreciation for our teachers, N. I. Akhiezer and M. G. Kreĭn, and also for A. Ja. Povzner; we once passed through *their* School of Functional Analysis.

Some of the problems and themes we owe to the suggestions of the following colleagues: G. R. Belickiĭ, V. I. Gurariĭ, M. I. Kadec, A. S. Markus, V. I. Macaev, Ju. L. Shmuljan. Various sections of the book were discussed with È. J. Žmud, I. S. Jokhvidov, A. V. Pogorelov, V. P. Potapov, V. A. Tkačenko, and B. I. Judovič. G. E. Shilov looked through the manuscript; his remarks helped us to improve the overall level of presentation. In the preparation of the manuscript we received great support from V. I. Kogan, V. I. Mitin, A. M. Rybalko, and Ju. F. Senčuk.

To all these persons we are sincerely grateful.

The authors

On May 30, 1968, when this book was already in production, Israil' Markovič Glazman died. The idea of this book, a finite dimensional "modelling" of functional analysis, belongs entirely to him. When somebody talked to him about complicated infinite-dimensional constructions, he usually asked: "And how does that look in the two-dimensional case?" Often enough that shocking question helped to a better understanding of the mathematical situation. All the mathematical activity of this unforgettable man of exceptional talent was directed toward recognizing the simple elements in complicated matters.

July 17, 1968 Ju. Ljubič

FOREWORD TO THE ENGLISH EDITION

The preparation of this edition has been aided by several communications from Professor Ljubič. The corrections made on the basis of his remarks and our own observations have resulted in numerous reformulations and also in the deletion of some problems. However, the original problem numbers were retained in order to avoid possible errors in the cross-references, whose number was greatly increased. This explains occasional gaps in the sequence of problem numbers.

The problems are numbered chapterwise. The accompanying text is distinguished from them by being shifted to the right everywhere with no exceptions by three "spaces". For the sake of clarity, since we avoided all indentations, paragraphs within a piece of text are only indicated by the "short end" of the preceding line. Where this could not be done, two dashes were inserted at the end of the line to show the end of the paragraph.

Originally all definitions were contained in the running text. We have extracted them. They are easily recognized, are numbered through each chapter, and their numbers appear also in the running heads of the pages.

Since the user of such a book will, as a rule, concentrate on certain chapters, we have inserted glossaries at the end of each of them, containing any new definition and new term in the same form in which it appears in the text.

We have adopted "iff" for "if and only if" almost everywhere. Nevertheless, all formal definitions have been stated in the customary conditional form, even though the biconditional form is often preferred today.

The bibliography, to which we added a few items, appears in its original sequence: books and papers in the cyrillic alphabet precede those in the Roman alphabet. References which are obtainable in English, French, or German translation have received that citation rather than the Russian one. We can only hope that we have not overlooked available translations. For the transliteration of Russian names we have used the code of "Mathematical Reviews" (disregarding a very few exceptions).

The index was compiled and re-alphabetized in the usual form from the glossaries. Since it contains all "named theorems" we felt we could omit a separate list of them.

G. P. Barker, Univ. of Missouri-Kansas City

G. Kuerti, Case Western Reserve Univ., Cleveland

CONTENTS

PUBLISHER'S NOTE

This format is intended to reduce the cost of publishing certain works in book form and to shorten the gap between editorial preparation and final publication. The time and expense of detailed editing and composition in print have been avoided by photographing the text of this book directly from the author's typescript.

ADDENDA

P. 249
Insert in Definition 4 after $E_1 \otimes \ldots \otimes E_p$: , but not all of its elements can be so obtained.

P. 271
Insert before problem 108: Applying the norm defined in D27 to $x_1 \otimes \ldots \otimes x_p$, we find:

P. 485
Add to General Literature in Linear Algebra: 22a. Markus, M. Finite Dimensional Multilinear Algebra. Part 1. Marcel Dekker, Inc., New York, 1973.

FINITE-DIMENSIONAL LINEAR ANALYSIS:

A SYSTEMATIC PRESENTATION IN PROBLEM FORM

CHAPTER I

THE COMPLEX LINEAR SPACE

Definition 1. *A complex linear space* is a set E, on which the operations of addition of elements and multiplication of elements by complex numbers are defined and the following axioms hold:

(1) E is closed under both operations;

(2) E is an *abelian group* relative to addition;

(3) multiplication by numbers (scalars) is associative, commutative, and connected with addition by distributive laws relative to numbers and elements;

(4) the product of the scalar unity with an arbitrary element equals that element.

The elements of the linear space[§] are called *vectors* (sometimes *points*) and are denoted by small Latin letters as distinguished from (complex) *numbers* or *scalars*, usually denoted by small Greek letters. The symbol 0 denotes either the null vector (the identity element of the group) or the number zero, depending on the context.

By virtue of axioms (1)-(4) the operations on vectors in a linear space satisfy the same rules as the corresponding

---

[§]Here and in the following, if no statement to the contrary is made, the term "linear space" means *complex* linear space, implying that the admissible scalars belong to the *field* of complex numbers. Sometimes we briefly say "space."

vector operations in the usual geometrical sense.

A trivial example of a linear space is the set consisting of the single element 0; the operations are then simply

$$0+0 = 0, \qquad \alpha\cdot 0 = 0\cdot\alpha = 0.$$

This space is called the zero-space; it is also denoted by the symbol 0.

Definition 2. Another, less trivial, example of a linear space is the set Φ_M of all scalar-valued functions on some abstract set M. Such functions are called *functionals*. For functionals addition and multiplication by scalars are defined naturally, i.e. in the same way as in the case of numerical functions of numerical arguments.

We obtain a special case by regarding M as the set of all natural numbers 1,2,... or some initial segment 1,2,...,n of it. Φ_M becomes now the set C^ω of all sequences of scalars $(\alpha_1,\alpha_2,\ldots)$ or, for fixed n, the set C^n of all possible systems of n scalars $(\alpha_1,\alpha_2,\ldots,\alpha_n)$ under termwise addition and scalar multiplication. The spaces C^k $(k=1,2,\ldots,n,\ldots,\omega)$ are called *arithmetic spaces*. As a special case we mention R^n, the *real* n-tuple space, where the scalars belong to the field of real numbers.

An important example of a linear space is the set Π of all polynomials $P(t)$ of one variable t or the set $\Pi^n \subset \Pi$ of polynomials of degree less than some natural number n with the usual definition of addition and scalar multiplication.

1. Linear Dependence and Independence. Rank of a System of Vectors.

Definition 3. A linear combination of vectors $x_1,\ldots,x_n$ (not necessarily distinct) with coefficients $\alpha_1,\ldots,\alpha_n$ is a vector of the form

$$u = \Sigma_{k=1}^{n} \alpha_k x_k .$$

A linear combination is called *trivial* if all its coefficients α_k are zero, and *nontrivial* otherwise. The trivial linear combination of any vectors is always zero.

Definition 4. The vectors $x_1, x_2, \ldots, x_m$ are called *linearly independent* if all their nontrivial linear combinations are different from zero; they are *linearly dependent*, if some nontrivial linear combinations vanish. We shall also speak of "a linearly independent (dependent) system of vectors $\{x_k\}_1^m$".

Definition 5. A system Γ *of vectors* is a nonempty *finite ordered* set of vectors. Γ_1 is a *subsystem* of system Γ, if set Γ_1 is a subset of set Γ.

We start with the elementary theorems on linear dependence and independence.

1. A system consisting of one vector x is linearly independent iff[§] $x \neq 0$.

2. A system $\{x_k\}_1^m$ $(m>1)$ is linearly independent iff none of the vectors of the system is a linear combination of the remaining ones.

3. Each subsystem Γ_1 of a linearly independent system of vectors Γ is linearly independent.

Speaking metaphorically, we may say that linear independence is a "hereditary" property.

Definition 6. Let Γ be some vector system. The *linear hull* of Γ is the set of all linear combinations of vectors of the system Γ; it is denoted by sp Γ (read "span").

Definition 7. A subsystem Γ_o of a system Γ is said to be a *complete subsystem of* Γ, if $\Gamma \subset \text{sp}\, \Gamma_o$, i.e. if each vector of the system Γ is a linear combination of the vectors of Γ_o.

4. Let Γ_1 be any subsystem of a system Γ. If Γ_1 contains a complete subsystem Γ_2 of Γ, then Γ_1 is also a complete subsystem of Γ.

[§]iff = if and only if.

Completeness is also a hereditary property (but relative to expansion rather than contraction of the system).

Let us now consider theorems 3 and 4 together. For this purpose we reformulate (3) in somewhat different terms: "Let Γ_1 be any subsystem of a system Γ. If Γ_1 is contained in a linearly independent subsystem Γ_2 of Γ, then Γ_1 is also a linearly independent subsystem of Γ."

Definition 8. If we replace in this formulation of theorem 3 "linearly independent" by "complete," and "is contained in" by "contains," then we evidently obtain theorem 4. Between theorems 3 and 4 there is a logical symmetry or, more precisely, a *duality*. It is based on a correspondence (or symmetry) of concepts and of relations. The concept of linear independence is dual to the concept of completeness. The relation between sets M and L, which is expressed by "M contains L" ($M \supset L$), is dual to "M is contained in L" ($M \subset L$).

Duality occurs frequently in a variety of mathematical theories. For example, in the general theory of sets duality is based on the correspondence between "to contain" and "to be contained in." Duality also holds between *intersection* and *union*, and this shows itself everywhere in the algebra of sets. The distributive laws are a good example:

$$(M \cup L) \cap N = (M \cap N) \cup (L \cap N), \qquad (M \cap L) \cup N = (M \cup N) \cap (L \cup N).$$

Clearly, the duality between some notions (or relations) can be a consequence of the duality between some other notions (relations). For example, the intersection of sets can be defined as that set which is contained in each of the given sets and contains every set with this property. The union of sets can be defined "dually" as that set which contains each of the given sets and is contained in every set with this property. Thus the duality between "contains" and "is contained in" is on a deeper level than that between "intersection" and "union."

In constructing a theory, the elucidation of dualities as they occur is very essential, since it permits a "doubling" of the number of theorems and strengthens structure and completeness of the theory. We shall therefore look out for duality wherever possible.

Definition 9. A complete and linearly independent subsystem of a system Γ of vectors, not all = 0, is called a *basic subsystem.* This concept is *self-dual.*

5. A subsystem of a system Γ of vectors, not all = 0, is basic iff it is a maximal linearly independent subsystem; i.e. if it is contained in no other linearly independent subsystem of Γ.

The dual assertion is:

6. A subsystem of a system Γ of vectors, not all = 0, is basic iff it is a minimal complete subsystem; i.e. if it contains no other complete subsystem of Γ.

Let us now formulate an existence theorem for a basic subsystem. Note first that any system Γ of vectors, not all = 0, contains a linearly independent subsystem (e.g. the subsystem which consists of one vector $x \in \Gamma$, $x \neq 0$), and a complete subsystem (e.g. the subsystem coinciding with the whole system Γ).

7. Any system of vectors, not all = 0, contains a basic subsystem. Moreover, any linearly independent subsystem can be extended to a basic subsystem and, dually, from any complete subsystem a basic subsystem can be extracted.

The role of basic subsystems is clarified by the following assertion:

8. If $x_{r_1}, x_{r_2}, \ldots, x_{r_p}$ is a basic subsystem of the system of vectors $\{x_k\}_1^m$, then any vector x_j $(j = 1,2,\ldots,m)$ can be represented as a linear combination

$$x_j = \Sigma_{k=1}^{p} \alpha_{kj} x_{r_k}$$

of vectors of the basic subsystem; the coefficients α_{kj} in this representation are unique.

Conversely:

9. If any vector of a system is uniquely representable as a linear combination of vectors of some subsystem, then that subsystem is basic.

The following lemma is essential to the fundamental theorem 11:

10. Let $\Gamma = \{x_k\}_1^m$ $(m>1)$ be some system of vectors and $\Gamma_1 = \{x_k\}_1^{m-1}$. If vectors u and $v \in \text{sp}\,\Gamma$, but $v \notin \text{sp}\,\Gamma_1$, then there exists a scalar α such that $u-\alpha v \in \text{sp}\,\Gamma_1$.

11. Any $m+1$ vectors of sp $\{x_k\}_1^m$ are linearly dependent.

One of the important consequences of (11) is:

12. All basic subsystems of a fixed vector system Γ contain the same number of vectors.

Definition 10. The number described in (12) is called the *rank* of the system Γ and is denoted by rk Γ. If all vectors in Γ equal zero, then rk $\Gamma = 0$.

Obviously, the rank of a system Γ does not exceed its cardinality (here the number of vectors $\in \Gamma$).

The rank of a system of vectors is "the upper measure" of its linear independence.

13. The rank of a system of vectors, not all = 0, equals the maximum cardinality of its linearly independent subsystems.

It is thus necessary and sufficient that for the linear independence of a system $\Gamma = \{x_k\}_1^m$, rk $\Gamma = m$.

The dual theorem to (13) is:

14. The rank of a system of vectors, not all = 0, equals the minimum cardinality of its complete subsystems.

Thus the rank of a system is "the lower measure" of its completeness.

Theorems (13) and (14) characterize the rank of a system as the extreme carbinality of the one or the other class of subsystems. We shall show that basic subsystems may be characterized as the "points" at which an extremum is attained. That the basic subsystems of Γ attain the extremum of cardinality is clear. Conversely:

15. Any linearly independent subsystem of a system Γ which consists of rk Γ vectors is basic; dually, any complete subsystem of a system Γ which consists of rk Γ vectors is basic.

Definition 11. Let Γ_1 and Γ_2 be any two systems of vectors. We shall write $\Gamma_1 \leq \Gamma_2$ if $\Gamma_1 \subset \mathrm{sp}\,\Gamma_2$. $\Gamma_1 \subset \Gamma_2$ implies of course $\Gamma_1 \leq \Gamma_2$.

16. For $\Gamma_1 \leq \Gamma_2$ it is necessary and sufficient that $\mathrm{sp}\,\Gamma_1 \subset \mathrm{sp}\,\Gamma_2$.

The relation $\Gamma_1 \leq \Gamma_2$ is a *quasiorder*[§] on the set of vector systems.

Definition 12. We shall say that the systems Γ_1 and Γ_2 are *equivalent* and write $\Gamma_1 \sim \Gamma_2$ if simultaneously $\Gamma_1 \leq \Gamma_2$ and $\Gamma_2 \leq \Gamma_1$. This relation satisfies the well-known requirements for equivalence relations.[§] We may thus speak of classes of equivalent systems.

Clearly, $\Gamma_1 \sim \Gamma_2$, iff $\mathrm{sp}\,\Gamma_1 = \mathrm{sp}\,\Gamma_2$.

Let us now investigate the behavior of the rank as a functional on the set of all vector systems.

17. If $\Gamma_1 \leq \Gamma_2$, then $\mathrm{rk}\,\Gamma_1 \leqq \mathrm{rk}\,\Gamma_2$.

Thus rank is a monotone nondecreasing functional (*Steinitz' Theorem*), and "monotone" refers to the quasiorder introduced. It may be considered as a functional on the *set of equivalence classes of systems*, since equivalent systems have equal ranks by Steinitz' theorem. The converse of (17) is false, but

18. If $\Gamma_1 \leq \Gamma_2$ and $\mathrm{rk}\,\Gamma_1 = \mathrm{rk}\,\Gamma_2$, then $\Gamma_1 \sim \Gamma_2$.

Thus in the case $\Gamma_1 \leq \Gamma_2$ the equality sign in $\mathrm{rk}\,\Gamma_1 \leqq \mathrm{rk}\,\Gamma_2$ holds iff $\Gamma_1 \sim \Gamma_2$.

Let us now note certain inequalities which easily follow from the preceding properties of the rank.

19. If the intersection of the systems $\Gamma_1, \Gamma_2, \ldots, \Gamma_q$ is not empty (hence is a system), then

---

[§]See the Dictionary of General Terms at the end of the book.

$$\operatorname{rk} \bigcap_{k=1}^{q} \Gamma_k \leq \min_{1\leq k\leq q} \operatorname{rk} \Gamma_k .$$

Here equality holds iff system $\bigcap_{k=1}^{q} \Gamma_k$ is equivalent to at least one of the systems $\Gamma_1,\ldots,\Gamma_q$.

20. For arbitrary systems of vectors $\Gamma_1,\Gamma_2,\ldots,\Gamma_q$

$$\operatorname{rk} \bigcup_{k=1}^{q} \Gamma_k \geq \max_{1\leq k\leq q} \operatorname{rk} \Gamma_k .$$

Here equality holds iff system $\bigcup_{k=1}^{q} \Gamma_k$ is equivalent to at least one of the systems $\Gamma_1,\Gamma_2,\ldots,\Gamma_q$.

Theorems (19) and (20) are dual.

The following self-dual theorem establishes the relation between the ranks of the intersection and of the union of two systems.

21. If the intersection of two systems Γ_1,Γ_2 is not empty, then

$$\operatorname{rk}(\Gamma_1\cap\Gamma_2) + \operatorname{rk}(\Gamma_1\cup\Gamma_2) \leq \operatorname{rk} \Gamma_1 + \operatorname{rk} \Gamma_2.$$

This inequality remains in force even for $\Gamma_1 \cap \Gamma_2 = \emptyset$ if we put $\operatorname{rk} \emptyset = 0$. Therefore in general

22. $\operatorname{rk}(\Gamma_1\cup\Gamma_2) \leq \operatorname{rk} \Gamma_1 + \operatorname{rk} \Gamma_2$.

In addition:

23. If $\operatorname{sp} \Gamma_1 \cap \operatorname{sp} \Gamma_2 = \{0\}$, then $\operatorname{rk}(\Gamma_1\cup\Gamma_2) = \operatorname{rk} \Gamma_1 + \operatorname{rk} \Gamma_2$.

Thus rank is an additive functional ("additive" refers here to union formation of vector systems).

Theorem (23) has a converse:

24. If $\operatorname{rk}(\Gamma_1\cup\Gamma_2) = \operatorname{rk} \Gamma_1 + \operatorname{rk} \Gamma_2$, then $\operatorname{sp} \Gamma_1 \cap \operatorname{sp} \Gamma_2 = \{0\}$.

Definition 13. Two systems of vectors Γ_1,Γ_2 which satisfy the condition

$$\operatorname{sp} \Gamma_1 \cap \operatorname{sp} \Gamma_2 = \{0\}$$

are called *mutually independent*.

In closing this section we shall characterize the rank by

listing its functional properties already established.

25. Let the functional $\rho(\Gamma)$, defined on the set of all vector systems, satisfy the following relations:

(1) If $\Gamma_1 \sim \Gamma_2$, then $\rho(\Gamma_1) = \rho(\Gamma_2)$.

(2) If Γ_1 and Γ_2 are mutually independent, then

$$\rho(\Gamma_1 \cup \Gamma_2) = \rho(\Gamma_1) + \rho(\Gamma_2).$$

(3) If Γ consists of one vector $x \neq 0$, then $\rho(\Gamma) = 1$. These relations imply $\rho(\Gamma) = \text{rk}\ \Gamma$ for any Γ.

Incidentally, no two of these relations are sufficient to characterize the rank. This is easily proved by constructing counterexamples.

2. Bases and Dimension of a Space. Isomorphic Spaces.

Definition 14. A nonempty set F of vectors of a space E is *linearly minimal* if each system $\Gamma \subset F$ is linearly independent.

Definition 15. The set F is *generating* (with respect to E) if for each $x \in E$ there is a system $\Gamma \subset F$ such that $x \in \text{sp}\ \Gamma$.

Definition 16. A space E is *finite dimensional* if it satisfies one of the two mutually dual conditions:

(1) Every linearly minimal set in E is finite.

(2) E contains a finite generating set.

With the aid of (11) it is easy to establish that

26. Conditions (1) and (2) are equivalent.

If a space is not finite dimensional, then it is said to be infinite dimensional. We give some examples.

27. The space Φ_M of all functionals on some set M is finite dimensional if the set M is finite; otherwise it is infinite dimensional. In particular, the arithmetic space C^n is finite dimensional for any natural number n, but the space C^ω is infinite dimensional.

28. The space Π^n of polynomials of degree less than n is finite dimensional, but the space Π of all polynomials is infinite dimensional.

In this book we shall only consider finite-dimensional spaces from now on.

If the space E is finite dimensional, then there exists a system of vectors $\Gamma \subset E$ such that sp $\Gamma = E$. A system of vectors which satisfies this condition is also called *complete*. (Note that "complete" now refers to a finite subsystem of E.)

Definition 17. A complete linearly independent system of vectors is called a *basis* of E.

29. In C^n the elements

$$e_1 = (1,0,\ldots,0),\ e_2 = (0,1,\ldots,0),\ldots,\ e_n = (0,0,\ldots,1)$$

form a basis.

Definition 18. The basis in (29) is called *canonical*.

30. In Π^n the polynomials $1,t,t^2,\ldots,t^{n-1}$ form a basis.

Definition 19. That basis is also called *canonical*.

Bases admit a characterization analogous to the characterization of a basic subsystem of a vector system (cf. (5) and (6)):

31. A system of vectors is a basis iff it is a maximal linearly independent system, i.e., if it is not contained in any other linearly independent system.

32. A system of vectors is a basis iff it is a minimal complete system, i.e., if it contains no other complete system.

The existence theorem for a basis is analogous to theorem (7):

33. In each space $E \neq 0$ there exists a basis. More precisely, any linearly independent system of vectors can be extended to a basis, and (dually) from any complete system of vectors a basis can be extracted.

The basis plays the role of a coordinate system in E (cf. (8) and (9)):

34. Let $\{e_k\}_1^n$ be a basis. Any vector $x \in E$ has a *unique decomposition* with respect to that basis: $x = \Sigma_{k=1}^n \xi_k e_k$.

Definition 20. The coefficients $\xi_1, \xi_2, \ldots, \xi_n$ are said to be the *coordinates* of the vector x *for the basis* $\{e_k\}_1^n$.

35. If the system of vectors $\{e_k\}_1^n$ is such that any vector x has a unique decomposition $x = \Sigma_{k=1}^n \xi_k e_k$, then it is a basis.

On the strength of (34) we shall introduce the matrix of a vector system. In general, a matrix ((n×m-matrix or matrix of size n × m) is a rectangular array of numbers

$$A = (\alpha_{jk}), \quad 1 \leq j \leq n, \quad 1 \leq k \leq m.$$

When n = m, the matrix A is square, and the number n is its order.

Definition 21. Let $\{u_k\}_1^m$ be some system of vectors and let

$$u_k = \Sigma_{j=1}^n \alpha_{jk} e_j \quad (k = 1,2,\ldots,m)$$

be the decompositions of the vectors for the basis $\{e_j\}_1^n$. Then

$$A = (\alpha_{jk}), \quad 1 \leq j \leq n, \quad 1 \leq k \leq m$$

is called the *matrix of the vector system* relative to the basis $\{e_j\}_1^n$.

In particular, the matrix of the basis relative to itself has the form $\{\delta_{jk}\}_{j,k=1}^n$, where δ_{jk} is the well-known Kronecker symbol

$$\delta_{jk} = \begin{cases} 0 & (j \neq k) \\ 1 & (j = k) \end{cases}.$$

The matrix (δ_{jk}) is the identity matrix.

For a fixed basis, the correspondence between all possible systems of vectors and all possible n-rowed matrices is bijective.

The following theorem, analogous to theorem (12), is fundamental:

36. All bases of a given space E have the same number of vectors.

Definition 22. The number of vectors in a basis is called the *dimension* of E and is denoted by dim E. If E = 0, then dim E = 0. A space whose dimension is n is called *n-dimensional*. Theorems (29) and (30) show that spaces C^n and Π^n are n-dimensional.

Dimension of a space is analogous to rank of a vector system and is closely connected with it:

37. The rank of any system of vectors $\subset E$ is not larger than dim E.

The rank of any complete system evidently equals the dimension of the space. Thus dim E is the maximum rank of all its possible vector systems. In addition, we have the analog to (13):

38. The dimension of a space $E \neq 0$ equals the maximum cardinality of all linearly independent vector systems $\subset E$.

The dual analog to (14) is:

39. The dimension of a space $E \neq 0$ equals the minimum cardinality of all complete vector systems $\subset E$.

From this point of view a basis can also be characterized as an extremal system of vectors:

40. Let dim $E = n > 0$. If the system $\{e_k\}_1^n$ is linearly independent or (dually) complete, then it is a basis of E.

Theorem (40) yields easily the following supplementary remark to (37):

41. If the rank of a system of vectors equals the dimension of the space, then this system is complete.

Fixing some basis Δ of the space E (dim $E = n$), we can describe all bases of E by their matrices relative to Δ. To each basis Γ will then correspond some n-th order square matrix $(\gamma_{jk})_{j,k=1}^n$, but not every such matrix will correspond to some basis. However,

42. If the matrix $G = (\gamma_{jk})_{j,k=1}^n$ corresponds to some basis Γ, then for any fixed pair of indices r,s, chosen from $1 \leq r \leq n$,

$1 \leq s \leq n$, there exists an $\varepsilon = \varepsilon_{r,s} > 0$ such that each matrix $\tilde{G} = (\tilde{\gamma}_{jk})_{j,k=1}^{n}$ with elements

$$\tilde{\gamma}_{jk} = \begin{cases} \gamma_{jk}, & j \neq r \text{ or } k \neq s \\ \gamma_{rs} + \eta, & j = r \text{ and } k = s \end{cases}$$

corresponds to some basis $\tilde{\Gamma}$ when $|\eta| < \varepsilon$.

We therefore can say that the set of bases in n-dimensional space depends on n^2 parameters; i.e. the elements of the matrix corresponding to it by (42).

Definition 23. A space E_1 is called *isomorphic* to a space E if there exists a *bijective* mapping j of the space E onto the space E_1 which satisfies the conditions

$$(1) \quad j(x_1+x_2) = jx_1 + jx_2,$$

$$(2) \quad j(\alpha x) = \alpha(jx).$$

The mapping j is then an *isomorphism* between E and E_1.-- Isomorphism is an equivalence relation, which permits us to speak of classes of isomorphic spaces. From the abstract point of view, isomorphic spaces are indistinguishable since any property that is formulated only in terms of the operations of addition and multiplication by scalars refers with equal force to all spaces in the same class.

If the spaces E and E_1 are isomorphic, then we write $E \simeq E_1$. The following theorem gives a criterion when two spaces are isomorphic.

43. $E \simeq E_1$ iff $\dim E = \dim E_1$.

Thus all n-dimensional linear vector spaces are isomorphic. E.g., the spaces C^n and Π^n are isomorphic.

In the following sections we shall, as a rule, fix some "fundamental" space E. However, in some important situations it will be necessary to consider simultaneously several "fundamental" spaces. The fundamental space will always be assumed $\neq 0$.

We reserve the letter n for the dimension of the space E.

3. Subspaces.

Definition 24. A nonempty set L of vectors in a space E is called a *subspace* if L is closed relative to the operations of addition and multiplication by scalars, i.e., if L itself is a linear space for those operations. The vector 0 is obviously contained in every subspace, and the set {0} containing only this vector is a subspace, the *zero subspace*. The space E itself is of course also a subspace. The subspaces 0 and E are the *trivial subspaces*. Since each subspace L is itself a linear space, dim L is a legitimate concept.

44. $\dim L \leq n$ is always true. Equality holds, iff L = E.

Each system of vectors generates some subspace:

45. The linear hull sp Γ of any vector system Γ is a subspace with $\dim \operatorname{sp} \Gamma = \operatorname{rk} \Gamma$.

Therefore:

46. E has subspaces of any dimension d, $0 \leq d \leq n$.

Consider now the inclusion relation for subspaces.

47. If L_1 and L_2 are subspaces and $L_1 \subset L_2$, then

$$\dim L_1 \leq \dim L_2.$$

Here the equality sign holds iff $L_1 = L_2$.

Definition 25. A set of subspaces L_ν is a *chain* if for each pair L_{ν_1}, L_{ν_2} either $L_{\nu_1} \subset L_{\nu_2}$ or $L_{\nu_2} \subset L_{\nu_1}$.

Thanks to the finite dimensionality of E we have the following principle of finiteness of chains.

48. Any chain of subspaces is finite.

Thus any chain of subspaces can be written in the form $L_o \subset L_1 \subset \ldots \subset L_m$ $(m \geq 0)$.

Definition 26. A chain is *maximal* if it is not a proper part of any other chain.

49. A chain $L_o \subset L_1 \subset \ldots \subset L_m$ is maximal iff m = n and $\dim L_k = k$ $(k = 0, 1, \ldots, n)$. Hence $L_o = 0$, $L_n = E$.

50. Any chain is contained in some maximal chain.

This maximal chain is in general not unique.

51. If $\{L_k\}_o^n$ is a maximal chain of subspaces, then there is a basis $\{e_k\}_1^n$ such that $\{e_k\}_1^m$ is a basis for the subspace L_m $(m = 1,2,\ldots,n)$.

Some operations can be performed on subspaces. Consider first set-theoretic operations.

The union of subspaces turns out to be a subspace only in exceptional cases:

52. A union of a finite set of subspaces is a subspace iff one of the subspaces contains all the others.

This obviously implies that the union of a finite set of subspaces distinct from E is also distinct from E. In contrast to (52) it is true, however, that

53. The intersection of any set of subspaces $\{L_\nu\}$ is a subspace and

$$\dim (\cap L_\nu) \leqq \min (\dim L_\nu).$$

Here the equal sign holds iff one of the subspaces L_ν is contained in all the others.

Definition 27. In the following it will often be necessary to consider the set $\mathbb{L}$ of all those subspaces that have some specified property. Let us distinguish between "maximal" and "largest" subspaces of the set $\mathbb{L}$ according to the following rule: If a subspace $L \in \mathbb{L}$ is not contained in any other subspace $\in \mathbb{L}$, then we shall call it maximal. If, however, a subspace $L \in \mathbb{L}$ contains every subspace $\in \mathbb{L}$, then we shall call it largest.

If a largest subspace exists in $\mathbb{L}$, then it is also maximal. But a maximal subspace in $\mathbb{L}$ is (the) largest subspace only if there is no other maximal subspace in $\mathbb{L}$. Thus one or more maximal subspaces may exist in a set $\mathbb{L}$, but only one largest subspace.

Minimal and smallest subspaces are obtained by the corresponding dual definitions.

As an example take the set $\mathbb{L}_F$ of all subspaces in E, which contain a certain set of vectors F. The intersection of all subspaces $\in \mathbb{L}_F$ also belongs to $\mathbb{L}_F$; it is thus the smallest subspace in the set $\mathbb{L}_F$. It is already known to us as the linear hull of the set F, denoted by sp F (D6). The intersection of a set of subspaces $\{L_\nu\}$ is the largest subspace contained in all the subspaces L_ν. But the corresponding dual notion is not the union, because we are operating in a class of sub<u>spaces</u> and not in a class of sub<u>sets</u> of the set E.

Definition 28. The dual required is called the *sum of the subspaces* $\{L_\nu\}$. It is the smallest subspace containing all L_ν. Hence this sum coincides with the linear hull $sp(\cup L_\nu)$. It is denoted by the usual symbols Σ or +. Since the terms of the sum are spaces, no misunderstanding should arise through this shorthand notation.

54. $\dim(\Sigma L_\nu) \geqq \max(\dim L_\nu)$. Here equality obtains iff one of the subspaces L_ν contains all the others.

Theorems (53) and (54) are analogous to theorems (19) and (20).

Intersection and sum of subspaces are subject to the following <u>finiteness principle</u> which ensues from the principle of finiteness for chains (D25):

55. Any set of subspaces contains a finite subset of subspaces with the same intersection.

Dually:

56. Any set of subspaces contains a finite subset of subspaces with the same sum.

Note: if a set of subspaces contains the intersection of any two of its subspaces, then it contains a smallest subspace. Dually: if a set of subspaces contains the sum of any two of its subspaces, then it contains a largest subspace.

The operation of addition of subspaces is associative and commutative, as is the operation of intersection. However, there is no distributive law. In the general case we have only:

57. For any set of subspaces $\{L_\nu\}$ and any subspace L the relations

$$L \cap \Sigma L_\nu \supset \Sigma(L \cap L_\nu), \quad L + \cap L_\nu \subset \cap(L + L_\nu)$$

are satisfied.

But in the case of <u>two</u> subspaces L_1 and L_2 we find:

58. If $L_1 \subset L$, then

$$L \cap (L_1 + L_2) = L \cap L_1 + L \cap L_2$$

and, dually, if $L_1 \supset L$, then

$$L + L_1 \cap L_2 = (L + L_1) \cap (L + L_2)$$

These are the so-called <u>modular laws</u>. They can be proved very simply if the following description of the sum of two subspaces is used:

59. The sum of two subspaces L_1, L_2 is the set of vectors of the form

$$x = x_1 + x_2 \quad (x_1 \in L_1,\ x_2 \in L_2).$$

Theorem (59) immediately carries over to the sum of any finite set of subspaces.

Let us now consider the system of two subspaces in more detail. We first establish the analog of theorem (21):

60. For any two subspaces L_1 and L_2 we have the equality

$$\dim(L_1 \cap L_2) + \dim(L_1 + L_2) = \dim L_1 + \dim L_2$$

(*Formula of Grassmann*).

From Grassmann's formula the analog to (22) follows:

61. $\dim(L_1 + L_2) \leq \dim L_1 + \dim L_2$.

We now formulate a sufficient condition for equality in (61).

Definition 29a. Reapplying the terminology of D13 to subspaces, we shall call subspaces L_1, L_2 *mutually independent* (and the system $\{L_1, L_2\}$ *independent*) if

$$L_1 \cap L_2 = 0.$$

Definition 30. The sum of two mutually independent subspaces L_1, L_2 is called their *direct sum* and is denoted by $L_1 \oplus L_2$.

62. Equality in (61), i.e.

$$\dim(L_1 + L_2) = \dim L_1 + \dim L_2$$

holds, iff the subspaces L_1 and L_2 are mutually independent (cf. D29 and (23), (24)).

Directness of a sum of two subspaces can be formulated as follows:

63. The sum of two subspaces L_1, L_2 is direct iff the representation

$$x = x_1 + x_2 \quad (x_1 \in L_1,\ x_2 \in L_2)$$

is unique for each $x \in L_1 + L_2$ (it would be enough to require unique representation for $x = 0$, i.e., $x = 0 \Rightarrow x_1 = 0,\ x_2 = 0$).

The notion of mutual independence is dual to the notion of completeness:

Definition 29b. A system of two subspaces L_1, L_2 is *complete* if

$$L_1 + L_2 = E,$$

i.e., if for each vector $x \in E$ there exists a representation

$$x = x_1 + x_2 \quad (x_1 \in L_1,\ x_2 \in L_2).$$

Definition 31. The subspaces L_1, L_2 are called *mutually complementary*, and the system $\{L_1, L_2\}$ is then *basic*, if

$$L_1 \oplus L_2 = E, \qquad (*)$$

i.e., if the system L_1, L_2 is complete and independent. Here either subspace is called the *complement* of the other.

Of course, equation (*) means that each vector x can be uniquely represented in the form

$$x = x_1 + x_2 \quad (x_1 \in L_1,\ x_2 \in L_2).$$

64. For any subspace L there exists a complement.

It is not unique if $L \neq E$, but

65. The dimension of any complement of the subspace L equals $n - \dim L$.

Definition 32. The quantity $n - \dim L$ is known as the *codimension* of the subspace L and is denoted by codim L. Thus by definition

$$\dim L + \operatorname{codim} L = n.$$

Codimension and dimension are dual concepts. This becomes evident from the following cycle of theorems.

66. If L_1, L_2 are subspaces and $L_1 \subset L_2$, then

$$\operatorname{codim} L_1 \geqq \operatorname{codim} L_2.$$

Equality occurs here iff $L_1 = L_2$.

67. For any set $\{L_\nu\}$ of subspaces

$$\operatorname{codim}(\cap L_\nu) \geqq \max(\operatorname{codim} L_\nu).$$

Equality occurs here iff one of the subspaces is contained in all the others.

68. For any set $\{L_\nu\}$ of subspaces the inequality

$$\operatorname{codim}(\Sigma L_\nu) \leqq \min(\operatorname{codim} L_\nu)$$

holds. Equality occurs here iff one of the subspaces L contains all the others.

69. For any two subspaces L_1, L_2 we have

$$\operatorname{codim}(L_1+L_2) + \operatorname{codim}(L_1 \cap L_2) = \operatorname{codim} L_1 + \operatorname{codim} L_2.$$

Hence

70. $\operatorname{codim}(L_1 \cap L_2) \leqq \operatorname{codim} L_1 + \operatorname{codim} L_2.$

Here

71. $\operatorname{codim}(L_1 \cap L_2) = \operatorname{codim} L_1 + L_2$

iff the system $\{L_1, L_2\}$ is complete.

From (61) and (70) it is easy to obtain a characterization of incompleteness and independence for a system of two subspaces $\{L_1, L_2\}$.

72. If $\operatorname{codim} L_1 > \dim L_2$, then the system $\{L_1, L_2\}$ is incomplete.

73. If $\dim L_1 > \operatorname{codim} L_2$, then the system $\{L_1, L_2\}$ is dependent.

Let $L_1, L_2, \ldots, L_m$ be any subspaces. Under what conditions will there exist a common complement for them? The answer can be obtained with the help of (52):

74. The subspaces $L_1, L_2, \ldots, L_m$ have a common complement iff their dimensions are equal.

To extend the notion of direct sum to a system of any number of subspaces $\{L_k\}_1^m$, proceed from (63) and define:

Definition 33. The sum of the subspaces $L_1, L_2, \ldots, L_m$ is a *direct sum* if for each $x \in \Sigma_{k=1}^m L_k$ the representation

$$x = \Sigma_{k=1}^m x_k \quad (x_k \in L_k;\ k = 1, 2, \ldots, m)$$

is unique, or, put more simply, if

$$\Sigma_{k=1}^m x_k = 0 \quad (x_k \in L_k;\ k = 1, 2, \ldots, m)$$

implies $x_k = 0$ $(k = 1, 2, \ldots, m)$. The direct sum will be denoted by the symbols $\Sigma^{\oplus}$ or $\oplus$.

75. The sum of subspaces $L_1, L_2, \ldots, L_m$ is direct, iff for any two disjoint subsystems of indices $\{j_1, \ldots, j_p\}$, $\{k_1, \ldots, k_q\}$ the subspaces

$$L_{j_1} + L_{j_2} + \ldots + L_{j_p} \quad \text{and} \quad L_{k_1} + L_{k_2} + \ldots + L_{k_q}$$

are mutually independent. It is even sufficient that the subspaces

$$L_1 + L_2 + \ldots + L_p \text{ and } L_{p+1}$$

be mutually independent for each $p = 1,2,\ldots,m-1$.

76. The sum of subspaces $L_1, L_2, \ldots, L_m$ is direct iff

$$\dim \Sigma_{k=1}^{m} L_k = \Sigma_{k=1}^{m} \dim L_k.$$

Definition 34. A system of subspaces $\{L_k\}_1^m$ not including the zero space is called *basic* if

$$\Sigma_{k=1}^{\oplus\, m} L_k = E.$$

One says then that the space E has been decomposed into the direct sum of the subspaces $L_1, L_2, \ldots, L_m$.

77. If $\{L_k\}_1^m$ is a basic system for the space E and Δ_k is some basis for the subspace L_k, then $\Delta = \cup_{k=1}^{m} \Delta_k$ is a basis for the space E.

Conversely:

78. Let Δ be a basis for the space E and $\{\Delta_1, \Delta_2, \ldots, \Delta_m\}$ an arbitrary decomposition of the vector system Δ into subsystems. Then the system of subspaces $\{\text{sp}\, \Delta_k\}_1^m$ is a basic system, and Δ_k is a basis for the subspace $\text{sp}\{\Delta_k\}$.

Definition 35. The *characteristic* of a basic system of subspaces $\{L_k\}_1^m$ is the name for the system of natural numbers $\dim\{L_k\}_1^m$.

79. A system of natural numbers $\{d_k\}_1^m$ is the characteristic of some basic system of subspaces, iff $\Sigma_{k=1}^{m} d_k = n$.

Of course, $m \leq n$. For $m = n$ the only possible characteristic is $\{1,1,\ldots,1\}$. It indicates the decomposition of E into a direct sum of one-dimensional subspaces.

In closing we characterize the dimension of a subspace by the set of its functional properties (cf. (25)):

80. Let the functional $d(L)$, defined on the set of all subspaces of a space E, possess the following two properties:
(1) if the subspaces L_1, L_2 are mutually independent, then $d(L_1 + L_2) = d(L_1) + d(L_2)$;

(2) if dim $L = 1$, then $d(L) = 1$.

Then $D(L) = \dim L$ for all L.

4. Factor Spaces. Homomorphisms. The Fredholm Alternative.

Definition 36. Let L be some subspace of E. We shall say that the vectors x and y are *congruent modulo* L,

$$x \equiv y \pmod{L},$$

if $x - y \in L$. In particular, the congruence $x \equiv 0 \pmod{L}$ means $x \in L$. To be congruent modulo L is clearly an equivalence relation. E is thus decomposed into classes so that the vectors in a single class are congruent modulo L (or, form a *congruence class modulo* L). The class generated by the vector x will be denoted by $[x]_L$ or simply by $[x]$ if L follows from the context.

81. $x \equiv y \pmod{L}$ and $x_1 \equiv y_1 \pmod{L}$ imply

$$x + x_1 \equiv y + y_1 \pmod{L}, \quad \alpha x \equiv \alpha y \pmod{L}.$$

By virtue of (81), addition and multiplication by a scalar can be introduced naturally in the set of classes relative to a fixed subspace by

$$[x] + [x_1] \underset{\text{def}}{=} [x + x_1], \quad \alpha[x] \underset{\text{def}}{=} [\alpha x].$$

Definition 37. The set of classes so becomes a linear space called the *factor space* of E with respect to L; it is denoted by E/L.

82. $E/0 \simeq E$, $E/E \simeq 0$ (notation in D23).

A more general assertion is:

83. If L' is a complementary subspace to L, then $E/L \simeq L'$.

Hence,

84. $\dim E/L = \operatorname{codim} L$.

This equation can be written in the complementary form

$$\dim E/L + \dim L = n.$$

In what follows we shall encounter a number of similar formulas.

Definition 38. The vectors $x_1, x_2, \ldots, x_m$ are called *linearly independent modulo* L if the classes $[x_1]_L, [x_2]_L, \ldots, [x_m]_L$ generated by them are linearly independent in the factor space E/L.

85. The vectors $x_1, x_2, \ldots, x_m$ are linearly independent modulo L if they are linearly independent, and if their linear hull and the subspace L are mutually independent (sufficient conditions).

On the other hand, vectors that are linearly independent modulo L are linearly independent a fortiori, since they are linearly independent mod 0 (necessary condition). More generally,

86. If the vectors $x_1, \ldots, x_m$ are linearly independent modulo L, then they are linearly independent modulo any $M \subset L$.

Definition 39. We now introduce the fundamental idea of homomorphism which generalizes the idea of isomorphism. Let E_1 be another linear space. A mapping $h: E \to E_1$ is a *homomorphism* from E into E_1 if

$$(1) \quad h(x_1 + x_2) = hx_1 + hx_2,$$

$$(2) \quad h(\alpha x) = \alpha hx.$$

Definition 40. If the relation between x and hx which establishes the homomorphism is one to one, i.e., if $hx_1 = hx_2 \Rightarrow x_1 = x_2$, then h is called a *monomorphism* (i.e., an *injective* homomorphism).

For example, if E is a subspace of E_1, then the homomorphism $i: E \to E_1$, defined by

$$ix = x \quad (x \in E \subset E_1)$$

is a monomorphism. It is called the *embedding* or the *insertion* of E into E_1.

Definition 41. If the homomorphism h maps the space E *onto* all of the space E_1, then it is called an *epimorphism* (i.e., a *surjective* homomorphism).

An example of an epimorphism will be given below.

Thus an isomorphism is a homomorphism, which is simultaneously a monomorphism and an epimorphism.

In the following propositions (87-90), h is a homomorphism from E into E_1.

87. The set of all solutions of the equation $hx = 0$ is a subspace of E.

Definition 42. This subspace of E is called the *kernel* of h and is denoted by Ker h.

88. The set of all vectors of the form $y = hx$ is a subspace of E_1.

Definition 43. This subspace of E_1 is called the *image* of h and is denoted by Im h.

Definition 44. The factor space E_1/Im h is known as the *cokernel* of the homomorphism h and denoted by Coker h. The factor space E/Ker h is the *coimage* of the homomorphism h and denoted by Coim h. The prefix "co" indicates the duality of the ideas (as already in "codimension").

89. h is a monomorphism iff Ker $h = 0$.

90. h is an epimorphism iff Coker $h = 0$.

Monomorphism and epimorphism are dual to each other. Isomorphism is self-dual.

Let us emphasize that a homomorphism h from E into[§] E_1 can be considered as an isomorphism of E onto Im h.

There is a close connection between homomorphisms and factor spaces:

91. Let M be some subspace of E. The mapping $h_M x = [x]_M$ is a homomorphism from E into E/M for which

$$\text{Ker } h_M = M, \quad \text{Im } h_M = E/M.$$

Definition 45. Thus h_M is an epimorphism. It is called

§"into" is the general term. It does not imply that Im h is a proper subspace of E_1.

the *contraction* of E with respect to M. Upon "contracting" E mod M we obtain the factor space E/M. The subspace M itself "contracts" to the point 0.

This result can be generalized. Let M and L be two subspaces of the space E, and $M \supset L$. $x \equiv y \pmod{L}$ evidently implies $x \equiv y \pmod{M}$, hence each class modulo L is entirely contained in exactly one class modulo M. Thus an <u>embedding of class into class</u> arises, which we shall denote by $h_{M/L}$. It operates from the factor space E/L to the factor space E/M. Obviously, $h_{M/O} = h_M$ (if E/O is identified with E); $h_{E/L} = 0$ for any subspace L, since $E/E = 0$.

92. The mapping $h_{M/L}$: $[x]_L \mapsto [x]_M$ is a homomorphism for which

$$\text{Ker } h_{M/L} = \{[x]_L \mid x \in M\}, \quad \text{Im } h_{M/L} = E/M.$$

Also Ker $h_{M/L} \simeq M/L$.

Definition 46. By (92), $h_{M/L}$ is an epimorphism. It is called the *contraction* of the space E by the subspace M *relative to the subspace* L.

93. Let h be a homomorphism from E into E_1. The mapping $\hat{h}$, defined by $\hat{h}[x]_{\text{Ker } h} = hx$, is an isomorphism of the spaces E/Ker h and Im h:

$$\text{Im } h \simeq E/\text{Ker } h. \qquad (*)$$

Thus by contracting E modulo Ker h we convert the homomorphism h into a monomorphism without changing its image. One may therefore say that the homomorphism h is a monomorphism "up to" or "not counting" a contraction modulo Ker h. Such a "normalization" of an arbitrary homomorphism plays an important role. We shall consider one of the more significant examples for formula (*), but for this we need two definitions:

Definition 47. The dimension of Im h is called the *rank* of the homomorphism h and is denoted by rk h; the dimension of Ker h is called its *defect* and denoted by def h.

Now, with $M \supset L$,

94. $\operatorname{rk} h_{M/L} = \operatorname{codim} M, \quad \operatorname{def} h_{M/L} = \dim M - \dim L.$

In particular, $\operatorname{rk} h_M = \operatorname{codim} M$, $\operatorname{def} h_M = \dim M$, and consequently

$$\operatorname{def} h_{M/L} = \operatorname{def} h_M - \operatorname{def} h_L.$$

If h is a homomorphism from E into E_1, then obviously

$$0 \leqq \operatorname{def} h \leqq n, \quad 0 \leqq \operatorname{rk} h \leqq n_1,$$

where $n = \dim E$ as always, and $n_1 = \dim E_1$.

95. For any homomorphism h from E into E_1 we have the complement formula

$$\operatorname{rk} h + \operatorname{def} h = n.$$

Definition 48. If $h \in \operatorname{Hom}(E,E_1) \underset{\text{def}}{=}$ set of all $h: E \to E_1$, then the difference of the dimensions

$$\dim E - \dim E_1$$

is called the *index* of h and is denoted by ind h.

Definition 49. If ind h = 0, then h is called a *Fredholm* homomorphism, of which the so-called *endomorphisms* of E ($h: E \to E$) are important particular cases.

Of course, a Fredholm homomorphism need not be an isomorphism. Its fundamental property is expressed in the following theorem:

96. A Fredholm homomorphism is an epimorphism iff it is a monomorphism (and so is an isomorphism).

Hence:

97. If h is a Fredholm homomorphism, then the inhomogeneous equation

$$hx = y$$

is globally solvable (i.e., solvable for any right-hand side y) then and only then, if the corresponding homogeneous equation

$$hx = 0$$

has only the trivial solution $x = 0$. In that case the

solution for given y is unique (first Fredholm theorem).

This result is often formulated as the Fredholm alternative: if h is a Fredholm homomorphism, then either the inhomogeneous equation is solvable for any right-hand side, or the corresponding homogeneous equation has nontrivial solutions.

Note here that the existence of nontrivial solutions of the homogeneous equation is equivalent to the non-uniqueness of the solution of the inhomogeneous equation whenever such a solution exists. And so either the inhomogeneous equation is globally solvable, or the solution is not unique in those cases when it exists.

The following theorem (98) is a generalization of (95) and will be the source of a number of further results.

Definition 50. Let h be a homomorphism from E into E_1, and let L be a subspace of E. The set of vectors

$$\{y \mid y = hx,\ x \in L\},$$

called the *image* of L, is a subspace $\subset E_1$. We shall denote it by hL. In particular, hE = Im h, hO = 0. Since $hL \subset \text{Im } h$,

$$\dim hL \leq \text{rk } h.$$

98. $\dim hL = \dim L - \dim(L \cap \ker h)$.

Hence $\dim hL \leq \dim L$, that is, a homomorphism never increases the dimension of a subspace. On the other hand,

$$\dim hL \geq \dim L - \text{def } h,$$

that is, a homomorphism cannot diminish the dimension of a subspace "too much". This estimate cannot be improved, because

99. There exists a subspace L such that

$$\dim hL = \dim L - \text{def } h.$$

Thus the defect of a homomorphism can be interpreted as a measure of the decrease of the dimension of certain subspaces. In particular it is clear that a homomorphism preserves the dimension of all subspaces if and only if it is a monomorphism.

Definition 51. Let us now introduce the object dual to hL. The set of vectors

$$\{x \mid x \in E,\ hx \in M\},$$

is called the *full inverse image* of $M \subset E_1$ and it is a subspace $\subset E$. We shall denote it by $h^{-1}M$. In particular $h^{-1}O = \operatorname{Ker} h$, $h^{-1}E_1 = E$. Since $h^{-1}M \supset \operatorname{Ker} h$, $\dim h^{-1}M \geqq \operatorname{def} h$.

100. In general

$$h^{-1}(hL) \supset L.$$

To make $h^{-1}(hL) = L$ for all L, h must be a monomorphism; this is also a necessary condition.

Dually:

101. $$h(h^{-1}M) \subset M$$

is always true, and $h(h^{-1}M) = M$ for all M iff h is an epimorphism.

102. $$\operatorname{codim} h^{-1}M = \operatorname{codim} M - \operatorname{codim}(M + \operatorname{Im} h) \quad \text{(D28)}.$$

The duality of formulas (98) and (102) shows up strikingly. From (102) it is evident that

$$\operatorname{codim} M \geqq \operatorname{codim} h^{-1}M.$$

Thus a homomorphism cannot lower the codimension of a subspace. On the other hand,

$$\operatorname{codim} M \leqq \operatorname{codim} h^{-1}M + \operatorname{codim}(\operatorname{Im} h),$$

that is, a homomorphism cannot increase the codimension of a subspace "too much". This estimate is again exact, because

103. There is a subspace M such that

$$\operatorname{codim} M = \operatorname{codim} h^{-1}M + \operatorname{codim}(\operatorname{Im} h).$$

The number codim(Im h) can be interpreted as a measure of

the increase of the codimension of a subspace. In particular, it is clear that a homomorphism conserves the dimension of all subspaces iff it is an epimorphism. Note that only for Fredholm homomorphisms

$$\operatorname{codim}(\operatorname{Im} h) = \operatorname{def} h,$$

that is, the maximal decrease of the dimension of h coincides here with the maximal increase of its codimension.-- Let us now consider the properties of a homomorphism h in regard to the operations of union and intersection of subspaces.

104. Let $\{L_\nu\}$ be some set of subspaces of E, and $\{M_\mu\}$ some set of subspaces of E_1. Then the relations

$$h(\Sigma L_\nu) = \Sigma h L_\nu, \quad h^{-1}(\Sigma M_\mu) \supset \Sigma h^{-1} M_\mu$$

hold and, dually,

$$h(\cap L_\nu) \subset \cap h L_\nu, \quad h^{-1}(\cap M_\mu) = \cap h^{-1} M_\mu.$$

The exact formulas are:

105. $$\cap h L_\nu = h(\,\cap(L_\nu + \operatorname{Ker} h)\,),$$

$$\Sigma h^{-1} M_\mu = h^{-1}(\,\Sigma(M_\mu \cap \operatorname{Im} h)\,).$$

Therefore, if $M_\mu \subset \operatorname{Im} h$ for all μ (which is true if h is an epimorphism), then

$$h^{-1}(\Sigma M_\mu) = \Sigma h^{-1} M_\mu.$$

Dually, if $L_\nu \supset \operatorname{Ker} h$ for all ν (which is true if h is a monomorphism), then

$$h(\cap L_\nu) = \cap h L_\nu.$$

106. Let the subspaces L_1 and L_2 form a complete system and let h be an epimorphism. Then hL_1 and hL_2 also form a complete system.

107. Let subspaces L_1 and L_2 be mutually independent, and h be a monomorphism. Then hL_1 and hL_2 are also mutually

independent.

A more general assertion is

108. If $L = \Sigma^{\oplus\, m}_{k=1} L_k$ and h is a monomorphism, then

$$hL = \Sigma^{\oplus\, m}_{k=1} hL_k.$$

5. Operations on Homomorphisms

Definition 52. Consider the set $\mathrm{Hom}(E,E_1)$ of all homomorphisms from E into E_1, define the sum of h_1 and h_2 by

$$(h_1 + h_2)x = h_1x + h_2x \quad (x \in E),$$

and the product αh by

$$(\alpha h)x = \alpha(hx) \quad (x \in E).$$

Thus *the set* $\mathrm{Hom}(E,E_1)$ *itself* becomes a *linear space*. The zero of this space is the homomorphism O: $Ox = 0$ $(x \in E)$.-- The following proposition indicates a general method of producing homomorphisms from E to E_1.

109. Let $\{e_k\}_1^n$ be some basis of the space E. For a given system of n vectors $\{v_k\}_1^n$ of E_1 there exists a unique homomorphism $h \in \mathrm{Hom}(E,E_1)$ satisfying

$$he_k = v_k \quad (k=1,2,\ldots,n).$$

If we now choose some basis $\{u_j\}_1^{n_1}$ of E_1, then the system $\{v_k\}_1^n = \Sigma_{j=1}^{n_1} \alpha_{jk}u_j$ uniquely determines and is determined by its matrix (D21).

Definition 53. The matrix (α_{jk}) is called the *matrix of the homomorphism* h relative to the pair of bases $\{e_k\}_1^n$, $\{u_j\}_1^{n_1}$. It has the dimensions $n_1 \times n$. If two homomorphisms are added, their matrices relative to a given pair of bases add:

$$(\alpha_{jk}) + (\beta_{jk}) = (\alpha_{jk} + \beta_{jk}).$$

For the multiplication of a homomorphism by a number λ we have analogously

$$\lambda(\alpha_{jk}) = (\lambda\alpha_{jk}).$$

110. Let $\{e_k\}_1^n$ be a basis of E and $\{u_j\}_1^{n_1}$ a basis of E_1. Define the homomorphism $I_{pq} \in \mathrm{Hom}(E,E_1)$ by putting

$$I_{pq}e_k = \delta_{qk}u_p \quad (q,k=1,2,\ldots,n;\ p=1,2,\ldots,n_1)^{\S}\ .$$

The system of homomorphisms $\{I_{pq}\}$ forms a basis for the space $\mathrm{Hom}(E,E_1)$.

It follows that

$$\dim \mathrm{Hom}(E,E_1) = \dim E \cdot \dim E_1.$$

In particular, the space Hom(E,E) is the space of all endomorphisms of the space E. Its dimension equals n^2.-- Let us observe one consequence of theorem (109) and of the results of Section 4 (cf. also (43)):

111. A monomorphism $h\colon E\to E_1$ exists iff $\dim E_1 \geqq \dim E$. An epimorphism $h\colon E\to E_1$ exists iff $\dim E_1 \leqq \dim E$.

Therefore, the simultaneous existence of a monomorphism and an epimorphism from E into E_1 implies $E \simeq E_1$, i.e., E and E_1 are isomorphic. It then follows from Fredholm's first theorem that every monomorphism or epimorphism from E into E_1 must be an isomorphism.

Definition 54. We shall now introduce the operation of multiplication of homomorphisms. Suppose yet another space E_2 is given and let $h_1 \in \mathrm{Hom}(E,E_1)$, $h_2 \in \mathrm{Hom}(E_1,E_2)$. We put

$$hx = h_2(h_1x) \quad (x\in E).$$

This relation defines a mapping $h\colon E\to E_2$ which is called

---

§ I_{pq} is not a matrix, but the symbol for a particular set of mappings h; the I_{pq} assign to $e_1 \in E$ the vectors $\{u_p\}_1^{n_1} \in E_1$ or $0 \in E_1$, if $q=1$ or $\neq 1$, respectively.

the *product of the homomorphisms* h_2 and h_1 and is denoted by h_2h_1 (watch the order, which is opposite to the order in speech). Homomorphisms are thus multiplied in the same way as mappings: the product mapping is the "resultant" mapping which arises from the successive execution of the factor mappings.

112. The product of two homomorphisms is a homomorphism. In particular, if $h_1 \in \mathrm{Hom}(E,E_1)$, $h_2 \in \mathrm{Hom}(E_1,E_2)$, then $h_2h_1 \in \mathrm{Hom}(E,E_2)$.

The associative law holds:

113. $$h_3(h_2h_1) = (h_3h_2)h_1.$$

The associative law is true for the product of any mappings, whenever successive composition is possible. In the multiplication of homomorphisms the matrices are multiplied according to the following rule: let $(\alpha_{jk}^{(1)})$ be the matrix of h_1 relative to the pair of bases Δ,Δ_1, and $(\alpha_{jk}^{(2)})$ the matrix of h_2 relative to Δ_1,Δ_2. Then the matrix (α_{jk}) of $h = h_2h_1$ relative to Δ,Δ_2 has elements

$$\alpha_{jk} = \Sigma_{s=1}^{n_1} \alpha_{js}^{(2)} \alpha_{sk}^{(1)}.$$

Let I_E denote the identity endomorphism of the space E, i.e., the mapping

$$I_E x = x \quad (x \in E),$$

(obviously a homomorphism). Identity homomorphisms play the role of units in multiplication.

114. If $h \in \mathrm{Hom}(E,E_1)$, then

$$hI_E = I_{E_1}h = h.$$

The matrix of the identity endomorphism with respect to the pair of coincident bases Δ,Δ is the unit matrix. Theorems (112)-(114) permit us to assert that the class of finite dimensional linear spaces and their homomorphisms are a category (see Dictionary of General Terms).

Let us also note the distributive laws:

115. If $h, h_1 \in \mathrm{Hom}(E,E_1)$ and $h_2 \in \mathrm{Hom}(E_1,E_2)$, then

$$h_2(h+h_1) = h_2h + h_2h_1;$$

and if $h \in \mathrm{Hom}(E,E_1)$ and $h_1, h_2 \in \mathrm{Hom}(E_1,E_2)$, then

$$(h_1+h_2)h = h_1h + h_2h.$$

Finally:

116. If $h_1 \in \mathrm{Hom}(E,E_1)$, $h_2 \in \mathrm{Hom}(E_1,E_2)$, and α is any scalar, then

$$h_2(\alpha h_1) = (\alpha h_2)h_1 = \alpha(h_2h_1).$$

For endomorphisms, multiplication is an internal operation: if $h_1, h_2 \in \mathrm{Hom}(E,E)$, then $h_2h_1 \in \mathrm{Hom}(E,E)$ also.

Definition 55. On the basis of the preceding theorems we may speak of the *algebra of endomorphisms*. We shall be concerned with this algebra in Chapter II. An algebra is a linear space, in which a multiplication is defined which satisfies the distributive laws

$$x(y_1+y_2) = xy_1 + xy_2, \quad (x_1+x_2)y = x_1y + x_2y$$

and commutes with scalar multiplication:

$$x(\alpha y) = (\alpha x)y = \alpha(xy).$$

If the multiplication is associative,

$$x(yz) = (xy)z,$$

the algebra is called *associative*; such, for example, is the algebra of endomorphisms.

We shall now consider the important problem of extending a homomorphism.

Definition 56. Let L be a subspace of E and $h \in \mathrm{Hom}(L,E_1)$. A homomorphism $\hat{h} \in \mathrm{Hom}(E,E_1)$ is an *extension* (or *continuation*) to E of the homomorphism h if

$$\hat{h}x = hx \quad (x \in L).$$

Conversely, if we are given a homomorphism $h \in \mathrm{Hom}(E,E_1)$, then the homomorphism $\check{h} \in \mathrm{Hom}(L,E)$ is a *restriction* to L of the homomorphism h if

$$\check{h}x = hx \quad (x \in L).$$

(Thus h is an extension of the homomorphism $\check{h}$.) The usual symbol for a restriction is a vertical bar: $\check{h} = h|L$. Obviously,

$$\mathrm{Im}(h|L) = hL, \quad \mathrm{Ker}(h|L) = (\mathrm{Ker}\ h) \cap L$$

The simplest theorem about extensions is the following:

117. Each homomorphism $h \in \mathrm{Hom}(L,E_1)$ has an extension $h \in \mathrm{Hom}(E,E_1)$.

More than that:

118. Let the vectors $e_1, e_2, \ldots, e_m$ of the space E be linearly independent modulo L, and $u_1, \ldots, u_m$ be arbitrary vectors of the space E_1. For each homomorphism $h \in \mathrm{Hom}(L,E_1)$ there is an extension $h \in \mathrm{Hom}(E,E_1)$ satisfying

$$he_k = u_k \quad (k = 1,2,\ldots,m).$$

Such an extension for given $\{e_k\}_1^m$ and $\{u_k\}_1^m$ is unique iff codim L = m (that is, if the linear hull of the system of vectors $\{e_k\}_1^m$ is a complement of the subspace L).

This result is not difficult to obtain from theorem (109). It has many applications, and we shall use it immediately to investigate the inversion of a homomorphism.

Definition 57. Let $h \in \mathrm{Hom}(E,E_1)$. A homomorphism $g \in \mathrm{Hom}(E_1,E)$ is called a *left inverse* of h if

$$gh = I_E;$$

it is a *right inverse* of h if

$$hg = I_{E_1}.$$

If the homomorphism has a left (right) inverse, then it is called *left (right) invertible*. If a homomorphism is both left and right invertible, then it is *two-sided invertible* or, briefly, *invertible*.

The terms "left" and "right" are here dual to each other.

119. A homomorphism h is left invertible iff it is a monomorphism (i.e., Ker h = 0).

In addition:

120. For a monomorphism h the left inverse is unique, iff it is an isomorphism (thus not only ker h = 0, but also Coker h = 0).

In dual fashion:

121. A homomorphism is right invertible iff it is an epimorphism (i.e., Coker h = 0).

In addition:

122. If h is an epimorphism, then the right inverse is unique iff it is an isomorphism (thus not only Coker h = 0, but also Ker h = 0).

Now consider immediate consequences of theorems (119)-(122):

123. A homomorphism is two-sided invertible iff it is an isomorphism.

124. If a homomorphism has a two-sided inverse, then its left and right inverses are not only unique but equal.

In this case each of them is denoted by h^{-1} and is simply <u>the inverse</u> homomorphism of h.

125. The homomorphism h^{-1} has a two-sided inverse and $(h^{-1})^{-1} = h$.

126. If a homomorphism is left invertible and its left inverse is unique, then it is also right invertible; its right inverse is unique and coincides with its left inverse.

For a Fredholm homomorphism (in particular, for an endomorphism) we can, with the aid of Fredholm's first theorem, strengthen theorem (126) considerably:

127. If a Fredholm homomorphism has either a left inverse or a right inverse, then it has a two-sided inverse.

Finally we note that

128. The left inverse to a monomorphism is an epimorphism, and the right inverse to an epimorphism is a monomorphism. The inverse of an isomorphism is an isomorphism.

Let us now investigate the image and kernel (respectively rank and defect) of the sum and product of homomorphisms.

129. If h_1 and $h_2 \in \mathrm{Hom}(E,E_1)$, then

$$\mathrm{Im}(h_1 + h_2) \subset \mathrm{Im}\, h_1 + \mathrm{Im}\, h_2, \quad \mathrm{Ker}(h_1 + h_2) \supset \mathrm{Ker}\, h_1 \cap \mathrm{Ker}\, h_2.$$

By the first of these relations

130. $$\mathrm{rk}(h_1 + h_2) \leq \mathrm{rk}\, h_1 + \mathrm{rk}\, h_2.$$

Hence

131. $$\mathrm{rk}(h_1 + h_2) \geq |\mathrm{rk}\, h_1 - \mathrm{rk}\, h_2|.$$

Further:

132. If $h_1 \in \mathrm{Hom}(E,E_1)$ and $h_2 \in \mathrm{Hom}(E_1,E_2)$, then

$$\mathrm{Im}(h_2 h_1) = h_2(\mathrm{Im}\, h_1), \quad \mathrm{Ker}(h_2 h_1) = h_1^{-1}(\mathrm{Ker}\, h_2).$$

This and theorem (98) imply the formula

133. $$\mathrm{rk}\, h_2 h_1 = \mathrm{rk}\, h_1 - \dim(\mathrm{Im}\, h_1 \cap \mathrm{Ker}\, h_2).$$

This formula can be written in the form

134. $$\mathrm{rk}\, h_2 h_1 = \mathrm{rk}\, h_2 - \mathrm{codim}(\mathrm{Im}\, h_1 + \mathrm{Ker}\, h_2),$$

and also in the form

135. $$\mathrm{def}\, h_2 h_1 = \mathrm{def}\, h_1 + \dim(\mathrm{Im}\, h_1 \cap \mathrm{Ker}\, h_2),$$

and finally as

136. $$\mathrm{def}\, h_2 h_1 = \mathrm{def}\, h_2 + \mathrm{codim}(\mathrm{Im}\, h_1 + \mathrm{Ker}\, h_2) + \mathrm{ind}\, h_1.$$

If in particular h_1 is a Fredholm homomorphism, then

$$\mathrm{def}\, h_2 h_1 = \mathrm{def}\, h_2 + \mathrm{codim}(\mathrm{Im}\, h_1 + \mathrm{Ker}\, h_2).$$

From (133) and (134) we get the inequality

137. $$\mathrm{rk}\, h_1 h_2 \leq \min(\mathrm{rk}\, h_2, \mathrm{rk}\, h_1).$$

Precisely:

138. If the system of subspaces $\{\mathrm{Im}\, h_1\ \mathrm{Ker}\, h_2\}$ of the space E_1 is independent (this happens in particular if h_2 is a monomorphism), then

$$\mathrm{rk}\, h_2 h_1 = \mathrm{rk}\, h_1 \qquad \text{(D29a)}.$$

If that system is complete (this happens in particular if h_1 is an epimorphism), then

$$\text{rk } h_2h_1 = \text{rk } h_2 \qquad \text{(D29b)}.$$

In the remaining cases

$$\text{rk } h_2h_1 < \min(\text{rk } h_2, \text{rk } h_1).$$

From theorem (135) follows the interesting <u>inequality of Sylvester</u> showing the <u>sublogarithmic property of the defect</u>:

139. $$\text{def } h_2h_1 \leqq \text{def } h_2 + \text{def } h_1.$$

Here the equality sign holds iff $\text{Ker } h_2 \subset \text{Im } h_1$.

This happens in particular if h_1 is an epimorphism or h_2 a monomorphism.

In connection with Sylvester's inequality let us note that the index has the <u>logarithmic property</u>:

140. $$\text{ind } h_2h_1 = \text{ind } h_2 + \text{ind } h_1.$$

The product of Fredholm homomorphisms is therefore a Fredholm homomorphism.

141. $$\text{def } h_2h_1 \geqq \text{def } h_1.$$

Here equality holds iff the system of subspaces $\{\text{Im } h_1, \text{Ker } h_2\}$ is independent (including the case that h_2 is a monomorphism).

By virtue of (136) we find:

142. If h_1 is a Fredholm homomorphism, then $\text{def } h_2h_1 \geqq \text{def } h_2$; here equality holds iff the system of subspaces $\{\text{Im } h_1, \text{Ker } h_2\}$ is complete.

Let us now clarify when the product $h = h_2h_1$ of two homomorphisms is a monomorphism or an epimorphism.

143. $h = h_2h_1$ is a monomorphism iff h_1 is a monomorphism and the system of subspaces $\{\text{Im } h_1, \text{Ker } h_2\}$ is independent.

As a particular case, the product of two monomorphisms is a monomorphism.

144. h_2h_1 is an epimorphism iff h_2 is an epimorphism and the system of subspaces $\{\text{Im } h_1, \text{Ker } h_2\}$ is complete.

In particular, the product of two epimorphisms is an epimorphism.

145. h_2h_1 is an isomorphism iff h_1 is a monomorphism, h_2 an epimorphism, and the system of subspaces $\{\text{Im } h_1, \text{Ker } h_2\}$ is basic.

In particular, the product of two isomorphisms is an isomorphism. In this case

146. $$(h_2h_1)^{-1} = h_1^{-1}h_2^{-1}.$$

Definition 58. For the following we shall need the idea of orthogonal homomorphisms. A homomorphism h_2 is said to be *orthogonal* to a homomorphism h_1 *from the left* if

$$h_2h_1 = 0. \qquad (*)$$

Right orthogonality is defined correspondingly.

Let us note that equation (*) is equivalent to the inclusion $\text{Im } h_1 \subset \text{Ker } h_2$. In modern algebra and topology an important role is played by such pairs of homomorphisms h_1, h_2 for which $\text{Im } h_1 = \text{Ker } h_2$. They are called exact pairs. A sequence of homomorphism in which any two neighboring homomorphisms form an exact pair is called an exact sequence.

147. Let $h_1 \in \text{Hom}(E,E_1)$. The set of homomorphisms $h_2 \in \text{Hom}(E_1,E_2)$, which are left orthogonal to the homomorphism h_1, is a subspace of $\text{Hom}(E_1,E_2)$ whose dimension is $(n_1 - \text{rk } h_1)n_2$ ($n_1 = \dim E_1$, $n_2 = \dim E_2$).

Consequently:

148. In order that there be no homomorphism other than zero which is left orthogonal to h_1, it is necessary and sufficient that h_1 be an epimorphism (i.e., right invertible).

149. Let $h_2 \in \text{Hom}(E_1,E_2)$. The set of homomorphisms $h_1 \in \text{Hom}(E,E_1)$, which are right orthogonal to the homomorphism h_2, is a subspace of $\text{Hom}(E,E_1)$ whose dimension is $n \cdot \text{def } h_2$.

Therefore:

150. In order that there be no homomorphisms other than zero which is right orthogonal to h_2, it is necessary and

sufficient that h_2 be a monomorphism (i.e., left invertible).

Definition 59. Let us now investigate the divisibility of homomorphisms. A homomorphism h_1 is called a *right divisor* of the homomorphism h if there is a homomorphism g (the *quotient* resulting from *dividing* h by h_1 *from the right*) such that

$$h = gh_1.$$

One also says that h is divisible by h_1 from the right.-- Correspondingly, one defines *left divisor* and *quotient* resulting from *dividing from the left.* These quotients are in general not unique. The special situation where h is the identity endomorphism was considered in (119)-(128).

We shall examine the right division.

151. Let the homomorphism $h \in \mathrm{Hom}(E,E_2)$ be divisible by the homomorphism $h_1 \in \mathrm{Hom}(E,E_1)$ from the right. Then the set of all quotients [h divided by h_1 from the right] is an equivalence class modulo $\mathbb{L}$ in $\mathrm{Hom}(E_1,E_2)$, where $\mathbb{L}$ is the subspace of homomorphisms which are left orthogonal to h_1.

Consequently:

152. Suppose the quotient [h divided by h_1 from the right] exists. It is unique iff h_1 is an epimorphism.

153. Let $h_1 \in \mathrm{Hom}(E,E_1)$ be a monomorphism, and let $h_1^{(-1)}$ be any of its left inverses. For any $h \in \mathrm{Hom}(E,E_2)$ the homomorphism $g = hh_1^{(-1)}$ is a quotient [h divided by h_1 from the right].

Thus a monomorphism is a right divisor of an arbitrary homomorphism. In the general case we find:

154. If h is divisible by h_1 from the right, then

$$\mathrm{Ker}\, h \supset \mathrm{Ker}\, h_1.$$

Let us establish that this necessary condition for right divisibility is also sufficient. We shall try to reduce this task to a problem of divisibility by a monomorphism, using contraction modulo Ker h_1 as in (93). With this aim in mind, we first consider contractions from the point of view of operations on homomorphisms.

155. Let L and M be two subspaces of a space E and $L \subset M$. Then

$$h_M = h_{M/L} \cdot h_L \quad \text{(notation in D45).}$$

Thus the contraction h_L is a right divisor of the contraction h_M, and the corresponding quotient (unique since h_L is an epimorphism) equals the relative contraction $h_{M/L}$.

This result can be reversed in the following sense:

156. If L and M are two subspaces of a space E and the contraction h_L is a right divisor of the contraction h_M, then $M \supset L$.

We can now give a new interpretation to theorem (93):

157. For any homomorphism h (acting on E) there exists a unique monomorphism $\hat{h}$ such that $h = \hat{h}f_h$, where f_h denotes the contraction modulo Ker h.

Now we are well prepared to obtain the basic result:

158. If $h \in \text{Hom}(E,E_2)$, $h_1 \in \text{Hom}(E,E_1)$, and

$$\text{Ker } h \supset \text{Ker } h_1,$$

then h is divisible by h_1 from the right. Any homomorphism g for which

$$g|\text{Im } h_1 = \hat{h}f_{h/h_1} \cdot \hat{h}_1^{(-1)} + r$$

may be considered as the quotient. Here f_{h/h_1} is the contraction of E modulo (Ker h/Ker h_1), $r \in \text{Hom}(\text{Im } h_1, E_2)$ is any homomorphism which is left orthogonal to h_1, and $.|.$ was introduced in D56. This result is an exhaustive description of all quotients [h divided by h_1 from the right].

The theory of left division is dual to the theory of right division.

159. Let the homomorphism $h \in \text{Hom}(E,E_2)$ be divisible from the left by the homomorphism $h_1 \in \text{Hom}(E_1,E_2)$. Then the set of all

quotients [h divided by h_1 from the left] is an equivalence class modulo $\mathbb{M}$ in Hom(E,E_1), where $\mathbb{M}$ is the subspace of homomorphisms which are right orthogonal to h_1.

Consequently:

160. Suppose the quotient [h divided by h_1 from the left] exists. Then it is unique iff h_1 is a monomorphism.

161. Let $h_1 \in \text{Hom}(E_1,E_2)$ be an epimorphism, and $h_1^{(-1)}$ any of its left inversions. For any $h \in \text{Hom}(E,E_2)$ the homomorphism $g = h_1^{(-1)}h$ is a quotient [h divided by h_1 from the left].

Thus an epimorphism h_1 is a left divisor of any homomorphism h. In the general case we have:

162. If h is left divisible by h_1, then $\text{Im}\, h \subset \text{Im}\, h_1$.

Conversely:

163. If $h \in \text{Hom}(E,E_2)$, $h_1 \in \text{Hom}(E_1,E_2)$, and $\text{Im}\, h \subset \text{Im}\, h_1$, then h is left divisible by h_1. One may take as quotient any homomorphism of the form

$$g = f_{h_1}^{(-1)}\hat{h}_1^{(-1)}h + r,$$

where $f_{h_1}^{(-1)}$ is any left inverse to the contraction f_{h_1}, and $r \in \text{Hom}(E,E_1)$ is any homomorphism which is right orthogonal to h_1. This result is an exhaustive description of all quotients [h divided by h_1 from the left].

6. Linear Functionals. Orthogonality. Biorthogonal Systems.

Definition 60. The *functional* f(x) defined on the space E is *linear*, if

$$(1) \quad f(x_1 + x_2) = f(x_1) + f(x_2),$$

$$(2) \quad f(\alpha x) = \alpha f(x)$$

This definition makes f a homomorphism of E into the arithmetical space C^1.

164. The linear functionals on E form a linear space with regard to the naturally defined operations of addition and multiplication by scalars.

Definition 61. This space of functionals is denoted by E' and is called the *conjugate* (or *dual*) of the space E.

165. Let E^1 be a one-dimensional space and e be any of its basis vectors (i.e., $e \in E^1$, $e \neq 0$). The formula

$$h_f x = f(x)e$$

defines a homomorphism from E to E^1. The mapping j: $E' \to \operatorname{Hom}(E,E^1)$, defined by the formula

$$jf = h_f \quad (f \in E'),$$

is an isomorphism.

166. dim E' = dim E.

Consequently,

167. $E' \simeq E$.

More generally, $E \simeq E' \simeq E'' \simeq \ldots$ where $E'' = (E')'$ etc. The isomorphism of the spaces E' and E can also be made evident by direct construction of bases:

168. Let $\Delta = \{e_k\}_1^n$ be some basis of E. The coordinates $\xi_k(x)$ $(k = 1,2,\ldots,n)$ of any vector $x \in E$ relative to the basis Δ may be considered as linear functionals $\in E'$. We denote them by f_k and have $f_k(x) = \xi_k$. In particular, $f_k(e_i) = \delta_{ik}$. $\{f_k\}_1^n$ is a basis in E'.

Definition 62. The basis $\{f_k\}_1^n$ is called the *conjugate* basis to Δ and is denoted by Δ'.

If $x = \Sigma_{k=1}^n \xi_k e_k$, then for any linear functional f

$$f(x) = \Sigma_{k=1}^n \alpha_k \xi_k,$$

where $\alpha_k = f(e_k)$ $(k = 1,2,\ldots,n)$. The right-hand side of this equation defines a linear functional in C^n, which is called a linear form.

The isomorphism of the spaces E and E" (in contrast to the isomorphism of E and E') can be given in a certain natural form without using an "incidentally" chosen basis.

169. Let $x \in E$. The formula

$$\phi_x(f) = f(x) \quad (f \in E'),$$

where f is variable and x is at first considered fixed, defines a linear functional ϕ_x on E' (i.e., $\phi_x \in E''$). With x now varying over E, the mapping ϕ: $E \to E''$, defined by the formula

$$\phi x = \phi_x \quad (x \in E),$$

is an isomorphism.

Definition 63. This isomorphism is the so-called *canonical isomorphism* of the spaces E and E".

What we have discovered is that the ideas of vectors and of linear functionals are dual: each linear functional on E is a vector in E' (167); each vector in E can be considered as a linear functional on E' (169).

(To avoid misunderstanding, let us emphasize that the term "vector" will denote, as before, an element of the "fundamental" vector space.)

Definition 64. Let L be a subspace of the space E and $g \in L'$, that is, g is a linear functional defined on L. In agreement with the general definition of extension of a homomorphism, the functional $\tilde{g} \in E'$ will be called an *extension* of the functional g, if

$$\tilde{g}(x) = g(x) \quad (x \in L).$$

The problem of extending linear functionals is one of the most important tasks in linear analysis. At present we shall consider its simplest aspect.

The following two theorems can be reduced to theorem (117) and (118):

170. Each functional $g \in L'$ has an extension $\tilde{g} \in E'$.

And stronger:

171. Let the vectors $x_1, x_2, \ldots, x_m$ be linearly independent modulo L and $\gamma_1, \gamma_2, \ldots, \gamma_m$ be arbitrary numbers. There exists

an extension $\tilde{g} \in E'$ of any functional $g \in L'$ satisfying the conditions

$$\tilde{g}(x_k) = \gamma_k \quad (k = 1,2,\ldots,m).$$

The extension is unique iff codim L = m.

In particular:

172. If the vector x_o does not belong to the subspace L, then there exists a functional $f_o \in E'$ such that

$$f_o(x) = 0 \text{ for all } x \in L, \quad f_o(x_o) \neq 0.$$

"The functional f_o separates the vector x_o from the subspace L" describes this situation.

Theorem (172) implies the following criterion for a system of vectors to be complete:

173. A system of vectors is complete iff the unique linear functional which vanishes on all vectors of the system is the zero functional.

Let us formulate the propositions dual to (172) and (173).

174. If M is a subspace in E', and the linear functional f_o does not belong to M, then there exists a vector $x_o \in E$ such that

$$f(x_o) = 0 \text{ for all } f \in M, \quad f_o(x_o) \neq 0.$$

175. A system of linear functionals is complete (in the space E') iff the unique vector on which all functionals of the system vanish is the zero vector.--Note here the related

Definition 65. A system of linear functionals all of which vanish only for $x = 0$ is called *total*.

Theorems (174) and (175) are obtained from theorems (172) and (173) by the canonical isomorphism.

Definition 66. A vector x and a linear functional f are (mutually) *orthogonal* if $f(x) = 0$ (i.e., $x \in \text{Ker } f$). This is denoted by $f \perp x$ (or $x \perp f$). The relation of orthogonality $f \perp x$ is bilinear:

176. If $f_1 \perp x$ and $f_2 \perp x$, then $(f_1 + f_2) \perp x$; if $f \perp x$, then

$\alpha f \perp x$ for any scalar α. Dually, if $f \perp x_1$ and $f \perp x_2$, then $f \perp (x_1 + x_2)$; if $f \perp x$, then $f \perp \alpha x$ for any scalar α.

Let L be a subspace in E and M a subspace in E'. They are (mutually) orthogonal if

$$f \perp x \quad (f \in M,\ x \in L).$$

This will be denoted by $M \perp L$.

177. The set of all functionals $f \in E'$ which are orthogonal to all vectors of some subspace $L \subset E$ is a subspace.

Definition 67. This subspace is called the *orthogonal complement*[§] of the subspace L and is denoted by $L^{\perp}$. In a dual manner we define the subspace $M^{\perp} \subset E$ as the orthogonal complement of the subspace $M \subset E'$. The orthogonal complement of any subspace N (in E or in E') is the largest subspace (in E' or in E, respectively) orthogonal to the subspace N.

Orthogonal complements and factor spaces are closely connected.

178. Let L be a subspace of the space E. $x \equiv y$ (modulo L) iff $f(x) = f(y)$ $(f \in L^{\perp})$.

Thus every functional $f \in L^{\perp}$ is constant on every class modulo L. Conversely:

179. If the functional $f \in E'$ is constant on every class modulo L, then $f \in L^{\perp}$.

Now we can establish the natural isomorphism between the spaces $(E/L)'$ and $L^{\perp}$:

180. Let $g \in (E/L)'$. The formula

$$\hat{g}(x) = g([x]_L) \quad (x \in E)$$

defines a functional $\hat{g} \in L^{\perp}$. The mapping $(E/L)' \to L^{\perp}$ so defined is an isomorphism.

Hence follows the complementation formula

[§]The sense of the word "complement" is here not the same as in D31: the orthogonal complement lies in E', but not in E; other than in "complementation," it is now unique, etc.

181. $\dim L^{\perp} + \dim L = n$.

Starting from this formula we shall investigate a mapping of the set of subspaces of E into the set of subspaces of E', which assigns to each $L \subset E$ its orthogonal complement $L^{\perp}$. The notation for this mapping will be $(\perp)$. We shall see that the duality in the geometry of subspaces, already well known to us, then follows automatically.

182. The $(\perp)$ mapping is an injection. More than that, it is an <u>involution</u>; i.e., $L^{\perp\perp} = L$.

183. The $(\perp)$ mapping is monotone decreasing: if $L_1 \subset L_2$, then $L_1^{\perp} \supset L_2^{\perp}$.

The self-dual formulae for the orthogonal complements of sum and intersection of subspaces are

184. $(L_1 + L_2)^{\perp} = L_1^{\perp} \cap L_2^{\perp}$, $(L_1 \cap L_2)^{\perp} = L_1^{\perp} + L_2^{\perp}$.

These formulae hold for sum and intersection of any set of subspaces.

185. A system of two subspaces $\{L_1, L_2\}$ is complete iff their orthogonal complements are mutually independent. The subspaces L_1 and L_2 are mutually independent iff the system of their orthogonal complements $\{L_1^{\perp}, L_2^{\perp}\}$ is complete.

As a consequence of this we have:

186. If $L_1 \oplus L_2 = E$, then $L_1^{\perp} \oplus L_2^{\perp} = E'$.

Theorem (186) cannot be extended to a system of more than two subspaces. This follows, for example, from the following inequality: If $\Sigma_{k=1}^{n} d_k = n$, then

$$\Sigma_{k=1}^{m} (n - d_k) = (m-1)n > n \quad (m > 2).$$

It is obvious, however, that if $\Sigma^{\oplus} L_{\nu} = E$, or merely $\cap L_{\nu} = 0$, then $\Sigma L_{\nu}^{\perp} = E'$, but this is not a direct sum.

In closing this section we shall consider so-called biorthogonal systems. These constructions are useful in many questions.

Definition 68. A system of vectors $\{x_k\}_1^m$ and a system of linear functionals $\{f_k\}_1^m$ are called (mutually) *biorthogonal* if

$$f_j(x_k) = \delta_{jk} \quad (j,k = 1,2,\dots,m),$$

i.e., $f_j \perp x_k$ $(j \neq k)$, $f_j(x_j) = 1$, $(j = 1,2,\dots,m)$.

187. Any basis Δ of E and its conjugate basis Δ' of E' are mutually biorthogonal.

Conversely, biorthogonal systems are always bases of their linear hulls, since

188. Each of two mutually biorthogonal systems is linearly independent.

Further:

189. For any linearly independent vector system $\Gamma \subset E$ there is a biorthogonal system $\Gamma' \subset E'$. The elements of the system Γ' are well-defined, disregarding arbitrary terms from $[\mathrm{sp}\ \Gamma]^{\perp}$.

Thus, if Γ is a basis, the system Γ' is uniquely defined and coincides with the conjugate basis.

190. If Γ is a linearly independent system, then the systems Γ and Γ' can be supplemented to form conjugate bases.

We shall now investigate the mutual relation of the linear hulls sp Γ and sp Γ'.

191. The subspace $[\mathrm{sp}\ \Gamma']^{\perp}$ is the complement of the subspace sp Γ.

Conversely:

192. If Γ is a linearly independent system and M is a subspace of E' such that $M^{\perp}$ is a complement to the subspace sp Γ, then there exists a biorthogonal system Γ' such that sp $\Gamma' = M$.

7. The Conjugate Homomorphism and Fredholm's Theory.

Let h be a homomorphism from E into E_1. We shall introduce the dual object h', the conjugate homomorphism. In doing so we shall significantly extend the apparatus of duality.

193. Let $f_1 \in E_1'$. The formula

$$f(x) = f_1(hx) \quad (x \in E)$$

defines a functional $f \in E'$. The mapping $h': E_1' \to E'$ defined by $h'f_1 = f$ is a homomorphism.

Definition 69. The homomorphism h' is called the *conjugate* of the homomorphism h. For example, $I_E' = I_{E'}$, $O' = O$.

194. With accuracy[§] up to the canonical isomorphisms of spaces E, E'' and E_1, E_1'', we have $h'' = h$.

Thus the conjugation mapping $\mathrm{Hom}(E,E_1) \to \mathrm{Hom}(E_1',E')$, which we denote by (') and define by $(')h = h'$, is an involution up to the canonical isomorphism.

195. The mapping (') is an isomorphism.

196. $(h_1h_2)' = h_2'h_1'$.

The following fundamental orthogonality relations hold for kernels and images.

197. $\mathrm{Ker}\ h' = (\mathrm{Im}\ h)^{\perp}$; $\mathrm{Ker}\ h = (\mathrm{Im}\ h')^{\perp}$, which follows by (194); and $(\mathrm{Ker}\ h')^{\perp} = \mathrm{Im}\ h$, $(\mathrm{Ker}\ h)^{\perp} = \mathrm{Im}\ h'$, which follows by (182).

This establishes at the same time $\mathrm{rk}\ h' = \mathrm{rk}\ h$.

In general, however, $\mathrm{def}\ h' \neq \mathrm{def}\ h$.

198. $\mathrm{def}\ h - \mathrm{def}\ h' = \mathrm{ind}\ h$.

Hence:

199. If h is a Fredholm homomorphism, then the homogeneous equations

$$hx = 0, \quad h'f_1 = 0$$

have the same maximal number of linearly independent solutions (second theorem of Fredholm).

[§]Since $h'' \in \mathrm{Hom}(E'',E_1'')$, and $h \in \mathrm{Hom}(E,E_1)$, the equation $h'' = h$, strictly speaking, makes no sense. But if we identify E'' with E and E_1'' with E_1 by means of the canonical isomorphism, then h'' coincides with h. In the following, the words "up to some isomorphism" will indicate an analogous identification, and we shall use the abbreviation [c.i.] for "up to a canonical isomorphism."

In the general case the maximal numbers of linearly independent solutions do not coincide, and their difference equals the index of the homomorphism.

The second theorem of Fredholm implies: if the Fredholm homomorphism h is a monomorphism, then h' is also a monomorphism. It is also obvious that h' is a Fredholm homomorphism together with h.

Let us note now that the relations of orthogonality (197) imply:

200. In order that a homomorphism h be an epimorphism (monomorphism) it is necessary and sufficient that the conjugate homomorphism h' be a monomorphism (epimorphism).

If, in particular, h is an isomorphism, then h' is also an isomorphism and

201. $(h')^{-1} = (h^{-1})'$.

By rephrasing theorem (200) we are able to say: for the global solvability of the inhomogeneous equation

$$hx = y$$

it is necessary and sufficient that the conjugate homogeneous equation

$$h'f_1 = 0$$

have only the trivial solution. This and the second theorem of Fredholm give a new confirmation of the first theorem of Fredholm (cf. (97)).

Theorem (197) reads in the language of equations:

202. For the solvability of the inhomogeneous equation

$$hx = y$$

it is necessary and sufficient that the vector y be orthogonal to all solutions of the conjugate homogeneous equation

$$h'f_1 = 0$$

(<u>third theorem of Fredholm</u>).

Theorem (197) can be generalized:

203. If $h \in \mathrm{Hom}(E,E_1)$, $L \subset E_1'$, and $M \subset E'$, then $h'L = (h^{-1}L^{\perp})^{\perp}$, $(h')^{-1}M = (hM^{\perp})^{\perp}$, corresponding to the last and first identity in (197), respectively.

Theorem (197) can be reobtained from these mutually dual formulae by substituting $L = E_1'$, $M = O$. Inversely, (203) can be derived from (197) by subjecting the homomorphism to a restriction. Let us use this as an occasion to study restricted homomorphisms within the scheme of duality.

We begin with a task that is interesting in itself: describing the subspace L' conjugate to the subspace $L \in E$.

204. Let i_L be the embedding of the subspace L into the space E and i_L' the homomorphism conjugate to i_L. Then

$$\mathrm{Im}\, i_L' = L', \quad \mathrm{Ker}\, i_L' = L^{\perp}.$$

Consequently (cf. (180)):

205. $$L' \simeq E'/L^{\perp}.$$

Of course, the natural isomorphism of the spaces $E'/L^{\perp}$ and L' is easy to perceive directly: All functionals $\in E'$ taken modulo $L^{\perp}$ coincide upon L (i.e., their restrictions to L are equal), and conversely. By virtue of theorem (170) each functional $\in L_1'$ can be considered as the restriction to L of some functional $\in E'$ (cf. (178)-(180)).

Now we can describe the homomorphism which is conjugate to the restriction of a given homomorphism. Recalling the notation h_M for the contraction modulo M, we claim

206. If $g \in \mathrm{Hom}(E,E_1)$ and L is a subspace of E, then

$$(g|L)' = h_{L^{\perp}} g'$$

up to the natural isomorphism of the spaces $E'/L^{\perp}$ and L'.

This formula is obtained from the representation $g|L = g i_L$.

We still want to take a quick look at some properties of the index.

207. If h is a monomorphism, then ind h = - def h', and conversely.

Thus, if h is a monomorphism, ind $h \leq 0$.

208. If h is an epimorphism, then ind h = def h, and conversely.

Thus, if h is an epimorphism, ind $h \geq 0$.

209. For any homomorphism h

$$\text{ind } h' = -\text{ind } h.$$

In closing this section we present the theorems of Fredholm translated from the language of homomorphisms to the language of functionals. First a preparatory remark:

210. If $\{f_k\}_1^m$ is a system of linear functionals defined on E, then the formulae

$$\alpha_k = f_k(x) \quad (k = 1,2,\ldots,m)$$

define a homomorphism from E to C^m.

211. The inhomogeneous system of equations

$$f_k(x) = \alpha_k \quad (k = 1,2,\ldots,n)$$

is solvable for x for any right-hand side, iff the corresponding homogeneous system

$$f_k(x) = 0 \quad (k = 1,2,\ldots,n)$$

has only the trivial solution $x = 0$.

212. The system of homogeneous equations

$$f_k(x) = 0 \quad \text{for } x \in E \ (k = 1,2,\ldots,n) \text{ and}$$

$$\Sigma_{k=1}^n \gamma_k f_k = 0 \quad \text{for } \{\gamma_k\}_1^n \in C^n$$

have the same maximal number of linearly independent solutions.

213. The inhomogeneous system of equations

$$f_k(x) = \alpha_k \quad (k = 1,2,\ldots,n)$$

is solvable iff for any numbers γ_k, satisfying the condition $\Sigma_{k=1}^n \gamma_k f_k = 0$, the relation $\Sigma_{k=1}^n \gamma_k \alpha_k = 0$ holds.

8. Bilinear Functionals and Tensor Products.

Definition 70. Let E_1, E_2 be two spaces and $E_1 \times E_2$ their *Cartesian product.* In some situations it is convenient to have in $E_1 \times E_2$ the operations of addition and multiplication by numbers. They are introduced naturally by

$$\{x,y\} + \{x_1,y_1\} = \{x + x_1, y + y_1\}, \quad \alpha\{x,y\} = \{\alpha x, \alpha y\}.$$

Thus the *Cartesian product* becomes a *linear space.*

214. $$\dim(E_1 \times E_2) = \dim E_1 + \dim E_2.$$

Definition 71. The functional $B(x,y)$ on $E_1 \times E_2$ is called *bilinear* if for each fixed $x \in E_1$ the functional

$$B_x(y) = B(x,y) \qquad (*)$$

on the space E_2 is linear (i.e., $B_x \in E_2'$), and (dually) if for each fixed y the functional

$$B_y'(x) = B(x,y) \qquad (**)$$

on the space E_1 is also linear (i.e., $B_y' \in E_1'$). Thus

(1) $B(x,y_1 + y_2) = B(x,y_1) + B(x,y_2)$,

(2) $B(x,\alpha y) = \alpha B(x,y)$,

(3) $B(x_1 + x_2,y) = B(x_1,y) + B(x_2,y)$,

(4) $B(\alpha x,y) = \alpha B(x,y)$.

In this section $B = B(x,y)$ denotes a bilinear functional on $E_1 \times E_2$.

There exists an intrinsic connection between bilinear functionals and homomorphisms, based on the definition of the bilinear functional. According to (*) each vector $x \in E_1$ corresponds to a linear functional $B_x \in E_2'$, i.e., to a certain mapping from E_1 into E_2'. We shall denote this mapping by h_B^{ℓ}. Analogously, (**) defines a mapping h_B^r

from E_2 into E_1' which assigns to each vector $y \in E_2$ the linear functional $B_y' \in E_1'$.

215. The functional

$$\Phi(f,x) = f(x) \quad (f \in E', x \in E)$$

is bilinear and $h^\ell_\Phi = I_{E'}$, $h^r_\Phi = I_E$ [c.i.].

216. The mappings h^ℓ_B and h^r_B are homomorphisms, and $h^r_B = (h^\ell_B)'$, [c.i.].

Definition 72. The homomorphisms h^ℓ_B and h^r_B will be called the *left* and *right generators* of the functional B. The spaces

$$S^\ell_B = \text{Im } h^\ell_B, \quad S^r_B = \text{Im } h^r_B$$

are called the *left* and *right supports* of the functional B.

217. $\dim S^\ell_B = \dim S^r_B$.

The common numerical value of these two dimensions is called the rank of the functional B and is denoted by rk B.

Definition 73. The spaces

$$K^\ell_B = \text{Ker } h^\ell_B, \quad K^r_B = \text{Ker } h^r_B$$

will be called the *left* and *right kernels* of the functional B, respectively. In general their dimensions are not equal. We shall call these dimensions the *left* and *right defects* of the functional B and denote them by def_ℓ B, def_r B, respectively. The difference

$$\text{ind } B = \text{def}_r B - \text{def}_\ell B$$

is called the *index* of the functional B.

Definition 74. If the index of a bilinear functional equals zero, then it is called a *Fredholm* functional. For a Fredholm functional the common value of the left and right defects is called the *defect* of the functional (notation: def B).

218. $\text{ind } B = \dim E_2 - \dim E_1$.

219. For a Fredholm functional B the complementation formula (95) holds:

$$\mathrm{rk}\ B + \mathrm{def}\ B = n = \dim E_1 = \dim E_2.$$

Clearly, we have here a certain variant of the Fredholm theory. The key to it is found in the duality relations (cf. (93), (197)):

220. $S_B^{\ell} \simeq E_1/K_B^{\ell}$; $S_B^r \simeq E_2/K_B^r$.

221. $S_B^{\ell} = (K_B^r)^{\perp}$; $S_B^r = (K_B^{\ell})^{\perp}$.

We stress here the relations

$$K_B^{\ell} \subset E_1,\quad K_B^r \subset E_2;\qquad S_B^{\ell} \subset E_2',\quad S_B^r \subset E_1',$$

which may serve as a check.

The structure of the set of all bilinear functionals on $E_1 \times E_2$ will now be investigated.

222. The bilinear functionals on $E_1 \times E_2$ form a linear space with the natural definitions of addition and multiplication by numbers.

We shall denote this space by $\mathbb{B}(E_1,E_2)$.

223. The mappings

$$h^{\ell}:\ \mathbb{B}(E_1,E_2) \to \mathrm{Hom}(E_1,E_2'),\quad h^r:\ \mathbb{B}(E_1,E_2) \to \mathrm{Hom}(E_2,E_1')$$

defined by the formulae $h^{\ell}B = h_B^{\ell}$, $h^r B = h_B^r$ are isomorphisms.

224. $\dim \mathbb{B}(E_1,E_2) = \dim E_1 \cdot \dim E_2$.

Bilinear functionals admit a matrix representation similar to homomorphisms.

Definition 75. Let $\Delta_1 = \{e_j\}_1^{n_1}$, $\Delta_2 = \{u_k\}_1^{n_2}$ be bases of the spaces E_1, E_2, respectively. The matrix

$$(B(e_j,u_k))$$

is called the matrix of the bilinear functional relative to the bases Δ_1 and Δ_2.

225. If (β_{jk}) is the matrix of the bilinear functional $B(x,y)$

relative to the pair of bases Δ_1, Δ_2 and $x = \Sigma_{j=1}^{n_1} \xi_j e_j$, $y = \Sigma_{k=1}^{n_2} \eta_k u_k$, then

$$B(x,y) = \Sigma_{j,k} \beta_{jk} \xi_j \eta_k$$

This establishes a bijection between the bilinear functionals on $E_1 \times E_2$ and the $(n_1 \times n_2)$ matrices.

Let us note still another way to describe the space $\mathbb{B}(E_1, E_2)$. It leads us to the fundamental idea of the tensor product of spaces.

226. Let f and g be (fixed) functionals $\in E_1', E_2'$, respectively. Then

$$B(x,y) = f(x)g(y) \quad (x \in E_1,\ y \in E_2)$$

is a bilinear functional.

Definition 76. This special kind of a bilinear functional, defined on $E_1 \times E_2$, is called the *tensor product of* the linear *functionals* f,g and is denoted by $f \otimes g$.

227. Let $\{f_j\}_1^{n_1}$ be a basis of E_1' and $\{g_k\}_1^{n_2}$ a basis of E_2'. Then the tensor products

$$f_j \otimes g_k \quad (j = 1,2,\dots,n_1;\ k = 1,2,\dots,n_2)$$

form a basis of the space $\mathbb{B}(E_1, E_2)$.

Definition 77. However, the expression

$$f(x)g(y) \quad (F \in E_1',\ g \in E_2';\ x \in E_1,\ y \in E_2)$$

can also be considered as a linear functional on $E_1' \times E_2'$ for fixed x and y (cf. (169)). By virtue of (226) and the canonical isomorphism of a space with its second conjugate this functional is bilinear. It is called the *tensor product of* the *vectors* x,y and is denoted by $x \otimes y$. By virtue of (227) the bases $\{e_j\}_1^{n_1}$ and $\{u_k\}_1^{n_2}$ generate the basis

$$e_j \otimes u_k \quad (j = 1,2,\dots,n_1;\ k = 1,2,\dots,n_2)$$

of the space $\mathbb{B}(E_1', E_2')$.

The space $\mathbb{B}(E_1', E_2')$ is called the *tensor product of* the *spaces* E_1 and E_2 and denoted by $E_1 \otimes E_2$. Obviously

228. $\dim (E_1 \otimes E_2) = \dim E_1 \cdot \dim E_2$.

We also note that $\mathbb{B}(E_1, E_2) = E_1' \otimes E_2'$ [c.i.].

229. Tensor multiplication has the linear properties

$$(1) \quad (x_1 + x_2) \otimes y = x_1 \otimes y + x_2 \otimes y,$$

$$(2) \quad x \otimes (y_1 + y_2) = x \otimes y_1 + x \otimes y_2,$$

$$(3) \quad \alpha x \otimes y = \alpha (x \otimes y),$$

$$(4) \quad x \otimes (\alpha y) = \alpha (x \otimes y).$$

From these properties and theorem (227) we easily prove that for any bilinear functional B on $E_1 \times E_2$ there exist two systems of linear functionals

$$\{f_k\}^{\nu},\ \{g_k\}^{\nu} \qquad (\nu = \min (n_1, n_2),$$

such that

$$B = \Sigma_{k=1}^{\nu} f_k \otimes g_k.$$

A more precise result is

230. Let $\rho = \mathrm{rk}\, B > 0$, $\{f_k\}_1^{\rho}$ be any basis for the subspace S_B^r, the right support of the functional B. Then there is a unique basis $\{g_k\}_1^{\rho} \in S_B^{\ell}$, the left support of B, such that

$$B = \Sigma_{k=1}^{\rho} f_k \otimes g_k.$$

One can of course operate inversely, i.e., give the basis of the left support and construct the corresponding basis of the right support.

Theorem (230) establishes the canonical form of an arbitrary bilinear functional. The task of reducing some object to canonical form occupies a central place in linear analysis. The words "reduce" and "canonical form" are used in a sense which depends upon the object and the situation in which it occurs. The study of an object in canonical form is significantly simpler than in general form. For example, from theorem (230) easily follows all the Fredholm theory for bilinear functionals (this theorem can be reformulated so that the result of (217) is not assumed known beforehand).

With the aid of (230) also the following fact is shown: Let $f \in E_1'$, $g \in E_2'$, and $F \in [\mathbb{B}(E_1,E_1)]'$. The last expression is the conjugate space to $\mathbb{B}(E_1,E_2)$, or the space of all linear functionals F that map $\mathbb{B}(E_1,E_2)$ into C^1. Thus F operates on the set of all $f \otimes g$ (cf. (226)) and may be written as $F(f \otimes g)$. Considered as a function of f and g, we denote it temporarily by $B_F(f,g)$.

231. $B_F(f,g)$ is a bilinear functional on $E_1' \times E_2'$. To every such element of $[\mathbb{B}(E_1,E_2)]'$ corresponds therefore an element of $E_1 \otimes E_2$. The mapping $[\mathbb{B}(E_1,E_2)]' \to E_1 \otimes E_2$ thus determined is an isomorphism.

This isomorphism is referred to as the "natural isomorphism" here and in (235) and (236). It can again be used to define an equality [n.i.] by identification of isomorphic structures.

This is a new and important interpretation of the tensor product of spaces: $E_1 \otimes E_2$ can be considered as the conjugate to the space $\mathbb{B}(E_1,E_2)$; one therefore can write

$$(E_1 \otimes E_2)' = E_1' \otimes E_2', \quad [\text{n.i.}].$$

Definition 78. We shall now introduce the *tensor* or *Kronecker product of homomorphisms*. Suppose we have two pairs of spaces $\{E_1,E_3\}$, $\{E_2,E_4\}$, and let $h_1 \in \mathrm{Hom}(E_1,E_3)$, $h_2 \in \mathrm{Hom}(E_2,E_4)$. The tensor product $h_1 \otimes h_2$ is a homomorphism from $E_1 \otimes E_2$ to $E_3 \otimes E_4$ which is defined in the following fashion. Let $B \in E_1 \otimes E_2$, i.e., B is a bilinear functional on $E_1' \times E_2'$; then

$$[(h_1 \otimes h_2)B](f,g) \underset{\text{def}}{=} B(h_1'f, h_2'g) \qquad (f \in E_3',\ g \in E_4').$$

Clearly, tensor multiplication of homomorphisms is linear in each factor.

We shall describe the kernel and the image of the tensor

product of homomorphisms and draw some conclusions from these relations. First a preparatory remark: $M_1 \otimes M_2$ can be considered as a subspace of $E_1 \otimes E_2$, if M_1 is a subspace of E_1, and M_2 is a subspace of E_2. Now

232. $\mathrm{Im}(h_1 \otimes h_2) = \mathrm{Im}\, h_1 \otimes \mathrm{Im}\, h_2$;

hence

233. $\mathrm{rk}(h_1 \otimes h_2) = \mathrm{rk}\, h_1 \cdot \mathrm{rk}\, h_2$.

Consequently:

234. The tensor product of epimorphisms is an epimorphism.

The formula for the kernel of the tensor product of homomorphisms looks more complicated:

235. $\mathrm{Ker}(h_1 \otimes h_2) = ((\mathrm{Ker}\, h_1)^{\perp} \otimes (\mathrm{Ker}\, h_2)^{\perp})^{\perp}$ [n.i.].

One obtains this formula easily from (232) if one makes use of the fact that

236. $(h_1 \otimes h_2)' = h_1' \otimes h_2'$ [n.i.].

237. $\mathrm{def}(h_1 \otimes h_2) = \mathrm{def}\, h_1 \cdot \mathrm{def}\, h_2 + \mathrm{rk}\, h_1 \cdot \mathrm{def}\, h_2 + \mathrm{rk}\, h_2 \cdot \mathrm{def}\, h_1$.

238. The tensor product of monomorphisms is a monomorphism.

239. The tensor product of isomorphisms is an isomorphism.

In addition, we note the formula for the index of a tensor product

240. $\mathrm{ind}(h_1 \otimes h_2) = n_2 \cdot \mathrm{ind}\, h_1 + n_3 \cdot \mathrm{ind}\, h_2$.

In particular:

241. The tensor product of Fredholm homomorphisms is a Fredholm homomorphism.

9. Complex Conjugation. Hermitian-linear Functionals. Hermitian Homomorphisms and Hermitian-bilinear Functionals.

Let f be a linear functional on the space E. Consider the functional $\bar{f}$ defined by

$$\bar{f}(x) = \overline{f(x)}$$

where the overbar on the right denotes the complex conjugate scalar. This functional is no longer linear;

although

(1) $\bar{f}(x_1 + x_2) = \bar{f}(x_1) + \bar{f}(x_2)$ is a linear relation,

(2) $\bar{f}(\alpha x) = \bar{\alpha}\bar{f}(x)$ is not.

Definition 79. A functional with these properties is called a *Hermitian-linear* functional.

242. The Hermitian-linear functionals on E form a linear space in regard to the natural definitions of addition and multiplication by numbers.

Definition 80. This space is called the *Hermitian-conjugate space* to E and denoted by E*.

Let us introduce the mapping j: E'→E* by the formula

$$jf = \bar{f} \quad (f \in E').$$

It is not a homomorphism since, again,

$$(1) \quad j(f_1 + f_2) = jf_1 + jf_2,$$

but

$$(2) \quad j(\alpha f) = \bar{\alpha} jf.$$

Definition 81. Any mapping with properties (1) and (2) is called a *Hermitian homomorphism*.

Definition 82. The mapping j: E'→E* is called the *canonical complex conjugation*.

For Hermitian homomorphisms the ideas of kernel and image can be introduced in the same way as for homomorphisms. By doing so, defect and rank are defined for Hermitian homomorphisms in a natural way. The classifications "epi-, mono-, iso-" are then maintained. The theory developed in Section 4 can be extended to Hermitian homomorphisms without any essential change.

243. The canonical complex conjugation is a Hermitian isomorphism.

We can say that the spaces E' and E* are Hermitian isomorphic. However, this assertion is in fact not different from the usual assertion of isomorphism (see (244)-(247)).

244. Let $\Delta = \{e_k\}_1^n$ be some basis of the space E. Then the

mapping j_Δ of E into itself defined by

$$x = \Sigma_{k=1}^{n} \xi_k e_k, \quad j_\Delta x = \Sigma_{k=1}^{n} \bar{\xi}_k e_k$$

is a Hermitian isomorphism.

245. The product of two Hermitian homomorphisms is a homomorphism.

246. The product of two Hermitian isomorphisms is an isomorphism.

247. If two subspaces are Hermitian isomorphic, then they are isomorphic.

Thus $E^* \simeq E$. Consequently, $E^{**} \simeq E$ and so forth.

Definition 83. For the spaces E and E** there exists even a *canonical isomorphism*:

$$\psi_x(g) = \overline{g(x)} \quad (x \in E,\ g \in E^*) \quad (\text{cf. } (169)).$$

The Hermitian isomorphism j permits the automatic transfer of the theory contained in Section 6 to the space E*. For example, if L is a subspace of E, then we may call the image $jL^\perp$ its orthogonal complement, etc.

However, it is also possible to construct an independent theory. For example:

248. The subspace $jL^\perp$ coincides with the set of Hermitian linear functionals g which are orthogonal to L in the sense that $g(x) = 0$ $(x \in L)$.

The role of the conjugate homomorphism h' $(h \in \text{Hom}(E_1, E_2))$ is now played by the Hermitian-conjugate homomorphism h*, defined in analogy to (193):

$$(h^* g_2)(x) = g(hx) \quad (x \in E_1,\ g_2 \in E_2^*).$$

Thus $h^* \in \text{Hom}(E_2^*, E_1^*)$. The analog to (195) is:

249. The mapping $(*)$: $\text{Hom}(E_1, E_2) \to \text{Hom}(E_2^*, E_1^*)$ defined by the formula $(*)h = h^*$ is a Hermitian isomorphism.

In particular, $(\alpha h)^* = \bar{\alpha} h^*$, while $(\alpha h)' = \alpha h'$.

250. The relation $h^* = j_1 h' j_2^{-1}$ holds, where $j_k\colon E_k' \to E_k^*$ $(k = 1, 2)$ are canonical complex conjugations.

Note here as a supplement to (245) that

251. The product of a Hermitian homomorphism and a homomorphism (in any order) is a Hermitian homomorphism.

We now introduce the notion of a Hermitian bilinear functional.

Definition 84. The functional $H(x,y)$ on $E_1 \times E_2$ is said to be *Hermitian-bilinear* if it is linear in x for each fixed y and Hermitian-linear in y for each fixed x.

For Hermitian-bilinear functionals the left and right generators $h_H^\ell\colon E_1 \to E_2{}^*$, $h_H^r\colon E_2 \to E_1{}'$ are defined similarly as for ordinary bilinear functionals.

252. The left generator h_H^ℓ is a homomorphism, the right generator h_H^r is a Hermitian homomorphism. Thus

$$h_H^r = j_1{}^{-1}(h_H^\ell)^* = (h_H^\ell)' j_2{}^{-1} \quad [\text{c.i.}].$$

Support, kernel, rank, defect, and index are defined for Hermitian exactly as for ordinary bilinear functionals.

The Fredholm theory holds in its previous form, but with the corresponding interpretation or orthogonal complements.

253. The Hermitian-bilinear functionals on $E_1 \times E_2$ form a linear space in regard to the natural definition of addition and multiplication by numbers.

254. Let $\{f_k\}_1^{n_1}$ be a basis of the space $E_1{}'$ and $\{g_j\}_1^{n_2}$ a basis of the space $E_2{}^*$. Then the tensor products

$$f_k \otimes g_j \quad (k = 1,2,\ldots,n_1;\ j = 1,2,\ldots,n_2),$$

defined by the same construction as in Section 8 form a basis of the space of Hermitian bilinear functionals on $E_1 \times E_2$ (cf. (226)).

By the same token this latter space has dimension $n_1 n_2$.

255. Let H be a Hermitian bilinear functional on $E_1 \times E_2$, $\rho = \text{rk}\ H > 0$, and $\{h_k\}_1^\rho$ be an arbitrary basis of the right support S_H^r. Then there is a unique basis $\{g_k\}_1^\rho$ in the left support S_H^ρ such that (similarly to (230))

$$H = \Sigma_{k=1}^{\rho} h_k \otimes g_k.$$

10. General Theory of Orthogonality.

Definition 85. Let E_1 and E_2 be two spaces, F a bilinear (or Hermitian bilinear) functional on $E_1 \times E_2$. The vectors $x \in E_1$, $y \in E_2$ are called *F-orthogonal* if

$$F(x,y) = 0,$$

and this is denoted by $x(F\perp)y$.
In the example

$$E_1 = E, \quad E_2 = E', \quad F(x,f) = f(x) \qquad (x \in E,\ f \in E'),$$

F-orthogonality $x(F\perp)f$ reduces to the ordinary orthogonality $x \perp f$ of a vector and a linear functional.
The relation of F-orthogonality is bilinear (cf. (176)).
Definition 86. The *subspaces* $L \subset E_1$, $M \subset E_2$ are called *F-orthogonal* if

$$x(F\perp)y \quad (x \in L,\ y \in M).$$

256. The set of those vectors $x \in E_1$ which are F-orthogonal to all vectors of an arbitrary subspace $M \subset E_2$ is a subspace.
Definition 87. The set described in (256) is the *left F-orthogonal complement* of M and is denoted by ${}^{F\perp}M$. The *right F-orthogonal complement* $L^{F\perp}$ of L is defined analogously. The F-orthogonal complement of a subspace N is the largest of the subspaces which are F-orthogonal to N. Below we shall write $(\perp)$ in place of $(F\perp)$.
257. If $M_1 \subset M_2$, then ${}^{(\perp)}M_1 \supset {}^{(\perp)}M_2$. If $L_1 \subset L_2$, then $L_1^{(\perp)} \supset L_2^{(\perp)}$.
The left and right kernels of the functional F can be written in terms of the F-orthogonal complements:
258. $K_F^{\ell} = {}^{(\perp)}E_2$, $K_F^{r} = E_1^{(\perp)}$.
But for arbitrary subspaces $L \subset E_1$, $M \subset E_2$ we find only:

259. $L^{(\perp)} \supset K_F^r$, ${}^{(\perp)}M \supset K_F^\ell$.

F-orthogonality reduces to the usual orthogonality by "inserting" the generators (D72):

260. For any subspace $M \subset E_2$, ${}^{(\perp)}M = (h_F^r M)^\perp$; for any subspace $L \subset E_1$, $L^{(\perp)} = (h_F^\ell L)^\perp$.

Therefore

261. $\dim\,({}^{(\perp)}M) = \operatorname{codim} M + \dim M \cap K_F^r - \operatorname{ind} F$.

262. $\dim\,(L^{(\perp)}) = \operatorname{codim} L + \dim L \cap K_F^\ell + \operatorname{ind} F$.

Definition 88. If F is a Fredholm (bilinear) functional (D74) and def F = 0, then F is called *regular*. For example, the functional $\Phi(f,x) = f(x)$ $(f \in E',\ x \in E)$ is regular.

263. For a regular functional F

$$\dim\,({}^{(\perp)}M) = \operatorname{codim} M, \quad \dim(L^{(\perp)}) = \operatorname{codim} L.$$

Let us now investigate to what extent the relation of F-orthogonality is an involution:

264. $({}^{(\perp)}M)^{(\perp)} \supset M$, ${}^{(\perp)}(L^{(\perp)}) \supset L$.

In addition, according to (259),

$$({}^{(\perp)}M)^{(\perp)} \supset K_F^r, \quad {}^{(\perp)}(L^{(\perp)}) \supset K_F^\ell.$$

It turns out that

265. $({}^{(\perp)}M)^{(\perp)} = M + K_F^r$, ${}^{(\perp)}(L^{(\perp)}) = L + K_F^\ell$;

so it is easy to establish the following proposition:

266. In order that $({}^{(\perp)}M)^{(\perp)} = M$, it is necessary and sufficient that $M \supset K_F^r$.

The criterion for the equality between ${}^{(\perp)}(L^{(\perp)})$ and L is formulated analogously.

267. If left and right defects of the functional F equal zero (in particular, if the functional is regular), then F-orthogonality is an involution:

$$({}^{(\perp)}M)^{(\perp)} = M, \quad {}^{(\perp)}(L^{(\perp)}) = L$$

for any subspaces $M \subset E_2$, $L \subset E_1$. This condition is not only sufficient, but also necessary.

268. ${}^{(\perp)}M_1 = {}^{(\perp)}M_2$, iff $M_1 + K_F^r = M_2 + K_F^r$.

The criterion for equality of $L_1^{(\perp)}$ and $L_2^{(\perp)}$ is formulated analogously.

Let us now consider the duality of sum and intersection from the viewpoint of F-orthogonality. Here the duality is incomplete. On the one hand

269. $^{(\perp)}(M_1 + M_2) = {}^{(\perp)}M_1 \cap {}^{(\perp)}M_2, \quad (L_1 + L_2)^{(\perp)} = L_1^{(\perp)} \cap L_2^{(\perp)}.$

But on the other hand, this does not imply in general that $^{(\perp)}(M_1 \cap M_2) = {}^{(\perp)}M_1 + {}^{(\perp)}M_2, \quad (L_1 \cap L_2)^{(\perp)} = L_1^{(\perp)} + L_2^{(\perp)}$, because F-orthogonality is not involutory. Indeed

270. $^{(\perp)}(M_1 \cap M_2) = {}^{(\perp)}M_1 + {}^{(\perp)}M_2$, iff

$$(M_1 + K_F^r) \cap (M_2 + K_F^r) = M_1 \cap M_2 + K_F^r.$$

In particular, it is sufficient that $M_1 \supset K_F^r$ or $M_2 \supset K_F^r$. Either condition is satisfied if the right defect equals zero (D73).

Analogous assertions are true for right orthogonal complements.

11. Topology

So far our considerations have been purely algebraic, but now we must touch that part of mathematics which forms the basis of analysis proper, that is to say topology.

As we are limiting ourselves to finite-dimensional spaces, topology will be allotted a comparatively small place. The "good" topology in a finite-dimensional space is in fact unique,§ so its nature must be algebraic after all.-- The topology in the space E can be given by defining the

§Here we mean topologies in which the operations of addition and multiplication by scalars defined in E are continuous. The precise formulation and the proof of the corresponding theorem, which requires somewhat deeper familiarity with general topology, can be found e.g. in G. Choquet's "Topology," Academic Press, 1966. Related results are considered below in (305) and in IV, (23).

<u>neighborhoods</u> of the elements of the space, i.e., of vectors. Open and closed sets are then defined in the usual fashion.

Definition 89. Let x_o be an arbitrary vector, $f_1, f_2, \dots, f_m$ some linear functionals, m some indefinite finite number, and $\varepsilon > 0$. The set of vectors x satisfying the inequalities

$$|f_k(x) - f_k(x_o)| < \varepsilon \quad (k = 1, 2, \dots, m),$$

denoted by $U(x_o; f_1, f_2, \dots, f_m; \varepsilon)$, is called a *neighborhood* of the vector x_o. The following assertions show that our definition of neighborhood satisfies the usual requirements.

271. The intersection of two neighborhoods of x_o contains a neighborhood of x_o.

272. If U is a neighborhood of x_o and $x_1 \in U$, then there exists a neighborhood of the vector x_1 contained in U.

273. Each neighborhood of x_o contains x_o.

274. Any two vectors $x_1 \neq x_2$ possess non-intersecting neighborhoods U_1 and U_2.

Definition 90. The neighborhoods U_1 and U_2 are said to *separate* x_1 and x_2. Proposition (274) expresses that E with the topology introduced is a *Hausdorff* (or *separated*, or T_2) space.

Definition 91. Neighborhoods of the form $U(x_o; f_1, f_2, \dots, f_n; \varepsilon)$, where $\Delta' = \{f_k\}_1^n$ is a fixed basis of the space E', are called *cubical neighborhoods* or *cubes* (relative to the basis Δ'). They form a *fundamental system of neighborhoods* in the sense that

275. Each neighborhood of a vector x_o contains some cubical neighborhood.

The topology of the space E is consistent with the operations of addition of vectors and multiplication of a vector by a scalar in the sense that these operations are continuous:

276. If $x_o + y_o = z_o$, then for each neighborhood W of z_o there are neighborhoods U and V of x_o and y_o, respectively, such that $x + y \in W$ if $x \in U$ and $y \in V$.

277. If $\alpha_o y_o = z_o$, then for each neighborhood W of the vector z_o there are neighborhoods U, V of the number α_o and of the vector y_o, respectively, such that $\alpha \in U$ and $y \in V$ imply $\alpha y \in W$.

Let us emphasize in this connection that the topology in E has been defined so that all linear functionals turn out to be continuous (this was done in the weakest possible way, i.e., the conditions in D89 cannot be further relaxed. In addition:

278. Each homomorphism $h \in \mathrm{Hom}(E,E_1)$ is a continuous mapping of the space E into the space E_1 (which is, of course, furnished with an analogous topology).

279. Each bilinear functional $B \in \mathbb{B}(E,E_1)$ is continuous.

Definition 92. We must now examine the so-called *induced topologies* in subspaces, factor spaces, etc. Each of these objects has already an autonomous topology which is analogous to that of the basic space E. We shall convince ourselves that the induced and the autonomous topologies coincide.

Let L be a subspace of the space E. The induced topology in L is defined in the following fashion: a subset $\Gamma \subset L$ is open if it has the form $\Gamma = G \cap L$, where G is an open set in E.

280. The induced and autonomous topologies of a subspace coincide.

In a factor space E/L the definition of the induced topology reads: a subset $\Gamma \subset E/L$ is open in E/L if it has the form $\Gamma = \{[x]_L\}_{x \in G}$ where G is an open set in E.

281. Induced and autonomous topologies of a factor space coincide.

Also in the space E' an induced topology can be introduced dually to D89. It is generated by some collection of vectors $\{x_k\}_1^m$ E and some $\varepsilon > 0$. The elements f in a meighborhood of f_o are characterized by

$$|f(x_k) - f_o(x_k)| < \varepsilon \quad (k = 1,2,\ldots,m).$$

The canonical isomorphism of the spaces E and E" implies that

282. The induced and autonomous topologies of the conjugate space coincide.

The situation is analogous for the topology of the Hermitian-conjugate space E*.

Generalizing this situation we shall consider the space of homomorphisms $\mathrm{Hom}(E,E_1)$. Let h_o be some homomorphism, take arbitrary vectors $x_1,x_2,\ldots,x_m$ in E, and arbitrary neighborhoods $U_1,U_2,\ldots,U_m$ of the vectors $h_ox_1,h_ox_2,\ldots,$ h_ox_m in E_1. The set of homomorphisms $h \in \mathrm{Hom}(E,E_1)$ for which

$$hx_k \in U_k \quad (k = 1,2,\ldots,m),$$

is then a neighborhood of the homomorphism h_o. This system of neighborhoods defines the induced topology in $\mathrm{Hom}(E,E_1)$.

283. The induced and autonomous topologies in the space of homomorphisms coincide.

Let us finally consider the space of bilinear functionals $\mathbb{B}(E,E_1)$ (this will also take care of the tensor product of the spaces E and E_1).

Let B_o be some bilinear functional. Take any vectors $x_1,x_2,\ldots,x_p$ in E, any vectors $y_1,y_2,\ldots,y_q$ in E_1, and some $\varepsilon > 0$. The set of bilinear functionals $B \in \mathbb{B}(E,E_1)$ which satisfy the inequality

$$|B(x_j,y_k) - B_o(x_j,y_k)| < \varepsilon$$
$$(j = 1,2,\ldots,p;\ k = 1,2,\ldots,q),$$

is called a <u>neighborhood of the functional</u> B_o. The system of these neighborhoods defines the induced topology in

$\mathbb{B}(E.E_1)$.

284. The induced and the autonomous topologies of the space of bilinear functionals coincide.

Definition 93. A set $M \subset E$ is called *bounded* if for any neighborhood U of the vector $x_o = 0$ there is an $\alpha > 0$ such that $\alpha x \in U$ $(x \in M)$.

The following proposition gives a simple criterion for boundedness.

285. A set M is bounded iff the set f(M) is bounded for all functionals $f \in E'$.

Hence:

286. If $\{e_k\}_1^n$ is an arbitrary basis of the space E and $x = \Sigma_{k=1}^n \xi_k(x) e_k$ $(x \in E)$, then for the boundedness of the set M it is necessary and sufficient that the numerical sets $\xi_k(M)$ $(k = 1,2,\dots,n)$ be bounded.

Thus the boundedness of a set of vectors is equivalent to coordinatewise boundedness. Hence:

287. If M is a bounded set in the space E, then each infinite subset of the set M has a limit point (Bolzano-Weierstrass Theorem).

Thus a bounded set in the space E is precompact[§]. Theorem (287) is easily inverted, but the converse theorem is not important for us.

We now consider the application of topology to certain problems of perturbation theory.

288. Let $r \geqq 0$ be an integer. The set of homomorphisms which satisfy $\text{rk}\, h \leqq r$ is closed; the set of homomorphisms which satisfy the inequality $\text{rk}\, h \geqq r+1$ is open.

For the proof the following lemma on the stability of linearly-independent systems may be useful:

289. If $\{x_k\}_1^m$ is a linearly independent system of vectors, then there exist neighborhoods $U_1, U_2, \dots, U_m$ of $x_1, x_2, \dots, x_m$, respectively, such that each system of vectors $\{y_k\}_1^m$, which

---

[§]See Dictionary of General Terms under "relative-compact".

is sufficiently close to the initial system and thus satisfies the conditions

$$y_k \in U_k \quad (k = 1,2,\ldots,m),$$

is linearly independent.

This lemma in its turn is conveniently proved by applying the technique of biorthogonal systems, (187)-(192).

As a consequence of (288),

290. The set of monomorphisms in the space $\mathrm{Hom}(E,E_1)$ is open.

291. The set of epimorphisms in the space $\mathrm{Hom}(E,E_1)$ is open.[§]

292. The set of isomorphisms in the space $\mathrm{Hom}(E,E_1)$ is open.

In contrast to (289) we have the following proposition that can be called a lemma on the <u>instability of linearly dependent systems</u>

293. Let $m \leq \dim E$, $\{x_k\}_1^m$ any system of vectors, and $U_1, U_2, \ldots, U_m$ some neighborhoods of the vectors $x_1, x_2, \ldots, x_m$, respectively. Then there exists a linearly independent system $\{y_k\}_1^m$ such that

$$y_k \in U_k \quad (k = 1,2,\ldots,m).$$

For the proof it is sufficient to apply induction, based on the fact that

294. Any subspace of the space E is closed.

295. If L is a proper subspace of E then L is nowhere dense in E.[§§]

We shall exhibit an application of lemma (293). Let us call a homomorphism <u>degenerate</u> if it is neither a monomorphism nor an epimorphism. It then follows:

296. The set of degenerate homomorphisms is nowhere dense in the space $\mathrm{Hom}(E,E_1)$.

Thus the set of nondegenerate homomorphisms is everywhere dense. The degenerate homomorphisms are an unstable

---

[§] Here it becomes evident that duality breaks down. Theorem (111) clarifies the situation.

[§§] A set is nowhere dense in E if its closure does not contain any nonvoid set that is open in E.

family, while the nondegenerate ones are a stable family by (290) and (291).

One more remark: from (289) follows the stability of bases (cf. (42)), and thus also the stability of complete systems. At the same time (293) implies the instability of incomplete systems of cardinality $m \geqq \dim E$.

The "qualitative" theorem (289) on the stability of bases will obtain a "quantitative" strengthening in Chapter IV.

12. Theory of Limits. Series. Elements of Infinitesimal Analysis.

From a topology on E we can construct a theory of limits.

Definition 94. Let $\{x_k\}_1^\infty$ be any sequence of vectors. Following the usual scheme, we call the vector x the *limit* of the sequence and denote it by $\lim_{k\to\infty} x_k$, if for any neighborhood U of x there exists a number $N = N_U$ such that

$$x_k \in U \quad (k \geqq N_U).$$

One also says that the sequence *converges* to x. The uniqueness of the limit is guaranteed by the separation axiom of the Hausdorff space, D90.

297. A sequence of vectors $\{x_k\}_1^\infty$ is convergent, iff the numerical sequence $\{f(x_k)\}_1^\infty$ converges for each $f \in E'$; and if it is known that $x = \lim_{k\to\infty} x_k$, then $f(x) = \lim_{k\to\infty} f(x_k)$.

298. Let $\{e_k\}_1^n$ be any basis of the space E and

$$x_k = \Sigma_{j=1}^n \xi_{jk} e_j \quad (k = 1,2,\dots,n).$$

The sequence $\{x_k\}_1^\infty$ converges to the vector $x = \Sigma_{j=1}^n \xi_j e_j$, iff

$$\lim_{k\to\infty} \xi_{jk} = \xi_j \quad (j = 1,2,\dots n).$$

Thus convergence in E coincides with coordinatewise convergence. Hence the basic theorems of the elementary theory of limits follow immediately:

299. If $\lim_{k\to\infty} x_k = x$, $\lim_{k\to\infty} y_k = y$, then $\lim_{k\to\infty}(x_k + y_k) = x + y$.

300. If $\lim_{k\to\infty} x_k = x$, $\lim \alpha_k = \alpha$, then $\lim_{k\to\infty}(\alpha_k x_k) = \alpha x$.

301. $\lim_{k\to\infty} x_k = x$, iff $\lim_{k\to\infty}(x_k - x) = 0$.

We now examine the connection between convergence and boundedness for sequences.

302. Each convergent sequence is bounded.

The converse, of course, is not true. Nevertheless, the Bolzano-Weierstrass theorem implies:

303. Each bounded sequence contains a convergent subsequence.

Note one consequence of this important theorem:

304. If all convergent subsequences of a bounded sequence have the same limit, then the sequence converges to that limit too.

Theorem (303) permits us to prove that it is impossible to define in E convergence to a unique limit differently from coordinatewise convergence. To define convergence means to give a class $\mathbb{L}$ of sequences $\{x_k\}_1^\infty$, called convergent, and a mapping function denoted by $\lim_{k\to\infty}$, which maps $\mathbb{L} \to E$, or $\{x_k\}_1^\infty \mapsto x$, such that the following axioms are satisfied:

(1) If $x_k = x$ $(k = 1,2,\ldots)$, then $\{x_k\}_1^\infty \in \mathbb{L}$ and $\lim_{k\to\infty} x_k = x$ (all stationary sequences $\in \mathbb{L}$)

(2) If $\{x_k\}_1^\infty \in \mathbb{L}$, then for any subsequence of indices $\{k_j\}_{j=1}^\infty$

$$\{x_{k_j}\}_{j=1}^\infty \in \mathbb{L} \text{ and } \lim_{j\to\infty} x_{k_j} = \lim_{k\to\infty} x_k.$$

305. If a convergence which satisfies conditions (299) and (300) is defined in E, then it coincides with coordinatewise convergence.

Definition 95. A sequence of vectors $\{x_k\}_1^\infty$ is called *fundamental* if for each functional $f \in E'$ the numerical sequence $\{f(x_k)\}_1^\infty$ is fundamental, i.e.,

$$\lim_{j,k\to\infty}\{f(x_k) - f(x_j)\} = 0.$$

306. A sequence of vectors converges iff it is fundamental (Cauchy's criterion).

This means that the space E is complete. (This notion of completeness of a space is quite different from the algebraic completeness introduced in D7.)

Let us pass on to the consideration of convergence in subspaces, factor spaces, etc. Here we shall obtain results which parallel theorems (280)-(284).

307. A sequence of vectors is convergent in a subspace iff it is convergent in the whole space. Thus the limits in either sense coincide.

308. In order that a sequence $\{[x_k]\}_1^\infty$ be convergent in the factor space E/L, it is necessary and sufficient that there exist a convergent sequence $\{x_k'\}_1^\infty$ in E such that

$$x_k' \equiv x_k \pmod{L} \qquad (k = 1,2,\ldots).$$

This implies

$$\lim_{k\to\infty}[x_k] = [\lim_{k\to\infty} x_k'].$$

In what follows we sometimes abbreviate $\lim_{k\to\infty} a_k = a$ by $a_k \to a$.

309. A sequence $\{f_k\}_1^\infty$ in E' is convergent, iff the numerical sequence $\{f_k(x)\}_1^\infty$ converges for all $x \in E$; and if it is known that $f_k \to f$, then $f_k(x) \to f(x)$ (cf. (297)).

310. A sequence $\{h_k\}_1^\infty$ in Hom(E,E_1) is convergent iff the sequence of vectors $\{h_k x\}_1^\infty$ is convergent for each $x \in E$; again, if $h_k \to h$, then $h_k x \to hx$.

311. A sequence $\{B_k\}_1^\infty$ in $\mathbb{B}(E,E_1)$ is convergent, iff the sequence $\{B_k(x,y)\}_1^\infty$ is convergent for each $x \in E$ and for each $y \in E_1$; again, if $B_k \to B$, then $B_k(x,y) \to B(x,y)$.

Theorems (309)-(311) can be strengthened in the following direction:

312. If the sequences $\{x_k\}_1^\infty \subset E$, $\{f_k\}_1^\infty \subset E'$ are convergent and $x_k \to x$, $f_k \to f$, then also the numerical sequence $\{f_k(x_k)\}_1^\infty$ is convergent and $f_k(x_k) \to f(x)$.

313. If the sequences $\{x_k\}_1^\infty \subset E$, $\{h_k\}_1^\infty \subset$ Hom(E,E_1) are

convergent and $x_k \to x$, $h_k \to h$, then also the sequence $\{h_k x_k\}_1^\infty \subset E_1$ is convergent and $h_k x_k \to hx$.

314. If the sequences $\{x_k\}_1^\infty \subset E$, $\{y_k\}_1^\infty \subset E_1$, $\{B_k\}_1^\infty \subset \mathbb{B}(E,E_1)$ are convergent and $x_k \to x$, $y_k \to y$, $B_k \to B$, then also the numerical sequence $\{B_k(x_k,y_k)\}_1^\infty$ is convergent and $B_k(x_k,y_k) \to B(x,y)$.

An important variant of the theory of limits is the theory of series. Here we shall be concerned with some special aspects of that theory, which arise in the vector case.

Definition 96. The formal vector series $S = \Sigma_{k=1}^{\infty} x_k$ is called *convergent* if the sequence of its partial sums $s_m = \Sigma_{k=1}^{m} x_k$ is convergent. By virtue of (297), the series S is convergent if and only if for every $f \in E'$ the formal numerical series $F = \Sigma_{k=1}^{\infty} f(x_k)$ is convergent; again, if the sum of the series S equals s, then the sum of the series F equals f(s).

Definition 97. Suppose now that S is some vector series. The set of those linear functionals f for which the series F converges is called the *domain of convergence* of the series S. For the *convergence* of a series it is thus necessary and sufficient that its domain of convergence coincide with *all* of the space E'.

315. The domain of convergence of an arbitrary vector series is a subspace of E'.

Definition 98. Let us call a vector series *fully divergent* if its domain of convergence equals zero (= the zero functional).

316. Let $\Sigma_{k=1}^{\infty} x_k$ be a vector series, $C \in E'$ be its domain of convergence, and L be some complement of the subspace $C^\perp$ in the space E (D67). If for all k

$$x_k = u_k + v_k \quad (u_k \in C^\perp,\ v_k \in L),$$

then the series $\Sigma_{k=1}^{\infty} u_k$ is fully divergent and the series $\Sigma_{k=1}^{\infty} v_k$ is convergent.

That property is also characteristic for the domain of convergence:

317. Let $M \subset E'$ be a subspace and L some complement of $M^{\perp}$ in E. If for the decomposition

$$x_k = u_k + v_k \quad (u_k \in M,\ v_k \in L)$$

the series $\Sigma_{k=1}^{\infty} u_k$ is completely divergent while the series $\Sigma_{k=1}^{\infty} v_k$ is convergent, then M is the domain of convergence of the series $\Sigma_{k=1}^{\infty} x_k$.

Definition 99. Theorem (316) leads us to the concept of the sum of a vector series relative to a subspace $L \subset E$, which is complementary to $C^{\perp}$. Under the conditions of (316) we shall consider the sum of the series $\Sigma_{k=1}^{\infty} v_k$ as the *sum of the series* $\Sigma_{k=1}^{\infty} x_k$ *relative to* L and denote it by

$$\Sigma_{k=1(L)}^{\infty} x_k.$$

There arises the question: how does that sum depend upon the subspace L?

For a convergent series, the subspace L cannot vary ($L = E$), and the relative sum equals the sum of the series. For a fully divergent series, L cannot vary either ($L = 0$), and the relative sum equals zero. In the general case we have:

318. For any vector series there is a vector s such that in the decomposition

$$s = u + v \quad (u \in C^{\perp},\ v \in L)$$

the component v always equals the sum of the series relative to any L that can be chosen as complement of $C^{\perp}$; thus

$$v = \Sigma_{k=1(L)}^{\infty} x_k.$$

The vector s is uniquely determined only modulo $C^{\perp}$.

For a convergent series $C^{\perp} = 0$, and the unique s equals the sum of the series. Let this be a hint for a possible proof of problem (318), viz. contraction modulo $C^{\perp}$.

Theorem (318) implies:

319. If the sum of a series relative to some subspace vanishes, then it must vanish relative to any subspace

complementary to $C^{\perp}$.

Theorem (319) is only a special case; the general situation is described by

320. If the sum of a series relative to some subspace does not vanish, then the set of the sums relative to all allowable subspaces is a class modulo $C^{\perp}$ (cf. (178), (179)).

A completely analogous theory of unconditional convergence can now be constructed.

Definition 100. The vector series $\Sigma_{k=1}^{\infty} x_k$ is called *unconditionally convergent* if the numerical series $\Sigma_{k=1}^{\infty} f(x_k)$ is unconditionally convergent for each $f \in E'$, i.e. $\Sigma_{k=1}^{\infty} |f(x_k)| < \infty$. Unconditional convergence clearly implies convergence. If a convergent vector series does not converge unconditionally, then it is *conditionally convergent*.

Now let S be some formal vector series. The set of linear functionals f, for which F in D96 is unconditionally convergent, will be called *domain of unconditional convergence* of the series S. Evidently, the domain of unconditional convergence is contained in the domain of convergence.

321. The domain of unconditional convergence of a vector series is a subspace in E'.

Let us call a series of vectors conditionally divergent if its domain of unconditional convergence equals zero.

322. Let $S = \Sigma_{k=1}^{\infty} x_k$ be a formal vector series, N its domain of unconditional convergence, and L any complementary subspace to $N^{\perp}$ in the space E. If

$$x_k = u_k + v_k \quad (u_k \in N^{\perp}, v_k \in L),$$

then the series $\Sigma_{k=1}^{\infty} u_k$ is conditionally divergent and the series $\Sigma_{k=1}^{\infty} v_k$ is unconditionally convergent.

323. Let the subspace $M \subset E'$ be such that for the decomposition

$$x_k = u_k + v_k \quad (u_k \in M^{\perp}, v_k \in L),$$

where L is some complementary subspace to $M^{\perp}$ in E, the series $\Sigma_{k=1}^{\infty} u_k$ turns out to be conditionally divergent while the series $\Sigma_{k=1}^{\infty} v_k$ is unconditionally convergent. Then M is the domain of unconditional convergence of the series $\Sigma_{k=1}^{\infty} x_k$.

Definition 101. We shall call the sum $\Sigma_{k=1}^{\infty} v_k$ defined in (323) the *unconditional sum* of the series S relative to the subspace $L \subset E'$ complementary to $N^{\perp}$.

324. For any vector series there is a vector $\tilde{s}$ such that in the decomposition $\tilde{s} = u + v$ the component v always equals the unconditional sum of the series relative to any subspace that can be chosen as complement of $N^{\perp}$. The vector $\tilde{s}$ is uniquely determined only modulo $N^{\perp}$.

If the series is unconditionally convergent, then one can take its sum for $\tilde{s}$. In the general case one can take for $\tilde{s}$ the sum of the series relative to any subspace $\tilde{L} \supset L$ complementary to $C^{\perp}$. (Recall that $N \subset C$, hence $N^{\perp} \supset C^{\perp}$.)

325. If the unconditional sum of the series relative to some subspace vanishes, then it vanishes relative to every subspace complementary to $N^{\perp}$.

326. If the unconditional sum of the series relative to some subspace is different from zero, then the set of all unconditional sums relative to all permissible subspaces is a class modulo $N^{\perp}$ (cf. (178), (179)).

327. A vector series is unconditionally convergent iff all series obtained by any reordering of the terms converge to the same sum.

Proposition (327) contains a partial generalization of the theorem of Riemann on conditional convergence of numerical series. In fact, Riemann's theorem in its full extent can be generalized to vector series. This generalization (a theorem of Steinitz) states: if $N \subset E'$ is the domain of unconditional convergence of a conditionally convergent vector series $\Sigma_{k=1}^{\infty} x_k$, then the sums of all possible convergent series, which are obtained from $\Sigma_{k=1}^{\infty} x_k$ by reordering its terms, form a class modulo $N^{\perp}$. We shall consider the theorem of Steinitz in Chapter VI.

In conclusion we make a short excursion into the theory of vector functions of a scalar argument.

Let I be some interval on the real axis and consider the function $x = x(t)$ on I with values in the space E (vector function). Since E has a topology we can in the usual manner introduce the notions of limit, continuity, derivative, and integral. The elementary theorems and formulas of analysis (with the exception of the theorems on mean values) immediately transfer to the vector case. At the same time some new variants of the theorems about continuity and differentiability of products do arise.

328. Let $\alpha(t)$ be a scalar function, $x(t)$ a vector function, and consider

$$u(t) = \alpha(t)x(t).$$

If $\alpha(t)$ and $x(t)$ are continuous at some point, then $u(t)$ is also continuous at that point. If $\alpha(t)$ and $x(t)$ are differentiable at a point, then $u(t)$ is differentiable there, and

$$u'(t) = \alpha'(t)x(t) + \alpha(t)x'(t).$$

329. Let $B(x,y)$ $(x \in E,\ y \in E_1)$ be a bilinear functional, $x(t)$ and $y(t)$ be functions with values in the spaces E and E_1, respectively, and put

$$\beta(t) = B(x(t),y(t)).$$

If $x(t)$ and $y(t)$ are continuous (differentiable) at some point, then $\beta(t)$ is also continuous (differentiable) at that point. One obtains for β':

$$\beta'(t) = B(x'(t),y(t)) + B(x(t),y'(t)).$$

This result applies in particular if the bilinear functional has the form

$$\beta(t) = f_t(x(t)),$$

where $x(t)$ is a function with values in E, f_t is a function with values in E':

330. Let $x(t)$ be a function with values in E, h_t be a function with values in $\mathrm{Hom}(E,E_1)$ so that

$$y(t) = h_t x(t).$$

If $x(t)$ and h_t are continuous (differentiable) at some point, then $y(t)$ is also continuous (differentiable) at that point. One obtains for y':

$$y'(t) = h_t' x(t) + h_t x'(t).$$

From (328)-(330) follow immediately the corresponding formulae for integration by parts.

We finally touch the theory of series of vector functions.

Definition 102. The series $\Sigma_{k=1}^{\infty} x_k(t)$ $(t \in I)$ is *uniformly convergent* on the set $K \subset I$ if this property is possessed by all series $\Sigma_{k=1}^{\infty} f(x_k(t))$ $(f \in E')$.

We shall quote one convenient test for uniform convergence, where we need the following

Definition 103. The sequence of vector functions $\{x_k(t)\}_1^{\infty}$ $(t \in I)$ is called *uniformly bounded* on the set K if the set

$$\{y \mid y = x_k(t),\ k = 1,2,\ldots;\ t \in K\}$$

is bounded.

331. If the numerical series $\Sigma_{k=1}^{\infty} \alpha_k$ converges unconditionally, and the sequence of vector functions $\{x_k(t)\}_1^{\infty}$ $(t \in I)$ is uniformly bounded on the set $K \subset I$, then the series $\Sigma_{k=1}^{\infty} \alpha_k x_k(t)$ is uniformly convergent on the set K.

The usual theorems on continuity, differentiability, and integrability for uniformly convergent series easily carry over to the vector case.

In linear analysis analytical vector functions of a complex variable play a very important role. Their theory is constructed in the standard form. In the finite-dimensional case (the only one we consider) polynomials and rational functions are especially important.

A vector polynomial has the form

$$P(\lambda) = \Sigma_{k=0}^{m} a_k \lambda^k,$$

where the coefficients a_k $(k = 0,1,\dots,m)$ are vectors. If $a_m \neq 0$, the polynomial $P(\lambda)$ has degree m.

A <u>rational</u> <u>vector</u> <u>function</u> is a function of the form

$$R(\lambda) = P(\lambda)/D(\lambda)$$

where $P(\lambda)$ is a vector polynomial, and $D(\lambda)$ is a <u>scalar</u> polynomial not identically zero.

332. In order that an analytic vector function $x(\lambda)$ be a polynomial (rational function), it is necessary and sufficient that this property be possessed by every scalar function $f(x(\lambda))$ $(f \in E')$, considered of course as function of λ.

The power series keeps its importance in the vector variant of the theory of analytic functions. It has the form $\Sigma_{k=1}^{\infty} a_k(\lambda - \lambda_o)^k$, where the coefficients a_k $(k = 0,1,\dots)$ are vectors.

333. Put $\ell(f) = \overline{\lim_{k\to\infty}} \sqrt[k]{|f(a_k)|}$ and define $\ell = \sup_{f \in E'} \ell(f)$. With $\rho = \ell^{-1}$, the power series

$$\Sigma_{k=0}^{\infty} a_k (\lambda - \lambda_o)^k \qquad (*)$$

converges (even unconditionally) everywhere in the circle $|\lambda - \lambda_o| < \rho$ and diverges everywhere outside of this circle (*Theorem of Cauchy-Hadamard*).

334. In each circle $|\lambda - \lambda_o| < r$ $(r < \rho)$ the convergence is uniform (*Theorem of Abel*).

The quantity ρ is called the <u>radius of convergence</u>, and the circle $|\lambda - \lambda_o| < \rho$ is the <u>circle of convergence</u> of the power series. The sum of the power series is analytic in the circle of convergence.

Another formulation of theorem (333) is: the radius of convergence of the vector power series (*) equals the largest lower bound of the radii of convergence of the scalar series $\Sigma_{k=0}^{\infty} f(a_k)(\lambda - \lambda_o)^k$ $(f \in E')$.

From power series one could go on to Laurent series and

develop the theory of singular points and residues. We shall not do this since nothing principally new emerges in comparison with the scalar case.

generating vector set D15
Grassmann's Formula (60)

Hausdorff space D90
Hermitian-bilinear functional D84
Hermitian-conjugate space (E^*) D80
Hermitian homom. D81
Hermitian-linear funct. D79
$Hom(E,E_1)$ D52
homomorphism D39

identity endomorphism (113)
identity matrix D21
iff (1)
image (Im) D43
image of subspace D50
independent system of two subspaces D29a
index of bilin. funct. (ind B) D73
index of homom. D48
induced topology D92
injective homom. D40
insertion D40
instability of lin. dependent system (293)
"into" (90)
inverse of a homom. (h^{-1}) (124)
involution (182)
isomorphic ($\simeq$) D23

kernel (Ker) D42
Kronecker product D78

largest subspace (w.r. to some property) D27
left defect D73
left divisor of homom. D59
left F-orthogonal complement D87
left generator (h_B^ℓ) D72
left inverse of homom. D57
left kernel (K_B^ℓ) D73
left orthogonality D58
left support D72
limit of vector sequence D94
linear combination of vectors D3
linear dependence D4

linear form D62
linear functional D60
linear hull D6
linear independence D4
linearly independent mod L D38
linearly minimal vector set D14

mapping (') (194)
matrix of a homom. D53
matrix of bilin. funct. D75
matrix of vector system D21
maximal chain of subspaces D26
maximal subspace (w.r. to some property) D27
modular laws (57)
monomorphism D40
mutually complementary subspaces D31
mutually independent subsp. D29a
mutually independent systems D13

neighborhood D89
neighborhood of bilin. funct. (283)
[n.i.] (231)
nowhere dense set (295)

orthogonal compl. ($L^\perp$) D67
orthogonal, vector to funct. ($x \perp f$) D66

polynomial space (28)
precompact set (287)
product of homom. D54

quasiorder ($\leq$) D11

rank of bilin. funct. (rk B) (217)
rank of homom. D47
rank of vector system D10
rational vector function (331)
regular Fredholm funct. D88
restriction of homom. ($h|L$) D56
right defect D73
right divisor of homom. D59
right F-orthogonal complement D87

right generator (h_B^r) D72
right inverse of homom. D57
right kernel (K_B^r) D73
right orthogonality D58
right support D72
rk D10

separating functional (172)
sp D6
stability of lin. indep. system (288)
Steinitz' Theorem (17)
Steinitz' Theorem on vector series (327)
subspace D24
subsystem of vectors D5
sum of subspaces D28
sum of vector series relative to a subspace D99
surjective homom. D41
Sylvester's Inequality (139)
system of vectors D5

tensor product of funct. ($f \otimes g$) D76
tensor product of vect. ($x \otimes y$) D77
total system of functionals D65

unconditionally convergent vector series D100
unconditional sum relative to subspace D101
uniform boundedness D103
uniform convergence D102

vector polynomial (331)

CHAPTER II

LINEAR OPERATORS IN COMPLEX LINEAR SPACES

Definition 1. An *operator* A acting in a space is a mapping of the space into itself. Such an operator is called *linear* if

$$(1) \quad A(x_1 + x_2) = Ax_1 + Ax_2,$$

$$(2) \quad A(\alpha x) = \alpha Ax.$$

Thus a linear operator is the same as an endomorphism, i.e., a homomorphism mapping the space into itself. What we know about homomorphisms applies therefore to linear operators.

A linear operator is a Fredholm homomorphism, hence the Fredholm theory applies completely.

Definition 2. If a linear operator A is an isomorphism (for this it is sufficient that it be a mono-or an epimorphism), then A is called a *regular* operator[§] (*automorphism*). For a regular A there exists the inverse isomorphism A^{-1}, which we shall now call the *inverse* operator. The identity endomorphism I_E will simply be called the *identity* (*operator*) and denoted by I.

Definition 3. For any linear operator A the conjugate

[§]or an invertible operator in agreement with the terminology established for homomorphisms (see I,(128)).

homomorphism (see I,D69) is the *conjugate operator* A', a linear operator in the space E'. Thus if A is regular, then so is A'. The same remarks apply to the *Hermitian conjugate operator* A*, defined in the space E* (see I, D80).

In the notation of Chapter I, the set of linear operators in the space E is Hom(E,E). Now we shall denote it by $\mathbb{M}$(E). It is a linear space of dimension n^2 by I,(110) and even an algebra (I,D55).

Below we shall say "operator" for short, meaning "linear operator in space E" unless another meaning is indicated.

1. The Algebra of Linear Operators.

Definition 4. The *operators* A and B are called *commutative* if

$$AB = BA.$$

This is denoted by A ∇ B.

Definition 5. For n>1 the algebra $\mathbb{M}$(E) is noncommutative, i.e., AB ≠ BA in general. A measure of noncommutativity of two operators is their *commutator*

$$[A,B] \underset{\text{def}}{=} AB - BA.$$

Definition 6. Operators of the form αI are called *scalar operators*. They commute with every operator of $\mathbb{M}$(E). Conversely:

1. If an operator commutes with all operators of $\mathbb{M}$(E), then it is a scalar operator (*Lemma of I. Schur*).

Definition 7. The center of the algebra $\mathbb{M}$(E) is the set of all those elements which commute with every element of the algebra. Schur's lemma shows that the center is the set {αI} of scalar operators in $\mathbb{M}$(E).

Note that the center is a <u>subalgebra</u>, i.e. a subspace closed relative to the operation of multiplication. There are many other subalgebras in $\mathbb{M}(E)$. A general method of constructing algebras consists in choosing some subset $\mathbb{A} \subset \mathbb{M}(E)$ and considering all possible products of the form

$$A_{i_1}, A_{i_2}, \dots, A_{i_s} \quad (A_{i_j} \in \mathbb{A}), \tag{*}$$

where the factors are not necessarily distinct.

2. The set of all linear combinations of products of the form (*) is the smallest subalgebra containing $\mathbb{A}$.

Definition 8. The set $\mathbb{A}$ in this construction is called a *base of the subalgebra*, and its elements are *generators*. (This base need not be a basis of a linear subspace.)

Of special interest are algebras with one generator. The center, which is generated by I, is one of them. In the general case, the subalgebra with the single generator A consists of all polynomials in A, i.e. of all elements of the form

$$\Sigma_{k=o}^{m} \alpha_k A^k.$$

The symbol A^k (k=0,1,2,...) denotes, as usual, the k-th power of the element A, and $A^o = I$, $A^1 = A$. Without enumerating the usual rules let us note only

3. If $A \nabla B$, then $(AB)^k = A^k B^k$ (k=0,1,2,...).
4. If $(AB)^2 = A^2 B^2$ with regular A and B, then $A \nabla B$.

Consider now subalgebras with one generator in more detail.

5. Each subalgebra with one generator is commutative. This, by the way, is a particular case of the following proposition: a subalgebra with base $\mathbb{A}$ is commutative, iff any two of its generators are commutative.

Denoting the subalgebra with one generator A by $\mathbb{P}(A)$, we now investigate its dimension.

6. The dimension of the subalgebra $\mathbb{P}(A)$, considered as a linear subspace, equals the largest integer m for which the powers $I, A, A^2, \dots, A^{m-1}$ are linearly independent.

This can be stated differently. Let

$$P(\lambda) = \Sigma_{k=o}^{r} \alpha_k \lambda^k$$

be some scalar polynomial in λ, and put

$$P(A) = \Sigma_{k=o}^{r} \alpha_k A^k.$$

Definition 9. Any polynomial $P(\lambda)$ not identically zero is called an *annihilator* of the operator A if the operator

$$P(A) = 0.$$

Here 0 denotes the *zero operator*, which maps all vectors into the zero vector. Now (6) may be reformualted:

7. The dimension of the subalgebra $\mathbb{P}(A)$ equals the smallest degree which the annihilating polynomial of the operator A can possess.

Definition 10. A polynomial of least degree annihilating the operator A is called a *minimum polynomial* of the operator A and is denoted by $M(\lambda;A)$. Its degree is called the *order of* A.

A minimum polynomial possesses several remarkable properties, the first of which is:

8. The set of polynomials annihilating the operator A consists exactly of those polynomials which are divisible by $M(\lambda;A)$.

An immediate consequence is:

9. For a given operator A the minimum polynomial is unique up to a scalar multiplier.

We can choose that multiplier so that the coefficient of the highest power of A becomes unity (monic polynomial): then the monic minimum polynomial of A is unique.

So far we have connected the dimension of the subalgebra $\mathbb{P}(A)$ with the annihilators of the operator A, but we have no estimate for its dimension other than the trivial

$$\dim \mathbb{P}(A) \le n^2.$$

This estimate can be considerably improved. To do this

we need

Definition 11. A polynomial $P(\lambda)$ different from zero is said to *annihilate the operator* A *for the vector* x (or briefly, $P(\lambda)$ is an annihilator of the vector x), if

$$P(A)x = 0.$$

10. For each vector $x \in E$ there exists an annihilating polynomial of degree $\leq n$.

Definition 12. A polynomial of least degree annihilating the vector x is called a *minimum polynomial of this vector*, and its degree is called the *order* of the vector x *relative to* the operator A. Thus the order of any vector relative to A does not exceed n. In analogy with (8) we have:

11. The set of polynomials annihilating a vector consists exactly of all those polynomials which are divisible by a minimal polynomial of the vector.

Consequently, the monic minimum polynomial of a vector is unique. Clearly, a minimum polynomial of the operator A is divisible by a minimum polynomial of each vector $x \in E$, and if $\Sigma_{\nu=0}^{m-1} \alpha_\nu \lambda^\nu + \lambda^m$ is the unique monic minimal polynomial of x, then the vectors $\{A^\nu x\}_{\nu=0}^{m-1}$ are linearly independent.

12. The set of minimum polynomials of all vectors $x \in E$, relative to A, contains the minimum polynomial of the operator A (cf. I,(52)).

Now it is clear that

13. $\dim \mathbb{P}(A) \leq n$.

This estimate cannot be improved: in the following section we shall prove that there is an operator $A \in \mathbb{M}(E)$ for which $\dim \mathbb{P}(A) = n$.

Annihilating polynomials are an effective tool in the <u>spectral theory of operators</u>. We shall now extend this technique somewhat, starting from an <u>arbitrary polynomial</u> $P(\lambda)$ and studying the kernel Ker $P(A)$.

14. Let $\{P_k(\lambda)\}_1^m$ be some system of nonzero polynomials, $\mathcal{D}(\lambda)$ their greatest common divisor, and $\mathcal{Q}(\lambda)$ their least common multiple. Then

$$\operatorname{Ker} \mathcal{D}(A) = \cap_{k=1}^{m} \operatorname{Ker} P_k(A), \quad \operatorname{Ker} \mathcal{Q}(A) = \Sigma_{k=1}^{m} \operatorname{Ker} P_k(A).$$

(Prove first $\operatorname{Ker} P_1(\lambda) \cap \operatorname{Ker} P_2(\lambda) \subset \operatorname{Ker} [P_1(\lambda) + P_2(\lambda)]$ and $\operatorname{Ker} P_1(\lambda) + \operatorname{Ker} P_2(\lambda) \subset \operatorname{Ker}[P_1(\lambda) P_2(\lambda)]$.)

15. If the polynomials $P_1(\lambda),\ldots,P_m(\lambda)$ are pairwise relatively prime, then

$$\operatorname{Ker} \mathcal{Q}(A) = \Sigma_{k=1}^{\oplus\, m} \operatorname{Ker} P_k(A)$$

(see I,D28 and D33 for notation).

Consequently:

16. If an <u>annihilating</u> polynomial $P(\lambda)$ of the operator A is decomposed into a product of pairwise relatively prime polynomials

$$P(\lambda) = \Pi_{k=1}^{m} P_k(\lambda),$$

then we obtain a direct-sum decomposition of the space

$$E = \Sigma_{k=1}^{\oplus\, m} \operatorname{Ker} P_k(A)$$

For completeness we also formulate the theorem dual to (14):

17. $\operatorname{Im} \mathcal{D}(A) = \Sigma_{k=1}^{m} \operatorname{Im} P_k(A), \quad \operatorname{Im} \mathcal{Q}(A) = \cap_{k=1}^{m} \operatorname{Im} P_k(A).$

The method of decomposing the annihilating polynomial into products will lead to a series of deeper results later, but we leave this subject for the present and consider an important class of subsets (indeed of subspaces) of the algebra $\mathbb{M}(E)$, the so-called ideals.

Definition 13. A subspace $\mathbb{J}$ of an algebra $\mathbb{A}$ is a *left ideal* if for each $j \in \mathbb{J}$ and every $a \in \mathbb{A}$ the product aj belongs to $\mathbb{J}$§. A *right ideal* is defined analogously. An

§For a <u>left</u> ideal the "roving" factor a is the <u>left</u> factor.

ideal is called *two-sided* if it is both a left and a right ideal. Each algebra $\mathbb{A}$ possesses the *trivial* two-sided ideals $\mathbb{A}$ and $\{0\}$. It should be clear that every ideal is a subalgebra.

18. An operator is regular iff it belongs to no ideal distinct from $\mathbb{M}(E)$.

It is even sufficient that the operator does not belong to any left ideal (or to any right ideal) distinct from $\mathbb{M}(E)$.-- We shall point out two lemmas which make the proof of the theorem (18) easier and at the same time present some independent interest.

19. An ideal is different from the algebra $\mathbb{M}(E)$ iff it does not contain the identity operator I.

20. The smallest left ideal containing the operator A is the set of all products of the form XA ($X \in \mathbb{M}(E)$).

The ideals of the algebra $\mathbb{M}(E)$ can be described in the following way:

21. Let M be some subspace of the space E. The set of all those operators A for which $\operatorname{Ker} A \supset M$ is a left ideal.

Denote this ideal by $\mathbb{J}^{\ell}_{M}$. With the aid of the theory of right divisors of homomorphisms (I, Sect. 5) we can establish the following proposition.

22. For any left ideal $\mathbb{J}$ there exists a unique subspace $M \subset E$ such that $\mathbb{J} = \mathbb{J}^{\ell}_{M}$.

Thus we have a bijective correspondence between the left ideals of the algebra $\mathbb{M}(E)$ and the subspaces of the space E. Under this correspondence, $\mathbb{J}_{O} = \mathbb{M}(E)$ and $\mathbb{J}^{\ell}_{E} = 0$.

For right ideals we have the dual criterion:

23. Let N be some subspace of the space E. The set of all those operators A for which $\operatorname{Im} A \subset N$ is a right ideal.

Denote this ideal by $\mathbb{J}^{r}_{N}$. The following proposition is established with the aid of the theory of right divisors of homomorphism.

24. For any right ideal $\mathbb{J}$ there exists a unique subspace $N \subset E$ such that $\mathbb{J} = \mathbb{J}^{r}_{N}$.

Thus we have a bijective correspondence between the right ideals of the algebra $\mathbb{M}(E)$ and the subspaces of the space E. This time $\mathbb{J}^r_O = 0$, $\mathbb{J}^r_E = \mathbb{M}(E)$.

Theorem (18) is reobtained as a simple corollary of theorems (21)-(24). Note also:

25. In the algebra $\mathbb{M}(E)$ there are no nontrivial two-sided ideals.

The duality of left and right ideals can be displayed directly by considering the conjugate mapping (cf. I,D69). It induces an involutory mapping of the algebra $\mathbb{M}(E)$ onto the algebra $\mathbb{M}(E')$, under which a left ideal becomes a right one, and a right ideal becomes a left one; this correspondence between ideals is also involutory. In this approach theorems (19) and (20) turn out to be immediate corollaries of theorems (17) and (18), and conversely.--

We close this section with the following remarks:

26. The set of regular operators (= automorphisms) form a multiplicative group. For $n>1$, this group is nonabelian.

For a regular operator A the power A^n makes sense not only with a nonnegative integer exponent, since we can define

$$A^{-m} = (A^{-1})^m \quad (m=1,2,3,\ldots).$$

The usual rules of operation with powers remain in force. In particular, powers of the same operator commute in multiplication, and all powers of a regular operator are obviously regular. Conversely, if some power of an operator A is regular, then the operator A is regular.

2. Eigenvalues and Eigenvectors of a Linear Operator. Invariant Subspaces.

Definition 14. The number λ is called an *eigenvalue*, often also a *characteristic* or *proper value*, of the operator A, if there exists a vector $x \neq 0$, such that

$$Ax = \quad x.$$

Definition 15. The vector x is called an *eigenvector* of the operator A *corresponding to the eigenvalue* λ.

This may be formulated differently: eigenvalues of the operator A are scalars λ for which

$$\mathrm{Ker}(A - \lambda I) \neq 0,$$

i.e., the operator $A - \lambda I$ is not regular; the eigenvectors corresponding to the eigenvalue λ are therefore those elements of the subspace $\mathrm{Ker}(A - \lambda I)$ which are different from zero.

Definition 15a. The subspace itself is called the *characteristic subspace* or the *subspace of eigenvectors* belonging to the eigenvalue λ.

Definition 15b. The set $\sigma(A)$ of eigenvalues of the operator A is called the *spectrum of the operator* A.

The study of the spectrum and of the geometrical structure of the operator connected with it is the subject of the spectral theory of operators.

27. Let $P(\lambda)$ be some polynomial. Then

$$P(\sigma(A)) \subset \sigma(P(A)).$$

(Here the right-hand side is the set of eigenvalues of the operator $P(A)$.)

Further, each eigenvector of the operator A corresponding to some eigenvalue μ is an eigenvector of the operator $P(A)$ corresponding to the eigenvalue $P(\mu)$.

Theorem (27) is a very preliminary formulation of the so-called spectral mapping theorem (cf. Section 4). For the present only the following consequence is important:

28. Let $P(\lambda)$ be some annihilating polynomial of the operator A. Then the spectrum of A is contained in the set of roots of the polynomial $P(\lambda)$.

In particular, the spectrum of A is contained in the set of roots of the minimum polynomial $M(\lambda;A)$. Hence it follows right away that

29. The spectrum of any operator contains no more than n points.

Can the spectrum of an operator be empty? To answer this question

30. Let $P(\lambda)$ be some annihilating polynomial of the operator A; let $\mathcal{D}(\lambda)$ be its greatest divisor in the class of polynomials which do not vanish on the spectrum $\sigma(A)$. Then the polynomial $P(\lambda)/\mathcal{D}(\lambda)$ is also an annihilating polynomial for the operator A.

Consequently:

31. Among the roots of any annihilating polynomial of the operator A there is at least one eigenvalue of the operator A.

This gives a first answer to our question:

32. No operator has an empty spectrum.

A stronger proposition follows from theorem (30):

33. The spectrum of the operator coincides with the set of roots of the minimum polynomial $M(\lambda;A)$.

Definition 16. Let $\Delta = \{e_k\}_1^n$ be some basis of the space E. The *matrix of the operator* A relative to the basis Δ (or, briefly, *in the basis* Δ) is the matrix of the homomorphism A relative to the pair of bases Δ,Δ (cf. I,D53) given by the decomposition of Ae_k:

$$Ae_k = \Sigma_{j=1}^n \alpha_{jk} e_j.$$

If $x = \Sigma_{k=1}^n \xi_k e_k$, then, clearly,

$$Ax = \Sigma_{j=1}^n (\Sigma_{k=1}^n \alpha_{jk}\xi_k) e_j. \qquad (*)$$

In order to formulate the rule for the transformation of the matrix of an operator under a change of basis we must introduce the idea of the inverse matrix:

Definition 17. Let some matrix A in a basis Δ define a regular operator A. Then it defines a regular operator in any basis and is called a *regular* (or *nonsingular*) *matrix.*

The same is true of the matrix of the operator A^{-1}, i.e., its regularity is also independent of the choice of the basis. This matrix is called the *inverse* to the matrix A and is denoted by A^{-1}. The inverse matrix satisfies the relation

$$AA^{-1} = A^{-1}A = E,$$

where E is the identity matrix, and is uniquely defined by either of these relations.

Now if A is the matrix of the operator A in the basis Δ, then its matrix in any basis Δ_1 equals

$$A_1 = T^{-1}AT,$$

where T is the matrix of the basis Δ_1 relative to the basis Δ.

One can try to change the basis of E in order to simplify the matrix of an operator so as to make its action more transparent. Let, for example, the basis be <u>characteristic</u>, i.e., consist of eigenvectors of the operator A so that

$$Ae_k = \lambda_k e_k \quad (k = 1,2,\ldots,n)$$

(the eigenvalues λ_k need not be pairwise distinct). Then (and only then) the matrix of the operator is <u>diagonal</u> ($\alpha_{jk} = 0$ for $j \neq k$), $\alpha_{kk} = \lambda_k$, and formula (*) takes the very simple form:

$$Ax = \Sigma_{j=1}^{n} \lambda_j \xi_j x_j.$$

A characteristic basis is also called a <u>basis of diagonal representation</u>. But not every operator possesses such a basis:

34. Let $\{e_k\}_1^n$ be some basis for the space E and let the operator D be defined by the formula:

$$De_1 = 0, \quad De_k = e_{k-1} \quad (k = 2,3,\ldots,n)$$

This operator has the unique eigenvalue $\lambda = 0$, and all of its eigenvectors have the form αe_1.

Thus for $n>1$ it is impossible to construct a basis from the eigenvectors of the operator D.

Definition 18. An operator A is called an *operator of scalar type* if there exists a characteristic basis for it, i.e., if its matrix can be reduced§ to diagonal form. The scalar operator αI (D6) serves as the simplest example; all vectors are eigenvectors for it (corresponding to the eigenvalue α); in this case all bases are characteristic.--We shall establish a simple geometric criterion for reducibility to diagonal form. Let us begin with a lemma:

35. If an operator A has a system of eigenvectors that correspond to pairwise distinct eigenvalues, then the system is linearly independent.

Consequently:

36. Let $\{E_k\}_1^m$ be a system of characteristic subspaces of the operator A which correspond to the points of the spectrum $\sigma(A) = \{\lambda_k\}_1^m$. Then the sum $\Sigma_{k=1}^m E_k$ is direct, and

$$\Sigma_{k=1}^m \dim E_k \leq n.$$

This, by the way, reestablishes (29).

Further:

37. In order that A be an operator of scalar type it is necessary and sufficient that

$$\Sigma_{k=1}^m \dim E_k = n.$$

From (37) follows

38. If the number of points in the spectrum of an operator equals n, then it is an operator of scalar type.

---

§Reducing a matrix to diagonal form means choosing a basis such that the matrix of the operator will be diagonal. Similarly, we shall also speak of the reduction of the matrix of an operator to some other special form.

But the number of points in the spectrum of an operator of scalar type can also be less than n. E.g., for a scalar operator the number of points equals 1. Yet it is nice to know that an "inverse" rule holds quite generally:

39. For any set σ of complex numbers containing no more than n elements, there exists an operator A such that $\sigma(A) = \sigma$.

Definition 19. If the number of points of the spectrum of the operator A equals n, then A is called an *operator with simple spectrum.*

According to (38), a simple spectrum implies that the operator is of scalar type.

Consider now the minimum polynomial of an operator of scalar type; it will lead us to an analytic criterion for the scalar type.

40. The minimum polynomial of an operator of scalar type has no multiple roots.

This result can be turned around with the aid of theorem (16) and the resolution of a polynomial into linear factors.

41. If the minimum polynomial $M(\lambda;A)$ has no multiple roots, then A is an operator of scalar type.

Still stronger:

42. If any annihilating polynomial of the operator A has no multiple roots, then A is an operator of scalar type.

For example:

43. If the operator A is an idempotent element of the algebra $\mathbb{M}(E)$, i.e., if $A^2 = A$, then it is an operator of scalar type.

44. If the operator A is a root of the identity, i.e., if $A^p = I$ for some integer $p \geqq 1$, then it is an operator of scalar type.

We note one consequence of theorems (39) and (40):

45. For each $m = 1,2,\ldots,n$ there exists an operator A such that

$$\dim \mathbb{P}(A) = m \quad \text{(notation in (6))}.$$

In the case $m = n$ this was also mentioned in the preceding section.

Definition 20. A subspace $L \subset E$ is called *invariant* relative to an operator A if A does not "remove" vectors from L, i.e., if

$$Ax \in L \quad \text{for} \quad x \in L.$$

The trivial subspaces E and {O} are invariant relative to any operator.

46. All characteristic subspaces of an operator are invariant.

A more general assertion is:

47. If $P(\lambda)$ is some polynomial, then the subspaces Ker $P(A)$, Im $P(A)$ are invariant with respect to the operator A.

Still more general is the assertion:

48. If $A \nabla B$, then the subspaces Ker B and Im B are invariant relative to the operator A.

Further:

49. If the subspace L is invariant relative to the operator A, then it is invariant relative to all polynomials $P(A)$.

50. If the subspace L is invariant relative to a regular operator A, then it is invariant relative to the inverse operator A^{-1}.

51. Sum (I,D28) and intersection of any set of invariant subspaces are invariant.

Definition 21. Let L be an invariant subspace of an operator A. The restriction $A|L$ is an operator in the space L. It is called the *part of the operator*, or the *partial operator*, *induced* by (or *lying* in) the subspace L. In addition, we can define the so-called *factor operator* A/L in the factor space E/L by putting

$$(A/L)[x] = [Ax].$$

52. Each annihilating polynomial of the operator A is an annihilating polynomial for all parts $A|L$ and for all factor operators A/L.

Consequently, the minimum polynomials of the partial and

factor operators are divisors of the minimum polymonial of the operator. Therefore the scalar character is *hereditary*:

53. All partial and factor operators belonging to operators of scalar type are also of scalar type.

There is a simple connection among the spectra of the operators A, $A|L$, A/L:

54. Let L be a nontrivial invariant subspace of the operator A. Then

$$\sigma(A) = \sigma(A|L) \cup \sigma(A/L).$$

We shall study the collection of invariant subspaces generated by some operator in E.

55. For $n>1$ each operator has a nontrivial invariant subspace.

56. If L is an invariant subspace of the operator A, and M is an invariant subspace of the factor operator A/L, then the set $N = \{x \mid [x]_L \in M\}$ is an invariant subspace of the operator A inbetween L and E, i.e.

$$L \subset N \subset E.$$

If M is nontrivial, then $N \neq L,E$.

57. If L_1, L_2 are invariant subspaces of the operator A for which

$$L_1 \subset L_2 \quad \text{and} \quad \dim L_2 - \dim L_1 > 1,$$

then there exists an invariant subspace inbetween L_1 and L_2:

$$L_1 \subset L \subset L_2, \quad L \neq L_1, L_2.$$

58. For each operator A there exists a maximal chain of invariant subspaces (I,D26)

$$0 = L_o \subset L_1 \subset L_2 \ldots \subset L_{n-1} \subset L_n = E.$$

In the language of matrices this theorem indicates the possibility of reducing the matrix of any operator to triangular form:

59. For any operator A there exists a basis $\{e_k\}_1^n$ such that

the matrix $(\alpha_{jk})_{j,k=1}^{n}$ of the operator is upper triangular in this basis:

$$\alpha_{jk} = 0 \quad (j>k) \qquad \text{(cf. I,(51)).}$$

Definition 22. The basis of (59) is called a *basis of triangular representation* of the operator A. The triangular representation is a generalization of the diagonal representation, and it *always* exists.

The connection between a triangular representation and a maximal chain of invariant subspaces is two-sided:

60. A basis $\{e_k\}_1^n$ is a basis of a triangular representation of an operator A, iff the linear hulls of the systems

$$\Gamma_m = \{e_k\}_1^m \quad (m = 1,2,\ldots,n)$$

are invariant subspaces of A.

As we shall see next, the triangular representation, like the diagonal one, completely discloses the spectrum of the operator (but not its geometric structure which that representation reflects rather crudely).

61. The diagonal elements α_{kk} $(k = 1,2,\ldots,n)$ of the matrix of an operator A, written in the basis of the triangular representation, are eigenvalues, and each eigenvalue of A is found among the diagonal elements (cf. (54)).

From theorem (61) ensues one useful corollary:

62. The set of operators with simple spectrum is everywhere dense in $\mathbb{M}(E)$.

By virtue of (38) the set of operators of scalar type is also everywhere dense. This fact is the basis of an often used method which transfers properties established for operators of scalar type to arbitrary operators through a limit process.

3. Root Subspaces. Basic Spectral Theorem.

Definition 23. Let the space E be decomposed into a direct sum

$$E = \Sigma_{k=1}^{\oplus\, m} L_k \qquad (*)$$

of invariant subspaces of the operator A, and let A_k be the part of the operator A lying in the subspace L_k. Then the operator A is called the *direct sum of the operators* A_k, and this is written in the form

$$A = A_1 \oplus A_2 \oplus \ldots \oplus A_m,$$

or

$$A = \Sigma_{k=1}^{\oplus\, m} A_k. \qquad (**)$$

If some vector x is represented in the form $x = \Sigma_{k=1}^{m} x_k$ corresponding to (*), then in correspondence with (**)

$$Ax = \Sigma_{k=1}^{m} A_k x_k.$$

Definition 24. A basis Δ in E can be obtained as the union of bases of the subspaces L_k; the matrix of the operator A in the basis Δ has then the *block-diagonal form*

$$\begin{bmatrix} A_1 & 0 & \cdots & 0 \\ 0 & A_2 & \cdots & 0 \\ \cdots & \cdots & \cdots & \cdots \\ 0 & 0 & & A_m \end{bmatrix}$$

where A_k is the matrix of the operator A_k in the corresponding basis and 0 is a rectangular matrix of zeros.-- Definition 18 of an operator of scalar type may now be reformulated: a scalar-type operator is a direct sum of scalar operators which act in the characteristic subspaces and effect multiplication by the corresponding eigenvalue:

$$A = \Sigma_{k=1}^{\oplus\, m} \lambda_k I_k \ . \qquad (***)$$

Here I_k is the identity operator in the k-th characteristic subspace and λ_k is the corresponding eigenvalue.-- The decomposition (***) is the simplest case of the spectral resolution of an operator. As we already know, it is not universal because it does not exist for all operators. Our aim is the generalization of the spectral resolution to arbitrary operators. (Note that the triangular representation does not solve this problem since it corresponds to a chain and not to a direct sum of invariant subspaces.)-- We shall generalize the notion of the spectrum of an operator, noticing that the spectrum of a scalar operator αI consists of a single point α.

Definition 25. The operator A will be called a *monosemeion* (from μόνον σημεῖον = single point) if its spectrum consists of one point. One says then that the operator is *concentrated* at this point. A nontrivial example of a monosemeion has occurred in (34).

63. The operator A is a monosemeion concentrated at the point α iff it satisfies the equation

$$(A - \alpha I)^r = 0$$

with some exponent $r > 0$.

The exponent r can be taken as the multiplicity of the root α in the minimum polynomial of the operator A. This exponent is the least possible one.

Definition 26. The multiplicity of an eigenvalue in the minimum polynomial will be called the *order of* that *eigenvalue* from now on. (Frequently "*index of* λ_k" is used.)

Definition 27. Let A be any operator, $\sigma(A) = \{\lambda_k\}_1^m$ its spectrum, and r_k the order of the eigenvalue λ_k. From now on the invariant subspace

$$W_k = \mathrm{Ker}(A - \lambda_k I)^{r_k}$$

will be called the *root subspace* of the operator A corresponding to the eigenvalue λ_k.

Theorem (16), which earlier led us to a criterion for the scalar type, now gives the possibility of proving the basic spectral theorem by the same method:

64. For every operator there is a *spectral decomposition* or *spectral resolution*

$$A = \Sigma^{\oplus\, m}_{k=1} A|W_k,$$

where $\{W_k\}_1^m$ is the set of root subspaces of the operator A.

Thus each operator A is decomposed into a direct sum of monosemeic operators concentrated at the pairwise distinct points of the spectrum of the operator A. This decomposition is unique up to the order of the terms:

65. If the operator A is the direct sum of monosemeions $A = \Sigma^{\oplus\, m}_{k=1} A_k$ concentrated at pairwise distinct points λ_k $(k = 1,2,\ldots,m)$, then $\sigma(A) = \{\lambda_k\}_1^m$ and $A_k = A|W_k$, where the sets W_k are the corresponding root subspaces.

This result follows from the two general theorems that follow:

66. If $A = \Sigma^{\oplus\, m}_{k=1} A_k$, then $\sigma(A) = \cup_{k=1}^m \sigma(A_k)$;

67. If $A = \Sigma^{\oplus\, m}_{k=1} A_k$, then the minimum polynomial $M(\lambda;A)$ coincides with the least common multiple of the minimum polynomials $M(\lambda;A_k)$ $(k = 1,2,\ldots,m)$.

Let us now study in greater detail the properties of root subspaces.

According to the basic spectral theorem the system of root subspaces $\{W_k\}_1^m$ for any operator is a basis, i.e. $\Sigma^{\oplus\, m}_{k=1} W_k = E$.

Definition 28a. Consider the characteristic $\{d_k\}_1^m$ of this decomposition. The number $d_k = \dim W_k$ is called the *multiplicity* of the eigenvalue λ_k. If $d_k = 1$, then the eigenvalue λ_k is *simple*; otherwise it is *multiple*. The polynomial

$$\mathcal{D}(\lambda;A) = \Pi_{k=1}^{m}(\lambda_k - \lambda)^{d_k}$$

is called the *characteristic polynomial*§ of the operator A.

The degree of the characteristic polynomial equals n and thus does not depend upon the operator, in contrast to the degree of the minimum polynomial.

Note that simplicity of the spectrum is equivalent with the assertion: all roots of the characteristic polynomial are simple.

68. If $A = \Sigma_{k=1}^{\oplus\, m} A_k$, then

$$\mathcal{D}(\lambda;A) = \Pi_{k=1}^{m} \mathcal{D}(\lambda;A_k) \ .$$

A more general fact is (cf. (54))

69. If L is a nontrivial invariant subspace of the operator A, then

$$\mathcal{D}(\lambda;A) = \mathcal{D}(\lambda;A|L)\mathcal{D}(\lambda;A/L)$$

The analogous generalization of theorem (67) is false.

70. If $(\alpha_{jk})_{j,k=1}^{n}$ is the matrix of the operator A in a basis of triangular representation, then

$$\mathcal{D}(\lambda;A) = \Pi_{k=1}^{n}(\alpha_{kk} - \lambda) \ .$$

Now theorem (61) can be strengthened:

71. Each eigenvalue occurs among the diagonal elements of the matrix of the operator in the basis of triangular representation as often as its multiplicity indicates.

Thus, for a given operator in a basis of triangular representation, the system of diagonal elements does not depend

§For the connection with the "characteristic equations of a matrix" see theorem (70).

upon the choice of that basis except for the order of the elements.

72. Let $\{\mu_j\}_1^n$ be a sequence of scalars representing the eigenvalues of the operator A and containing each eigenvalue as often as its multiplicity indicates. For any permutation $\{\mu_{j_k}\}_{k=1}^n$ of this system exists then a basis of triangular representation of A such that the diagonal elements of the matrix of the operator are

$$\alpha_{kk} = \mu_{j_k} \quad (k = 1,2,\ldots,n) .$$

The minimum polynomial of any operator A can now be written in the form§

$$M(\lambda;A) = \Pi_{k=1}^m (\lambda_k - \lambda)^{r_k} ,$$

where $\{\lambda_k\}_1^m = \sigma(A)$ and r_k is the order of the eigenvalue λ_k.

What is the connection of the multiplicity r of an eigenvalue with its order d? An operator of scalar type is a sufficiently typical example. According to (40), $r_k = 1$ $(k = 1,2,\ldots,m)$, while the multiplicity $d_k \geqq 1$ can be arbitrary, provided $\Sigma_{k=1}^m d_k = n$.

73. For any operator A we have the inequality

$$d_k \geqq r_k \quad (k = 1,2,\ldots,m) .$$

Hence,

74. The characteristic polynomial is an annihilator:

$$\mathcal{D}(A;A) = 0$$

(*Cayley-Hamilton Theorem*).

(73) implies again that the degree of the minimum polynomial does not exceed n.

The easiest way to obtain theorem (73) is by considering

---

§Recall that the minimum polynomial is defined up to a multiplicative constant (cf. (9)).

the <u>truncated</u> <u>root</u> <u>subspaces</u>

$$W_k^s = \mathrm{Ker}(A - \lambda_k I)^s \quad (s = 0,1,2,\dots) .$$

Obviously, $W_k^o = \{0\}$. The subspace W_k^1 coincides with the characteristic subspace E_k which corresponds to the eigenvalue λ_k. All W_k^s are of course invariant. The general picture is this:

75. The inclusions

$$W_k^o \subset W_k^1 \subset W_k^2 \subset \dots \subset W_k^{r_k}$$

are strict and $W_k^s = W_k^{r_k}$ for $s > r_k$.

Hence inequalities stronger than (73) follow:

76. $d_k \geqq \dim E_k + (r_k - 1) \quad (k = 1,2,\dots,m)$.

Let us note now that

77. $(A - \lambda_k I)^j \; W_k^{s+j} \subset W_k^s \quad (j,s = 0,1,2,\dots)$.

Therefore (cf. I,(98)):

78. $\mathrm{Dim}\; W_k^{s+j} \leqq \dim W_k^s + \dim W_k^j \quad (j,s = 0,1,2,\dots)$.

In particular:

79. $\dim W_k^{s+1} \leqq \dim W_k^s + \dim E_k \quad (s = 1,2,\dots)$.

Hence, as a supplement to (76):

80. $d_k \leqq r_k \dim E_k \quad (k = 1,2,\dots,m)$.

Definition 28b. An operator A is called *nonderogatory*§ or of *simple structure* (the Russian text uses "simple operator"), if its characteristic polynomial coincides with its minimum polynomial, i.e., if

$$d_k = r_k \quad (k = 1,2,\dots,m).$$

81. An operator is nonderogatory iff all of its characteristic subspaces are one-dimensional.

In particular:

82. An operator with simple spectrum is nonderogatory.

The converse assertion is false, as is shown by example

---

§The term "derogatory" was introduced by Sylvester (1884).

(34). However:

83. If an operator of scalar type is nonderogatory, then its spectrum is simple.

Another criterion (distinct from (81)) for an operator to be nonderogatory is:

84. An operator is nonderogatory iff the degree of its minimum polynomial is equal to n.

In other words, an operator A is nonderogatory iff $\dim \mathbb{P}(A) = n$ (cf. (7)).

Using theorem (52), we can go from this criterion to the following one:

85. An operator A is nonderogatory iff there exists a vector x for which the system

$$x, Ax, \ldots, A^{n-1}x$$

is a basis of E.

In other words, the vector x must have the order n relative to the operator A (see D12).

Definition 29. A vector x (as in (85)) is called a *generating vector* of the operator A, and the corresponding basis is a *Frobenius basis*. The matrix of the nonderogatory operator in this basis has *Frobenius' canonical form*:

$$\begin{bmatrix} \alpha_o & 1 & 0 \ldots 0 \\ \alpha_1 & 0 & 1 \ldots 0 \\ \alpha_2 & 0 & 0 \ldots 0 \\ \cdot & \cdot & \cdots\cdots \\ \alpha_{n-2} & 0 & 0 \ldots 1 \\ \alpha_{n-1} & 0 & 0 \ldots 0 \end{bmatrix}$$

in which α_k are the coefficients of the characteristic or, what is here the same thing, the minimum polynomial written in the form

$$\mathcal{D}(\lambda;A) = (-1)^n (\lambda^n - \Sigma_{k=0}^{n-1} \alpha_k\lambda^k).$$

Among other things, this gives an obvious method of constructing an operator with a given characteristic polynomial.

In connection with theorem (85) note that

86. for <u>any</u> vector x the linear hull of the system

$$x, Ax, \ldots, A^{n-2}x, A^{n-1}x$$

is an invariant subspace.

Definition 30. This subspace is called the *cyclic subspace generated by the vector* x, and by virtue of (85) it is characterized by the fact that $A|sp\{xA^k\}_{k=0}^{n-1}$ is nonderogatory.

87. The dimension of the cyclic subspace generated by the vector x equals the order of the vector x relative to the operator A.

4. Jordan's Theorem. Classification of Operators.

Further development of the basic spectral theorem brings us to the analysis of the <u>fine structure</u> of an operator.

Definition 31. First let us introduce a new idea. A nonderogatory monosemeic operator (cf. always the same example (34)) is called a *unicellular*§ *operator* (a translation of the authors' term). We shall see immediately that unicellular operators possess, in a definite sense, the simplest structure.

88. If a is a unicellular operator, then the chain of its truncated root subspaces§§

---

§The reason for the choice of this term will become clear in what follows.

§§The index in the notation for the truncated root subspace is here omitted since there is only one eigenvalue.

$$W^o \subset W^1 \subset W^2 \subset \ldots \subset W^r$$

is maximal (I,D26).

89. The set of invariant subspaces of a unicellular operator is exactly its set of truncated root subspaces.

Definition 32. An invariant subspace of an operator A is called *reducing* if there exists a complement M E of it which is also invariant relative to A. An operator which possesses nontrivial reducing subspaces is called *reducible*.

An irreducible operator is necessarily monosemeic by virtue of the spectral decomposition. In the following it will become clear that it necessarily is also a unicellular operator. In the meantime we note the converse assertion:

90. A unicellular operator is irreducible.

The high point of the spectral theory is *Jordan's Theorem*, which we now formulate. It consists of two parts. The first part establishes a canonical form for the matrix of a unicellular operator, which is closely connected with Frobenius' canonical form:

91. If A is a unicellular operator concentrated at the point α, then there exists a basis $\{e_k\}_1^n$ such that

$$Ae_k = \alpha e_k + e_{k-1} \qquad (k = 1,2,\ldots,n;\ e_o = 0).$$

In this basis the matrix of the operator has the form

$$Z = \begin{bmatrix} \alpha & 1 & 0 & \ldots & 0 \\ 0 & \alpha & 1 & \ldots & 0 \\ \ldots & \ldots & \ldots & \ldots & \ldots \\ 0 & 0 & 0 & .\alpha & 1 \\ 0 & 0 & 0 & \ldots & \alpha \end{bmatrix}.$$

We will call such a matrix a *Jordan cell* or a *Jordan block*. The basis vector e_1 is an eigenvector. The remaining vectors e_k are said to be <u>associated</u> to the vector e_1, and e_k is called the <u>associated vector</u> of the (k-1)st order. Obviously, the entire basis is uniquely defined by the vector e_n. This is the key to the proof of theorem (91).-- Incidentally, theorem (91) has the converse

92. If the operator A possesses a basis $\{e_k\}_1^n$ such that

$$Ae_k = \alpha e_k + e_{k-1} \quad (k = 1,2,\dots,n;\ e_o = 0),$$

then it is a unicellular operator concentrated at the point α.

The second and principal part of Jordan's theorem asserts that

93. Each operator is a direct sum of unicellular operators.

The proof of theorem (93) will be sketched in (96), but let us first note some of its consequences.

94. If an operator is irreducible, then it is unicellular.

Now we are able to characterize unicellular operators purely geometrically:

95. An operator is unicellular iff its invariant subspaces are pairwise <u>comparable</u>: of any two invariant subspaces one is contained in the other.

By the spectral resolution theorem (64), it is sufficient to prove theorem (93) for monosemeic operators. The proof can proceed on the basis of the following general lemma (cf. (77)):

96. If the vectors $u_j \in W_k^{s+1}$ $(j = 1,2,\dots,\ell)$ are linearly independent mod W_k^s, then the vectors

$$v_j = (A - \lambda_k I)u_j \in W_k^s \quad (j = 1,2,\dots,\ell)$$

are linearly independent mod W_k^{s-1} $(s = 1,2,\dots)$.

This lemma implies among other things the interesting amplification of inequality (78):

97. The sequence $\{\dim W_k^s\}_{s=o}^{\infty}$ is concave, i.e.

$$\dim W_k^{s+1} - 2 \dim W_k^s + \dim W_k^{s-1} \le 0 \quad (s = 1,2,\dots)$$

(*Frobenius' Inequality*).

For $s = r_k$ the inequality is strict.

Definition 33. Jordan's theorem (93), (91) means that the matrix of any operator can be reduced to the following *Jordan canonical form*

$$\begin{bmatrix} Z_1 & 0 & 0 & \dots & 0 \\ 0 & Z_2 & 0 & \dots & 0 \\ \dots & \dots & \dots & \dots & \dots \\ 0 & 0 & 0 & \dots & Z_p \end{bmatrix},$$

where Z_j is a Jordan cell. We shall call the basis, in which the matrix of the operator has the Jordan canonical form the *Jordan basis* of that operator.

The Jordan basis is clearly a basis of triangular representation, and Jordan's theorem makes precise theorem (59) on triangular representations. It is final in the sense that it guarantees the decomposition of any operator into a direct sum of operators which do not permit further decomposition, hence the term "irreducible operator" suggested by D32. Moreover, Jordan's theorem makes an exhaustive classification of linear operators possible. This will be carried out right away.

Definition 34. We shall say that the operators A and B are *similar* and write $A \approx B$ if there exists an automorphism T such that

$$B = TAT^{-1} .$$

Similarity of operators is an equivalence relation, so one can speak of the class of similar operators.

We also define similarity for matrices. The defining relation is

$$B = TAT^{-1} ,$$

so that similar operators in the same base have similar matrices. On the other hand, similar matrices can be considered as matrices of the same operator in distinct bases (cf. D17).

98. The operators A and B are similar iff for each basis Δ there exists a basis Δ_1 such that the matrix of the operator B in the basis Δ_1 equals the matrix of the operator A in the basis Δ.

We may think of the operator as an object to be constructed by choosing a basis and giving a matrix in that basis. Then we can say that the notion of similarity removes the distinction between operators caused by the "accidental" choice of a basis ("reference system", the physicists would say). A class of similar operators consists of all operators with the same geometric properties In distinction from the individual operators it is invariant under the group of automorphisms (cf. (26)) of the basic space.

We shall now describe in more detail a collection of properties common to operators of the same similarity class.

99. If $B = TAT^{-1}$, then for any polynomial $P(\lambda)$ we have the equality

$$P(B) = TP(A)T^{-1}.$$

100. The minimal polynomials of similar operators coincide.

Consequently, the spectra of similar operators coincide.

101. If one of two similar operators is of scalar type, then so is the other. More exactly, if $B = TAT^{-1}$ and $\{e_k\}_1^n$ is a characteristic basis of the operator A, then $\{Te_k\}_1^n$ is a characteristic basis of the operator B, and the eigenvalues corresponding to the vectors e_k and Te_k are equal.

The dimensions of corresponding characteristic subspaces are thus equal.

Generalizing (101) we can assert:

102. If $B = TAT^{-1}$, then

$$TW_k^s(A) = W_k^s(B) \quad (k = 1,2,\ldots,m;\ s = 0,1,2,\ldots).$$

And hence:

103. The characteristic polynomials of similar operators coincide.

In other words, the multiplicities of the corresponding equal eigenvalues are equal.

Now we can formulate the *basic classification theorem.*

104. Two operators A and B are similar iff a Jordan basis Δ for A and a Jordan basis Δ_1 for B exists such that the matrix of the operator A in Δ coincides with the matrix of B in Δ_1 (cf. (98)).

It is also not difficult to point out a full system of independent numerical invariants which determine a class of similar operators (in other words: it determines an operator up to similarity). Let λ be an eigenvalue of the operator A. Denote by $\kappa_A(\lambda;\nu)$ the number of those Jordan cells that correspond to λ and have the same dimension ν. Then

105. For the similarity $A \approx B$ it is necessary and sufficient that

$$\sigma(A) = \sigma(B) \text{ and } \kappa_A(\lambda;\nu) = \kappa_B(\lambda;\nu) \quad (\nu = 1,2,\ldots)$$

for each eigenvalue λ.

An equivalent system of invariants is the system of dimensions

$$\{\dim W_k^s\}_{1 \leq k \leq m,\quad 1 \leq s \leq r_k} .$$

106. $\kappa_A(\lambda_k;\nu) = -(\dim W_k^{\nu+1} - 2\dim W_k^{\nu} + \dim W_k^{\nu-1}) \quad (\nu = 1,2,\ldots)$

The Frobenius inequality (97) follows anew from (106), which makes it clear that a strengthening of that inequality is impossible. We note in passing that

107. $\kappa_A(\lambda_k;r_k) > 0,\ \kappa_A(\lambda_k;\nu) = 0$ for $\nu > r_k$.

Thus the order r_k of the eigenvalue λ_k coincides with the largest order of a Jordan cell that belongs to that eigenvalue.

The following relation is, in a sense, reciprocal to (106):

108. $\dim W_k^s = \Sigma_{\nu=1}^{s} \nu\kappa_A(\lambda_k;\nu) + s\Sigma_{\nu=s+1}^{r_k} \kappa_A(\lambda_k;\nu) \quad (s = 1,2,\ldots,r_k).$

In particular

$$\dim W_k^{r_k} = \Sigma_{\nu=1}^{r_k} \nu\kappa_A(\lambda_k;\nu).$$

Definition 35. We shall say that the operators A and B have the same *Jordan structure* if their spectra can be brought into a bijective relation such that

$$r_k(A) = r_k(B),\ \dim W_k^s(A) = \dim W_k^s(B)$$
$$(k = 1,2,\ldots,m;\ s = 1,2,\ldots,r_k).$$

The criteria of similarity can be greatly simplified in the presence of suitable supplementary information about the operators.

109. If two nonderogatory (D28b) operators have the same spectrum and the multiplicities of the corresponding eigenvalues are equal, then the operators are similar. In particular, if two operators with simple spectrum have the same spectrum, then they are similar.

110. If two operators of scalar type (D18) have the same spectrum and the multiplicities of corresponding eigenvalues are equal, then they are similar.

In either case the criterion of similarity is simply the coincidence of the characteristic polynomials.

In closing let us consider duality in spectral theory i.e., the relation between the spectral properties of the operators A and A' (I,D69).

111. For any polynomial $P(\lambda)$ we have

$$P(A') = [P(A)]'.$$

112. The minimum polynomials of the operators A' and A coincide.

Consequently, the spectra of the operators A and A' coincide (this is also clear from the Fredholm theory); and if A is an operator of scalar type, then so is A'. The latter assertion can be made more precise:

113. If $\{e_k\}_1^n$ is a characteristic basis for the operator A,

then the biorthogonal system $\{e_k'\}_1^n$ is a characteristic basis for the operator A', and the eigenvalues that correspond to the vectors e_k and e_k' are equal.

Consequently the dimensions of the corresponding characteristic subspaces are also equal.

Let us note now that the notion of similarity immediately transfers to operators acting in distinct spaces.

Definition 36. The operators A and A_1 acting, respectively, in the spaces E and E_1 are called *similar* $(A \approx A_1)$ if $E \approx E_1$. There exists then an isomorphism $h \in \mathrm{Hom}(E, E_1)$ such that $A_1 = hAh^{-1}$.

The criteria (104), (105), (109), (110) remain valid for operators acting in distinct spaces.

For any operator it is true that

114. $A' \approx A$.

One of the possible proofs of this theorem consists in passing to the limit with sequences of operators of scalar type, but we shall establish it by algebraic methods as before. Theorem (114) will follow from (115) and (116).

115. If $\sigma(A) = \{\lambda_k\}_1^m$ and $j \neq s$, then the root subspace of the operator A' that corresponds to the eigenvalue λ_j is orthogonal to the root subspace of the operator A that corresponds to the eigenvalue λ_s:

$$W_j(A') \perp W_s(A) \quad \text{(cf. I,(176))}.$$

Let us consider now the truncated root subspaces. According to Fredholm's theory

116. $\dim W_k^s(A') = \dim W_k^s(A)$ $(k = 1,2,\ldots,m;\ s = 0,1,2,\ldots)$.

Note that in particular

117. The characteristic polynomials of the operators A' and A coincide.

We could study the connection between A and A* in a similar way, but we can transfer the theorems established for A and A' to A and A*, using the canonical complex conjugation j (I,D82):

118. $A^* = jA'j^{-1}$.

Hence it follows that:

119. $\sigma(A^*) = \overline{\sigma(A)}$.

Consequently, the operators A and A* are in general not similar, but their Jordan structures (D35) are the same. But, of course,

120. If the spectrum of the operator A is real, then

$$A^* \approx A.$$

5. Resolvent and Operational Calculus.

Definition 37a. Let A be any operator. The operator

$$R_\lambda(A) = (A - \lambda I)^{-1},$$

which exists for $\lambda \notin \sigma(A)$, is called the *resolvent of the operator* A. We shall often write R_λ for $R_\lambda(A)$ in what follows.

In this section the behavior of the resolvent as a function of λ will be studied. We start with a trivial but useful remark:

121. If $\lambda \notin \sigma(A)$, then the equation

$$Ax - \lambda x = y$$

has a unique solution x for any right-hand side y. This solution is

$$x = R_\lambda y.$$

122. If $Ax = \lambda x + y$, then

$$A^k x = \lambda^k x + \Sigma_{j=0}^{k-1} \lambda^{k-j-1} A^j y \quad (k = 1,2,\ldots).$$

123. For <u>any</u> polynomial $P(\lambda) = \Sigma_{k=0}^{m} \alpha_k \lambda^k$ we have
$(P(A) - P(\lambda)I)R_\lambda = \Sigma_{j=0}^{m-1} P_j(\lambda)A^j \quad (\lambda \notin \sigma(A))$, where
$P_j(\lambda) = \Sigma_{k=0}^{m-j-1} \alpha_{k+j+1}\lambda^k \quad (j = 0,1,\ldots,m-1)$.

124. If $P(\lambda)$ is now some <u>annihilating</u> polynomial of A and W

is the set of its roots, then

$$R_\lambda = -\frac{1}{P(\lambda)} \Sigma_{j=0}^{m-1} P_j(\lambda)A^j \qquad (\lambda \notin W).$$

Thus the resolvent can be characterized by

125. $R_\lambda(A)$ is a rational function of λ. Its poles belong to the spectrum $\sigma(A)$. It is regular at ∞ and $R_\infty(A) = 0$.

The behavior of the resolvent at the points of the spectrum and at ∞ can be described more precisely:

126. Each point $\lambda_k \in \sigma(A)$ is a pole for the resolvent; its order equals the order of the eigenvalue λ_k.

Thus all poles of the resolvent of A are simple iff A is an operator of scalar type (cf. (40)).

It is now easy to guess the role of the resolvent in spectral theory. The point is (and we have seen this already) that the resolvent makes it possible to describe the spectrum in terms of analytic-function theory. Hence function-theoretical methods become applicable in operator theory.

127. The Laurent expansion of the resolvent about ∞ has the form

$$R_\lambda = -\Sigma_{k=0}^{\infty} A^k/\lambda^{k+1}.$$

This series converges for $|\lambda| > \rho(A)$, where $\rho(A)$ is the maximum of the moduli of the eigenvalues of A.

Definition 37b. $\rho(A) = \max |\lambda_i|$ is called the *spectral radius* of A.

The Cauchy-Hadamard theorem (I,(333)) for power series applied to (127) yields the following formula for the spectral radius:

128. $\rho(A) = \sup_{f\in E', x\in E} \left\{ \overline{\lim} |f(A^k x)|^{1/k} \right\}$.

The following functional relation, known as *Hilbert's Equation* or resolvent equation proves quite useful in the subsequent study of the resolvent:

129. $R_\lambda - R_\mu = (\lambda-\mu)R_\lambda R_\mu \qquad (\lambda,\mu \notin \sigma(A)).$

From (129) the differential equation for the resolvent follows

130. $dR_\lambda/d\lambda = R_\lambda^2 \quad (\lambda \notin \sigma(A))$,

and (130) yields for all derivatives of R_λ

131. $d^k R_\lambda/d\lambda^k = k!R_\lambda^{k+1} \quad (k = 1,2,\ldots;\ \lambda \notin \sigma(A))$.

Hence it follows:

132. The Taylor series for the resolvent in a neighborhood of λ_o is

$$R_\lambda = \Sigma_{k=o}^{\infty} R_{\lambda_o}^{k+1} (\lambda - \lambda_o)^k, \quad \lambda_o \notin \sigma(A).$$

It is noteworthy that Hilbert's functional equation, under suitable restrictions, has no other solutions than the resolvent:

133. If the <u>operator function</u> $R(\lambda)$ is given on some point set M of the complex plane and satisfies

$$R(\lambda) - R(\mu) = (\lambda-\mu)R(\lambda)R(\mu) \quad \text{for } \lambda,\mu \notin M,$$

and if the <u>operator</u> $R(\lambda_o)$ is regular for some $\lambda_o \notin M$, then there exists an operator A such that $R(\lambda)$ coincides with the resolvent $R_\lambda(A)$ at all points of M.

We shall now take the so-called *operational* or *operator calculus of F. Riesz and N. Dunford* as an example that demonstrates the usefulness of the notion of the resolvent. The following result reveals the methodical idea.

134. For any polynomial $P(\lambda)$ the integral representation

$$P(A) = -\frac{1}{2\pi i}\int_C P(\lambda)R_\lambda d\lambda \qquad (*)$$

is valid. Here C is any circuit[§] enclosing the spectrum of the operator A.

$$-\frac{1}{2\pi i}\int_C R_\lambda d\lambda = I \quad \text{and} \quad -\frac{1}{2\pi i}\int_C \lambda R_\lambda d\lambda = A$$

[§]A circuit is a simple, closed, rectifiable curve or a finite system of such curves without mutual intersections. The integration proceeds in the positive sense on all curves.

are particular cases of formula (*), which is the exact operator analog of the Cauchy integral formula for analytic functions. $(\lambda I - A)^{-1}$ plays the role of the Cauchy kernel $(\lambda-\alpha)^{-1}$, yet $P(\lambda)$ in (*) is not just some analytic function but a polynomial. No other possibility exists so far, since the symbol $P(A)$ makes sense only for polynomials. It would be a natural extension, however, if we take the Cauchy formula as a definition.

With this in mind, let the scalar function $\varphi(\lambda)$ be holomorphic on some open set G_φ containing the spectrum of A. (G_φ need not necessarily be connected; this will enable us to admit "piecewise analytic" functions φ). Denoting the class of the functions φ of λ by ϕ_A, we put down as definition of $\varphi(A)$

$$\varphi(A) = -\frac{1}{2\pi i}\int_C \varphi(\lambda) R_\lambda d\lambda, \qquad (**)$$

where $C \subset G_\varphi$ is any circuit enclosing $\sigma(A)$. This is a correct definition in the sense that the right-hand side of (**) does not depend on C.§

The class ϕ_A is an algebra with respect to the usual operations (addition, multiplication by scalars, and multiplication of functions). Consider the mapping $\mathbb{F}_A : \phi_A \to \mathbb{M}(E)$, defined by $\mathbb{F}_A\varphi = \varphi(A)$. This mapping is a homomorphism between algebras:

135. Let φ and $\psi \in \phi_A$, and $\vartheta = \alpha\varphi + \beta\psi$, where α and β are scalars. Then $\vartheta(A) = \alpha\varphi(A) + \beta\psi(A)$.

---

§If the set G_φ contains the closed disk $|\lambda| \leq \rho(A)$, then R_λ in (**) may be replaced by (127). It may also be possible to use only one power series for $\varphi(\lambda)$ along C. The resulting expansion of $\varphi(\lambda)R_\lambda(A)$ will have polynomials in A as coefficients, which reappear in the result after the termwise integration. In general, one may have to use, along C, several expansions like (132) for $R_\lambda(A)$ and several Taylor expansions for $\varphi(\lambda)$, but the algebraic structure of the final result is the same as before.

136. Let φ and $\psi \in \Phi_A$ again, and $\vartheta = \varphi\psi$. Then $\vartheta(A) = \varphi(A)\psi(A)$. Now turn to the image of the homomorphism $\mathbb{F}(A)$: using expansion (127) or proceeding as in the last footnote, we obtain the rather unexpected result

137. $$\operatorname{Im} \mathbb{F}_A = \mathbb{P}(A) \qquad \text{(notation in (6))}.$$

In other words, all possible operators $\varphi(A)$, assigned to the analytic function $\varphi(\lambda)$ by $\mathbb{F}_A$ are polynomials in A, whose degrees are not larger than r-1, where r is the degree of the minimal polynomial $M(\lambda;A)$ (D10). We have even theorem

138. For each function $\varphi \in \Phi_A$ there exists only one polynomial $P_\varphi(\lambda)$ of degree $\leq$r-1 such that

$$P_\varphi(A) = \varphi(A).$$

It should be clear that the operator A in this formula is fixed in the sense that there is no mapping between analytic functions and polynomials without giving an A; A "creates" the mapping, i.e., the polynomial-function P_φ depends on the choice of A (see (140)).

There is no difficulty in characterizing the kernel of the homomorphism $\mathbb{F}_A$ (i.e., those φ for which $\varphi(A)$ equals the zero operator for given A).

139. Ker $\mathbb{F}_A$ consists of exactly those $\varphi \in \Phi_A$ that are "divisible" by the minimal polynomial $M(\lambda;A)$, i.e.,

$$\varphi(\lambda) = M(\lambda;A)\psi(\lambda), \quad \text{where } \psi \in \Phi_A \quad \text{(cf. (8))}.$$

This also leads easily to an explicit algebraic construction of the polynomial $P_\varphi(A)$:

140. Let $\sigma(A) = \{\lambda_k\}_1^m$. For any $\varphi \in \Phi_A$ the polynomial $P_\varphi(\lambda)$ is the solution of the Hermite interpolation problem

$$P_\varphi^{(j)}(\lambda_k) = \varphi^{(j)}(\lambda_k) \qquad (j = 1,2,\ldots,r_{k-1};\ k = 1,2,\ldots,m)$$

Here r_k denotes the order of the eigenvalue λ_k as usual (D26); the superscript denotes differentiation.

141. Let $\varphi \in \Phi_A$, set $\psi = \frac{1}{\varphi}$ and assume $\psi \in \Phi_A$, i.e., $\varphi(\lambda) \neq 0$ for $\lambda \in \sigma(A)$. Then the operator $\varphi(A)$ is regular and its inverse

$[\varphi(A)]^{-1} = \psi(A)$.

Of course, if $\varphi(\lambda) \equiv \lambda$, then (141) becomes the trivial assertion: if $0 \notin \sigma(A)$, then A is regular.

Let us now show that (141) can be inverted.

142. Every eigenvector of A, belonging to some eigenvalue μ, is an eigenvector of $\varphi(A)$ belonging to the eigenvalue $\varphi(\mu)$.

143. If $\varphi \in \phi_A$ and $\varphi(A)$ is regular, then $\frac{1}{\varphi} \in \phi_A$.

These theorems imply the very important *Spectral Mapping Theorem*:

144. If $\varphi \in \phi_A$, then the set $\varphi(\sigma(A)) = \sigma(\varphi(A))$.

This theorem can also be derived from Jordan's theorem (D33). Starting from the remark

145. If $A = \Sigma^{\oplus\, m}_{k=1} A_k$, then $\varphi(A) = \Sigma^{\oplus\, m}_{k=1} \varphi(A_k)$ $(\varphi \in \phi_A)$;

we can prove:

146. If A is a monosemeic operator and its matrix in a Jordan basis is

$$\begin{bmatrix} \alpha & 1 & 0 & \dots & 0 \\ 0 & \alpha & 1 & \dots & 0 \\ \dots & & & & \\ 0 & 0 & \dots & \alpha & 1 \\ 0 & 0 & \dots & 0 & \alpha \end{bmatrix}$$

then the matrix of the operator $\varphi(A)$ in the same basis is

$$\begin{bmatrix} \varphi & \frac{1}{1!}\varphi' & \dots & \frac{1}{(n-1)!}\varphi^{(n-1)} \\ 0 & \varphi & \dots & \frac{1}{(n-2)!}\varphi^{(n-2)} \\ \dots & & & \\ 0 & 0 & \dots & \frac{1}{1!}\varphi' \\ 0 & 0 & \dots & \varphi \end{bmatrix},$$

φ and its derivatives being taken at α.

The spectral mapping theorem follows directly from (145)

and (146). In addition, we can now formulate an improved version of it, which takes account of the multiplicity of eigenvalues (cf. (71)).

147. Let A be any operator, $\sigma(A) = \{\lambda_k\}_1^m$, and d_k the multiplicity of the eigenvalue λ_k $(k = 1,2,\ldots,m)$. With $\varphi \in \Phi_A$ and $\mu \in \sigma(\varphi(A))$, we obtain for the multiplicity of the eigenvalue μ of the operator $\varphi(A)$

$$d(\mu) = \Sigma d_k,$$

the sum to be taken over those k for which $\varphi(\lambda_k) = \mu$.

Here follows an important result of operational calculus:

148. If $\varphi \in \Phi_A$ and $\psi \in \Phi_{\varphi(A)}$, then the composite function $\vartheta(\lambda) = \psi(\varphi(\lambda))$ is an element of Φ_A and

$$\vartheta(A) = \psi(\varphi(A)).$$

An easy consequence of this is

149. Let $\varphi \in \Phi_A$ be univalent. Then the inverse function $\psi \in \Phi_{\varphi(A)}$, and $\psi(\varphi(A)) = A$.

Theorem (149) is an efficient tool for the solution of operator equations of the form

$$\psi(X) = A,$$

where the function ψ and the operator A are given.

150. Let ψ be holomorphic and univalent on some open set G and $\sigma(A) \subset \psi(G)$. Then $\psi(X) = A$ has a unique solution in the class of operators satisfying $\sigma(X) \subset G$.

Special cases such as the determination of the root and the logarithm of an operator illustrate well theorem (150):

151. $$X^p = A,$$

where p is a natural number and A regular, has a unique solution in the class of those operators whose spectrum lies in the angular domain $0 \leq \arg \lambda < 2\pi/p$.

152. $$e^X = A,$$

where A is regular, has a unique solution in the class of the operators with $\sigma(X)$ on the strip $0 \leq \operatorname{Im} \lambda < 2\pi$.

Definition 38. The solution of (151) is called the

arithmetical root of degree p of the operator A and is denoted by $A^{1/p}$.

Definition 39. The solution of (152) is called the *principal value of the logarithm* of the operator A and is denoted by log A.

So far we have considered $\varphi(A)$ with fixed operator A and variable $\varphi \in \Phi_A$. Let us now take the dual viewpoint: φ is fixed and A is variable. Here functions given by their power series expansions are particularly interesting. So let $\varphi(\lambda) = \Sigma_{k=0}^{\infty} \alpha_k \lambda^k$ converge on some disk $|\lambda| < \rho$ with ρ = radius of convergence and ask: When can we assert the convergence of the operational power series $\Sigma_{k=0}^{\infty} \alpha_k A^k$?

153. If the spectral radius $\rho(A)$ of the operator A satisfies $\rho(A) < \rho$, then $\Sigma_{k=0}^{\infty} \alpha_k A^k$ converges and has the sum $\varphi(A)$.

When φ is an entire function (a special case), then $\varphi(A) = \Sigma_{k=0}^{\infty} \alpha_k A^k$ for <u>all</u> $A \in \mathbb{M}(E)$. $e^A = \Sigma_{k=0}^{\infty} A^k/k!$ is an example.

But $\rho(A) < \rho$ is not a necessary condition, although

154. Convergence of $\Sigma_{k=0}^{\infty} \alpha_k A^k$ implies $\rho(A) \leq \rho$.

The complete answer to this question is:

155. Necessary and sufficient conditions for the convergence of $\Sigma_{k=0}^{\infty} \alpha_k A^k$ are: (1) $\rho(A) \leq \rho$; (2) if $\rho(A) = \rho$, then for each eigenvalue λ on the boundary of the disk of convergence $|\lambda| = \rho$ the differentiated series $\Sigma_{k=j}^{\infty} \alpha_k k(k-1)\ldots(k-j+1)\lambda^{k-j}$ converge for $j = 0,1,\ldots,r-1$. Here r is again the order of the eigenvalue λ.

The following corollary ensues from (155):

156. Suppose $\Sigma_{k=0}^{\infty} \alpha_k \lambda^k$ diverges for all λ with $|\lambda| = \rho$. Then $\Sigma_{k=0}^{\infty} \alpha_k A^k$ converges iff $\rho(A) < \rho$.

An example for this situation is the *Neumann series* $\Sigma_{k=0}^{\infty} A^k$, an operational geometric progression:

157. The Neumann series converges iff $\rho(A) < 1$; and $\Sigma_{k=0}^{\infty} A^k = (I - A)^{-1}$.

Note also that

158. The Neumann series converges iff $\lim_{k\to\infty} A^k = 0$.

Definition 40. The operator A is *nilpotent* (or a *Volterra operator*), if $A^r = 0$ for some positive integer r. The smallest r satisfying this condition coincides with the order of A (cf. D10).

If $\lambda \in \sigma(A)$ and W is the corresponding root subspace, then the operator $(A - \lambda I)|W$ is clearly nilpotent and its order coincides with the order of the eigenvalue λ.

159. An operator is nilpotent iff its spectral radius is zero.

160. For any operator there is a decomposition $A = S + N$, where S is of scalar type and N is nilpotent and commutes with S.

Such a decomposition is called a *Dunford decomposition* of the operator A; its study is deferred to the next section.

6. Commuting Operators. Functions of an Operator.

Definition 41. Let us agree to call the operator B a *function of the operator* A, if there exists a function $\varphi \in \phi_A$ such that $B = \varphi(A)$, i.e., if $B \in \operatorname{Im} \mathbb{F}_A$.

161. If B is a function of A, then $B \nabla A$ (D4).

The converse is false as one sees e.g., if one sets $A = I$. Investigating the situation that arises here, we find:

162. If $B \nabla A$, then the characteristic subspaces of A are invariant relative to B (cf. (48)).

163. If A is an operator with simple spectrum (D19), and $B \nabla A$, B is a function of A.

But B need not have a simple spectrum! The scalar-type property of A is preserved however:

164. If A is of scalar type, then so are all functions of A (cf. (142)).

Spelled out in detail: if $A = \Sigma_{k=1}^{\oplus\, m} \lambda_k I_k$ with I_k the unit operator in the kth characteristic subspace, then (cf.145)

$$\varphi(A) = \Sigma_{k=1}^{\oplus\, m} \varphi(\lambda_k) I_k \qquad (\varphi \in \phi_A).$$

If A is of scalar type, (162) has a converse:

165. Let A be of scalar type and the characteristic subspaces of A be invariant relative to B, then $B\nabla A$.

We could also say: if $A = \Sigma_{k=1}^{\oplus\, m} \lambda_k I_k$ (of course, $\lambda_j \neq \lambda_k$ for $j \neq k$), then the general form of an operator commuting with A is $B = \Sigma_{k=1}^{\oplus\, m} B_k$, where the operators B_k can be arbitrarily chosen in their respective subspaces.

From this result ensues the converse of (163), but again only for operators of scalar type:

166. If A is of scalar type, and if all operators of scalar type that commute with A are functions of A, then A has a simple spectrum.

Theorem (165) also provides easy access to the Dunford decomposition $A = S + N$, introduced at the end of the last section.

167. If $S = \Sigma_{k=1}^{\oplus\, m} \lambda_k I_k$ ($\lambda_j \neq \lambda_k$ for $j \neq k$), then $N = \Sigma_{k=1}^{\oplus\, m} N_k$, where all N_k are nilpotent, and $A = \Sigma_{k=1}^{\oplus\, m} (\lambda_k I_k + N_k)$.

168. The spectra of the operators S and A coincide; the characteristic subspaces of A coincide with the corresponding root subspaces of A.

This implies that

169. The Dunford decomposition of the operator A is unique.

Definition 42. The operator S in the Dunford decomposition is called the *scalar part* of the operator A and the operator N its *nilpotent part*.

It is now obvious that

170. The operator A is of scalar type iff its nilpotent part vanishes.

Another proposition worth mentioning is

171. B commutes with A iff it commutes with the scalar and nilpotent parts of A individually.

This means in particular that the root subspaces of A are

invariant relative to B (cf. (162)). Actually a more general formulation is valid:

172. If $B\nabla A$, then the truncated root subspaces of A (cf. (75)) are invariant relative to B.

We now try to describe the class $\mathbb{K}$ of those operators A for which the set of operators commuting with A is the set of functions of A. The results (163) and (164) show that the intersection of $\mathbb{K}$ with the set of scalar-type operators is the class of operators with simple spectrum (D19) or, what is the same, the class of nonderogatory operators (D28b) of scalar type (cf. (82) and (83)). This suggests that the following theorem is true.

173. The set of operators commuting with A coincides with the set of functions of A iff A is nonderogatory.

Thus $\mathbb{K}$ defined above is the nonderogatory class. Necessity for (173) can easily be established on the basis of (81); sufficiency follows from (174) and (175).

174. Let $A = \Sigma^{\oplus\, m}_{k=1} A_k$, and let the minimum polynomials of A_k be pairwise relatively prime. Then for every selection $\{B_k\}_1^m$ of operators B_k that are functions of A_k, the operator $B = \Sigma^{\oplus\, m}_{k=1} B_k$ is a function of A.

175. Let A be a unicellular (D31) operator and $\{e_k\}_1^n$ its Jordan basis. Then B commutes with A, iff a system of scalars $\{\alpha_k\}_o^{n-1}$ exists such that

$$Be_k = \Sigma_{j=o}^{k-1} \alpha_j e_{k-j} \qquad (k = 1,2,\ldots,n).$$

This means that $B = \Sigma_{j=o}^{n-1} \alpha_j A^j$.

Definition 43. Let us write $B\nabla\nabla A$, if B commutes with every operator that commutes with A.

176. If $B\nabla\nabla A$, then $B\nabla\varphi(A)$ $(\varphi\in\phi_A)$. In particular, $B\nabla A$.

177. If $B\in\varphi(A)$, then $B\nabla\nabla A$. $(\varphi\in\phi_A)$.

Is every operator B satisfying $B\nabla\nabla A$ a function of A? In clarifying the situation we shall first make the simplifying assumption that A is of scalar type. Using (165) and

Schur's lemma (1) we can establish

178. If A is of scalar type and $B\nabla\nabla A$, then B is a function of A.

For the general case we use the reducing subspace (D32) as a tool:

179. If the subspace L reduces the operator A, then any operator $X:L\to L$ that commutes with $A|L$ can be extended to become an operator $Y:E\to E$ that commutes with A.

180. If $B\nabla\nabla A$, then every reducing subspace of A also reduces B, and $(B|L)\nabla\nabla(A|L)$.

181. If A is a monosemeion (D25), then each of its maximal cyclic subspaces (D30 and I,D27) is reducing.

182. If A is a monosemeion and $B\nabla\nabla A$, then each maximal cyclic subspace L of the operator A can be associated with a polynomial $P_L(\lambda)$ such that

$$B|L = P_L(A|L).$$

Now we are able to formulate the remarkably general answer to the question posed before:

183. For any operator A, every operator B satisfying $B\nabla\nabla A$ is a function of A.

We note a corollary of (183) that follows right away by (171):

184. The scalar and nilpotent parts of A are functions of A.

It is not difficult to find these functions with the aid of the Jordan canonical form of A.

Let us now apply (183) to study again a question answered in the last section: Is the solution of the operator equation

$$\psi(X) = A \qquad (*)$$

unique?

We presuppose as in (150) that the function ψ is holomorphic and univalent in some region G and that $\sigma(A) \subset \psi(G)$. Equation (*) has then a solution in the class of operators with $\sigma(X) \subset G$ by (150).

As a preparatory step, let us note two general formulas dealing with commutators (D5): For any two operators X and B let $Q = [X,B]$. Then

185. $[X^k,B] = \Sigma_{j=1}^{k-1} X^{k-j}QX^j \quad (k = 2,3,\ldots)$.

186. If $\psi \in \phi_X$ and u is an eigenvector of the operator X belonging to the eigenvalue μ, then

$$[\psi(X),B]\mu = \psi(X,\mu)Qu.$$

Here $\psi(\lambda,\mu) \equiv (\psi(\lambda) - \psi(\mu))/(\lambda - \mu)$ (and $= \psi'(\mu)$ for $\lambda = \mu$).

187. Equation (186) remains valid for the solution of the inhomogeneous equation

$$Xu - \mu u = v,$$

if v satisfies $QX^m v = 0 \quad (m = 0,1,\ldots)$.

188. If X solves equation (*) and $B\nabla A$, then $Qu = 0$ for all eigenvectors and associated vectors (cf. (91)) of the operator X.

It follows that

189. All solutions of equation (*) are functions of A.

Now we can prove again that the solution of (150) is unique.

190. The equation $\psi(X) = A$ has a unique solution.

In closing we return to theorem (162) that initiates this section and note that it implies

193. If the operators A and B of scalar type commute, then they possess a common characteristic basis.

This formulation can be strengthened:

194. If A_ν is any set of pairwise commuting operators of scalar type, then there exists a common characteristic basis for all operators A_ν.

Note also the following variant:

195. If A_ν is any set of pairwise commuting operators containing an operator with simple spectrum, then there exists a common characteristic basis for all operators A_ν.

The existence of a common characteristic basis for A and B

does not mean that one of them is a function of the other.

However:

196. If two operators A and B have a common characteristic basis, then there exists an operator T such that A and B are functions of T.

Theorems (193) and (196), applied to operators A and B of scalar type imply: from $A\nabla B$ ensues the representability of A and B as functions of the same operator T. For general operators, $A\nabla B$ is only a necessary condition for the existence of such a representation.

7. The Trace of an Operator.

Definition 44. Let $\{\lambda_k\}_1^m$ be the spectrum and $\{d_k\}_1^m$ the characteristic of an operator A. The scalar

$$\mathrm{tr}\, A \underset{\mathrm{def}}{=} \Sigma_{k=1}^{m} d_k \lambda_k$$

is called the *trace of the operator* A, i.e. the sum of the eigenvalues of A, each taken as often as its multiplicity indicates.

The traces of similar operators are obviously equal.

197. If $S+N$ is the Dunford decomposition of A, then $\mathrm{tr}\, A = \mathrm{tr}\, S$.

198. If $\mathcal{D}(\lambda;A) = (-1)^n(\lambda^n - \Sigma_{k=o}^{n-1} \alpha_k\lambda^k)$ (cf. D29), then

$$\mathrm{tr}\, A = \alpha_{n-1}.$$

If the matrix of an operator is taken in a basis of triangular representation, then the sum of its diagonal elements is clearly the trace. It is quite important that this method of calculating the trace is valid in an arbitrary basis. We shall establish an even more general fact:

199. Let $\{u_k\}_1^q$ be some system of vectors and $\{f_k\}_1^q$ some system of linear functionals. Then

$$Ax = \Sigma_{k=1}^{q} f_k(x)u_k$$

determines a linear operator A with the trace

$$\text{tr } A = \Sigma_{k=1}^{q} f_k(u_k).$$

Note here that

200. For every operator A and every complete vector system $\{u_k\}_1^q$ there is a system of linear functionals $\{f_k\}_1^q$ such that

$$Ax = \Sigma_{k=1}^{q} f_k(x)u_k.$$

The system $\{f_k\}_1^q$ is unique iff $\{u_k\}_1^q$ is a basis.

As a consequence of (199) we obtain

201. If $(\alpha_{jk})_{j,k=1}^{n}$ is the matrix of A in any basis, then

$$\text{tr } A = \Sigma_{k=1}^{n} \alpha_{kk}.$$

Thus the sum of the diagonal elements of the matrix is independent of the choice of the basis.

Let us now look at tr X as a functional defined on the space of linear operators $\mathbb{M}(E)$.

202. tr X is a linear functional.

By the way, this property of the trace, which has become so obvious after theorem (201), need not be viewed as a mere consequence of D44.

Using the trace, we can obtain the general form of a functional defined on $\mathbb{M}(E)$:

203. For every $\tau \in (\mathbb{M}(E)0'$ there is a unique operator $T \in \mathbb{M}(E)$ such that the representation

$$\tau(X) = \text{tr }(TX) \qquad (X \in \mathbb{M}(E))$$

is valid. The mapping $\tau \mapsto T$, or $(\mathbb{M}(E))' \to \mathbb{M}(E)$, is an isomorphism.

There is a set of interesting problems connected with

$$\text{tr}(X) = 0. \qquad (*)$$

We first point out a certain class of solutions of that equation.

204. For any two operators A and B

$$\mathrm{tr}[A,B] = 0, \quad \text{or} \quad \mathrm{tr}\, AB = \mathrm{tr}\, BA.$$

But is it possible to write <u>any</u> solution of (*) as a commutator? Theorem (210) contains the answer to that question.

205. If $Q = [A,B]$ and $C\nabla A$, then $\mathrm{tr}\, QC = 0$.

Using the Fredholm theory and the general form of a linear functional on $\mathbb{M}(E)$, we can prove the converse assertion:

206. Let Q be some fixed operator. If $\mathrm{tr}\, QC = 0$ for every C that commutes with some fixed A, then there is an operator B such that $Q = [A,B]$; in other words, the equation $[A,X] = Q$ is soluble in X.

207. If A is nonderogatory (D28b), then $[A,X] = Q$ is soluble iff

$$\mathrm{tr}\, QA^k = 0 \quad (k = 0,1,\ldots,n-1).$$

208. Let A be an operator with simple spectrum (D19), $\{e_k\}_1^n$ its characteristic basis, and $(\kappa_{jk})_{j,k=1}^n$ the matrix of an operator Q in that basis. The equation $[A,X] = Q$ is soluble in X, iff $\kappa_{kk} = 0$ for $k = 1,2,\ldots,n$.

209. If $\mathrm{tr}\, Q = 0$, then there is a basis, such that all diagonal elements of the matrix of Q, referred to that basis, vanish.

The result of the steps carried out in the foregoing is the theorem

210. If $\mathrm{tr}\, Q = 0$, then Q can be represented in the form $Q = [A,B]$, where A can be selected from the class of nonderogatory operators.

Note one interesting consequence of theorem (210):

211. Let $\{A_j\}_1^s$ and $\{B_j\}_1^s$ be any two systems of operators. Then the sum of the commutators

$$\Sigma_{j=1}^s [A_j,B_j]$$

can be written in the form of a commutator [A,B].

Another series of theorems derives from (205):

212. If A and B are such that $[A,B] \nabla A$, then [A,B] is nilpotent.

And conversely:

213. If N is a nilpotent operator, then it can be represented in the form $N = [A,B]$ with $A \nabla N$.

Thus the conclusion of (212) cannot be strengthened. However,

214. Let A be an operator of scalar type and B arbitrary. Put

$$Q_{k+1} = [A,Q_k] \quad (k = 1,2,\ldots) \text{ and } Q_1 = [A,B].$$

If $Q_m = 0$ for some $m \geqq 1$, then $Q_1 = 0$, hence $A \nabla B$.

Forming the commutator [...] may be considered as a multiplication. It is linear in the first "factor" since

$$[A_1+A_2,B] = [A_1,B] + [A_2,B], \quad [\alpha A,B] = \alpha[A,B]$$

and it is anticommutative, since

$$[B,A] = -[A,B],$$

but it is not associative. A substitute for the associative law is *Jacobi's Identity*

215. $[A,[B,C]] + [B,[C,A]] + [C,[A,B]] = 0$.

Definition 45. An algebra generated in a linear space by a multiplication concept with the properties enumerated is called a *Lie algebra*.

8. Projectors and the Decomposition of the Identity.

Definition 46. An operator P is a *projector*, if it satisfies

$$P^2 = P.$$

According to (43), every projector is thus an operator of

of scalar type.

216. For every projector

$$\sigma(P) \subset \{0,1\}$$

For $P \neq 0$ and $\neq I$, the spectrum $\sigma(P) = \{0,1\}$.

The spectral decomposition in that case has the form

$$P = 1.I_1 \oplus 0.I_o = I_1 \oplus 0 .$$

217. If P is a projector, then Im P coincides with the characteristic subspace belonging to the eigenvalue $\lambda = 1$.

Ker P coincides with the characteristic subspace belonging to the eigenvalue $\lambda = 0$ (trivial).

218. The formulae Im P = L and Ker P = M establish a bijective relation between projectors and pairs of subspaces $\{L,M\}$ which satisfy $L \oplus M = E$

Definition 47. If Im P = L and Ker P = M, then one says that P *projects* every $x \in E$ *upon L parallel* to M, or simply: P projects on L, etc. This is an intuitive description of: "if $x = x_1 + x_2$ ($x_1 \in L$, $x_2 \in M$), then $Px = x_1$".

Definition 48. Theorem (218) assigns to each projector P the *complementary* projector $\overline{P}$ defined by

$$\text{Im } \overline{P} = \text{Ker } P \text{ and Ker } \overline{P} = \text{Im } P.$$

Obviously $\overline{\overline{P}} = P$.

Between complementary projectors we have the obvious relation

219. $$P + \overline{P} = I.$$

This is the simplest form of the so-called decomposition of the identity. We now formulate the general definition, requiring two auxiliary definitions and theorem (221).

Definition 49. The projectors P and Q are *mutually orthogonal* (cf. I,D58 and what follows), if

$$PQ = 0 \text{ and } QP = 0,$$

or also, if

$$\text{Im } Q \subset \text{Ker } P \text{ and Ker } Q \subset \text{Im } P.$$

For instance:

220. The complementary projectors P and $\overline{P}$ are mutually orthogonal.

Definition 50. A system of pairwise mutually orthogonal $\{P_k\}_1^m$ is called an *orthogonal system of projectors.*

221. If $\{P_k\}_1^m$ is an orthogonal system of projectors, then the operator $P = \Sigma_{k=1}^m P_k$ itself is a projector, with

$$\text{Im } P = \Sigma_{k=1}^{\oplus\, m} \text{ Im } P_k \text{ and Ker } P = \bigcap_{k=1}^{m} \text{Ker } P_k.$$

Definition 51. An orthogonal system of projectors $\{P_k\}_1^m$ is a *decomposition of the identity,* if

$$\Sigma_{k=1}^m P_k = I.$$

If $\{P_k\}_1^m$ is such a decomposition, then obviously $\text{Im}\{P_k\}_1^m$ is a basic system of subspaces (I,D34).

222. The equations $\text{Im } P_k = L_k$ $(k = 1,2,\ldots,m)$ determine a bijective relation between the decompositions of the identity and the basic systems of subspaces.

223. If $\{P_k\}_1^m$ is a decomposition of the identity, then

$$\Sigma_{k=1}^{\oplus\, m} I_k = I, \qquad (*)$$

where $I_k = P_k|\text{Im } P_k$. This again determines a bijective relation between the decompositions of the identity and the decompositions of the form (*). (Also the form (*) could have been named a decomposition of the identity, but we want to affix only one sense to that term).

Definition 52. A decomposition of the identity by the operator A is generated by the spectral decomposition of E in the sense of Section 3; that is, the basic system of subspaces in (222) is identified with the root subspaces of D27. This decomposition is well-defined up to the order of succession of the projectors, which are called *root projectors.*

224. In order that the subspace L be invariant under the operator A, it is necessary that $\overline{P}AP = 0$ for each P that projects on L, and sufficient that $\overline{P}AP = 0$ for any P whatever.

225. The subspace L reduces the operator A (D32), iff $A\nabla P$ for some P that projects on L.

226. Let $\{P_k\}_1^m$ be the decomposition of the identity by the operator A. Then

$$A = \Sigma_{k=1}^m A_k,$$

where $A_k = AP_k = P_k A \quad (k = 1,2,\ldots,m)$.

The operators A_k are pairwise mutually orthogonal, i.e., $A_j A_k = O$ $(j \neq k)$.

In (formal) distinction from the decomposition (64), the operators A_k act on the entire space E. But, of course, $A_k|W_k = A|W_k$, and $A_k|W_j = O$ for $j \neq k$. Thus the term *spectral decomposition* will also be used for (226). The simple form, whichit assumes for a scalar-type operator, is given by

227. If A is of scalar type and $\{P_k\}_1^m$ is the decomposition of the identity which it effects, then

$$A = \Sigma_{k=1}^m \lambda_k P_k,$$

where λ_k is the corresponding eigenvalue. For all $\varphi \in \phi_A$ we have, moreover,

$$\varphi(A) = \Sigma_{k=1}^m \varphi(\lambda_k) P_k.$$

Note also the *decomposition of the resolvent*

$$R_\lambda = \Sigma_{k=1}^m P_k/(\lambda_k - \lambda),$$

which is simply the decomposition of the rational function R_λ of λ into partial fractions (cf.(125),(126)).

228. The root projectors P_k $(k = 1,2,\ldots,m)$ of <u>any</u> operator A are functions of A in the sense of D41, and we can even exhibit their explicit analytic form:

229. Let $\sigma(A) = \{\lambda_k\}_1^m$ and Γ_k be a simple circuit separating the point λ_k from the remaining spectral points. Then the *Formulae of F. Riesz* hold:

$$P_k = -(1/2\pi i) \int_{\Gamma_k} R_\lambda d\lambda.$$

These formulae show that a root projector always equals the

negative residue of the resolvent at the corresponding pole. For a scalar-type operator the Riesz formulae follow directly by the residue theorem from the decomposition of the resolvent above.

230. Let $H_k(\lambda)$ (k = 1,2,...,m) be the Hermite-interpolation polynomial satisfying $H_k(\lambda_k) = 1$, $H_k^{(j)}(\lambda_k) = 0$ $(j = 1,2,..,r_k-1)$, r_k being the order of the eigenvalue λ_k of the operator A. Then (cf.(140))

$$P_k = H_k(A).$$

The decomposition of the identity is closely related to the Dunford expansion (160).

231. Let $\{P_k\}_1^m$ be the decomposition of the identity by A and $\sigma(A) = \{\lambda_k\}_1^m$. Then the scalar part of A (cf.(184)) is given

$$S = \Sigma_{k=1}^m \lambda_k P_k = \Sigma_{k=1}^m \lambda_k H_k(A).$$

Another interesting consequence of the Riesz formulae is

232. In whatever way the operator A and the vector $x \neq 0$ are chosen, $R_\lambda(A)x$ is never an entire function of λ.

The breaking up of the resolvent into partial fractions, obtained above for scalar-type operators, can be extended to arbitrary operators:

233. For any operator A with $\sigma(A) = \{\lambda_k\}_1^m$

$$R_\lambda(A) = -\Sigma_{k=1}^m \Sigma_{j=o}^{r_k-1} P_k^{(j)}/(\lambda-\lambda_k)^{j+1},$$

where $P_k^{(j)} = (A - \lambda_k I)^j P_k$ for $j \neq 0$, and $P_k^{(0)} = P_k$, the kth root projector of A.

Note that $(A - \lambda_k I)P_k^{(r_k-1)} = 0$.

From (233) follows the *Interpolation formula of Lagrange-Sylvester:*

234. $$\varphi(A) = \Sigma_{k=1}^m \Sigma_{j=o}^{r_k-1} \varphi^{(j)}(\lambda_k) \cdot P_k^{(j)}/j! \quad \text{for all}$$

$\varphi(\lambda) \in \phi_A$ (cf.(227),(146), and also (138),(140)).

We now leave these spectral questions and return to the general theory of projectors.

235. Let P be a projector, $\{e_k\}_1^r$ some basis of the subspace Im P, $\{e\}_{r+1}^n$ some basis of the subspace Ker P, and

$x = \Sigma_{k=1}^{n}\xi_k(x)e_k$ the decomposition of x in the basis $\{e_k\}_1^n$. Then

$$Px = \Sigma_{k=1}^{r}\xi_k(x)e_k.$$

Note that the system of functionals $\{\xi_k\}_1^r$ is biorthogonal to the vector system $\{e_k\}_1^r$ (I,D68). Conversely:

236. If Γ $\{e_k\}_1^r$ is a linearly independent vector system, and $\Gamma' = \{f_k\}_1^r$ a system of linear functionals biorthogonal to Γ, then the operator P, acting on x according to the formula

$$Px = \Sigma_{k=1}^{r}f_k(x)e_k ,$$

is a projector and

$$\text{Im } P = \text{sp } \Gamma, \text{ Ker } P = (\text{sp } \Gamma')^{\perp}.$$

Another converse:

237. If some projector P acts on x according to

$$Px = \Sigma_{k=1}^{r}f_k(x)e_k ,$$

where $\Gamma = \{e_k\}_1^r$ and $\Gamma' = \{f_k\}_1^r$ are linearly independent systems of vectors and linear functionals respectively, then Γ and Γ' are biorthogonal.

Now one can easily obtain (see (199)) that

238. $\text{rk } P = \text{tr } P$ for any projector P.

This result can be applied to prove a general theorem on the sum of projectors:

239. The sum of projectors $\Sigma_{k=1}^{m}P_k$ is a projector, iff the system $\{P_k\}_1^m$ is orthogonal (D50).

Necessity is here non-trivial, but sufficiency is containe in (221). And (239) will convince us that orthogonality must not be removed from the postulates that define the decomposition of the identity (D51).

Let us now consider other operations on projectors.

240. The difference of two projectors $P_1 - P_2$ is a projector, iff $\overline{P}_1$ and P_2 are mutually orthogonal. With this condition satisfied, the following relations hold for $P = P_1 - P_2$:

$$\text{Im } P = \text{Im } P_1 \cap \text{Ker } P_2, \text{ Ker } P = \text{Ker } P_1 \oplus \text{Im } P_2.$$

Note that the mutual orthogonality of $\overline{P}_1$ and P_2 is equivalent with

$$\text{Im } P_1 \supset \text{Im } P_2 \text{ and Ker } P_1 \subset \text{Ker } P_2 \quad \text{(cf. D49)}.$$

41. The product P_1P_2 of two projectors is a projector, iff he commutator $[P_2, P_1]$ maps the subspace Im P_2 into the subpace Ker P_1. If this condition is satisfied and $P = P_1P_2$, hen

$$\text{Im } P = \text{Im } P_1 \cap (\text{Im } P_2 + \text{Ker } P_1 \cap \text{Ker } P_2),$$
$$\text{Ker } P = \text{Ker } P_2 + \text{Ker } P_1 \cap (\text{Im } P_1 + \text{Im } P_2).$$

Note the special case

42. If P_1 and P_2 commute, then their product P is a proector, and

$$\text{Im } P = \text{Im } P_1 \cap \text{Im } P_2, \ \text{Ker } P = \text{Ker } P_1 + \text{Ker } P_2.$$

This result can be immediately extended to any system $\{P_k\}_1^m$ of pairwise commuting projectors. In fact, there is the more general theorem

43. If the product of projectors $P_1P_2 \ldots P_m$ remains unchanged n all cyclic permutations of the factors, then it is also a rojector, P, and

$$\text{Im } P = \bigcap_{k=1}^{m} \text{Im } P_k, \ \text{Ker } P = \Sigma_{k=1}^{m} \text{Ker } P_k \quad \text{(recall I,D28)}.$$

And a final remark:

44. If P is a projector, then P' is also a projector in the onjugate space E', and

$$\text{Im } P' = (\text{Ker } P)^{\perp}, \ \text{Ker } P' = (\text{Im } P)^{\perp} .$$

9. Elements of Perturbation Theory.

The question which we shall only touch in this section is: how does the structure of an operator change, if the operator changes slightly (i.e., suffers a perturbation). We have already obtained one important result in this direction (I,(292)). In the language of the present chapter its formulation would be

245. The set $\mathbb{R}$ of regular operators in $\mathbb{M}(E)$ is open.

Moreover:

246. The set $\mathbb{R}$ is connected.

Considering the inversion of the operator A as a mapping $\mathbb{R} \to \mathbb{R}$ and denoting it by $\Omega A = A^{-1}$, we find that

247. The mapping Ω is continuous.

In other words, the operator inverse to A depends continuously on A. Let us here properly stress that the same is true for the multiplication of operators:

248. The mapping $\Pi(A,B) = AB$ with $A, B \in \mathbb{M}(E)$ is continuous.

We can apply (245) and (246) to investigate the dependence of the spectrum $\sigma(A)$ on the operator A.

249. If the point $\lambda_o \notin \sigma(A_o)$, then there exists a neighborhood $\mathcal{U}_o$ of λ_o and a neighborhood $\mathbb{U}_o$ of A_o such that $\lambda \notin \sigma(A)$ for all $\lambda \in \mathcal{U}_o$ and all $A \in \mathbb{U}_o$.

250. The resolvent $R_\lambda(A) = (A = \lambda I)^{-1}$ is continuous in the pair of arguments A and λ, wherever $\lambda \notin \sigma(A)$

251. $\sigma(A)$ depends continuously on the operator A.

For any operator A_o and any neighborhood $\mathcal{V}_o$ of the set $\sigma(A_o)$ in the complex plane this means that there is a neighborhood $\mathbb{U}_o$ of the operator A_o such that $\sigma(A) \subset \mathcal{V}_o$ for all $A \in \mathbb{U}_o$.

Theorem (249) implies in particular:

252. The set of operators with a simple spectrum is open (cf. (62)).

It is also true that

253. The set of operators with a simple spectrum is connected.

Starting from the following remark, we can make (250) considerably more precise.

254. Suppose $\lambda_o \in \sigma(A_o)$ and no other spectral points of $\sigma(A_o)$ are on the disk $|\lambda - \lambda_o| \leqq \varepsilon$. Then the Riesz projector

$$P_A = -(1/2\pi i) \int_C R_\lambda d\lambda \qquad (C \text{ is the circle } |\lambda - \lambda_o| = \varepsilon).$$

is a continuous function of A in some neighborhood of the operator A_o.

Consequently,

255. There is a neighborhood $\mathbb{U}_o$ of the operator A_o such that for $A \in \mathbb{U}_o$

$$\mathrm{rk}\, P_A = \mathrm{rk}\, P_{A_o} \qquad (\text{cf.I,(288)}).$$

In other words, the sum of the multiplicities of those eigenvalues of an operator $A \in \mathbb{U}_o$, which lie in the disk $|\lambda-\lambda_o|<\varepsilon$, is equal to the multiplicity of λ_o. This has as a corollary:

256. The characteristic polynomial $\mathcal{D}(\lambda;A)$ depends continuously on the argument A, and even on the <u>pair</u> of arguments A and λ.

For the minimal polynomial this assertion would be false. This is connected with the fact that the set of operators of scalar type is not open, as can be easily seen. (That set is not closed either, see (62)). Yet, incidentally:

257. The set of nonderogatory operators is open (see (85) and I,(289)).

Note also (cf.(245),(247)):

258. If the scalar function $\varphi(\lambda)$ is holomorphic on some open set G, then the set $\mathbb{M}_\varphi$ of those operators A for which $\varphi\in\phi_A$ is open, and the mapping $\mathbb{M}(E) \to \mathbb{M}(E)$, which assigns the operator $\varphi(A)$ to the operator A, is continuous.

Let us now briefly review the elements of analytic perturbation theory. The subject of investigation is here the holomorphic operator function A_ζ of the scalar argument ζ.

259. If the operator function A_ζ is holomorphic on the open set H, then A_ζ^{-1} is holomorphic on that subset of H where $A_\zeta \in \mathbb{R}$ and

$$dA_\zeta^{-1}/d\zeta = -A_\zeta^{-1}(dA_\zeta/d\zeta)A_\zeta^{-1}\ .$$

260. The resolvent $R_\lambda(A_\zeta) = (A_\zeta-\lambda I)^{-1} \quad (\lambda\notin(A_\zeta))$ of the holomorphic operator function A_ζ is holomorphic in the <u>pair</u> of arguments ζ and λ.

261. Let the scalar function $\varphi(\lambda)$ be holomorphic on some open

open set G, and the operator function A_ζ be holomorphic on some open set H. Then the operator function $\varphi(A_\zeta)$ is holomorphic on that subset of H where $A_\zeta \in \mathbb{M}_\varphi$.

The following result has fundamental importance:

262. The characteristic polynomial $\mathcal{D}(\lambda;\zeta) \underset{\text{def}}{=} \mathcal{D}(\lambda;A_\zeta)$ of a holomorphic operator function A_ζ is holomorphic in the pair of arguments ζ and λ.

The proof of (262) can be reduced to that of (261) by the formula

263. $\mathcal{D}(\lambda;A) = \exp \operatorname{tr}[\log(A-\lambda I)] \qquad (\lambda \notin \sigma(A))$

10. The Determinant Operator. The Group Commutator.

Definition 53. The *determinant operator* is the multiplicative counterpart of the trace. Let A be some operator, $\{\lambda_k\}_1^m$ its spectrum, and $\{d_k\}_1^m$ the characteristic of its spectral resolution. The determinant of A is the scalar

$$\det A = \prod_{k=1}^{m} \lambda_k^{d_k} ,$$

that is the product of its eigenvalues, each taken as often as its multiplicity indicates.

264. $\det A = \mathcal{D}(0;A)$

Hence, and from the results of the foregoing section, ensues

265. det A is a continuous functional on $\mathbb{M}(E)$.

266. If A_ζ is a holomorphic operator function in some region G, then $\det A_\zeta$ is also holomorphic in G.

By (263)

267. $\det A = \exp \operatorname{tr}(\log A) \qquad (A \in \mathbb{R})$.

Supplementing the foregoing, one verifies that

$A \notin \mathbb{R} \Rightarrow \det A = 0$.

The operator A is therefore regular iff $\det A \neq 0$.

Formula (264) can be generalized:

268. $$\mathcal{D}(\lambda;A) = \det(A - \lambda I) .$$

Thus determinants of similar operators are clearly equal, and hence

269. $$\det BA = \det AB \quad \text{(cf. (204))}.$$

Moreover,

270. $$\mathcal{D}(\lambda;AB) = \mathcal{D}(\lambda;BA)\ .$$

This identity can be derived from the following assertion:

271. If at least one of the operators A and B is regular, then $AB \approx BA$ (notation in D34).

In general, however, $AB \not\approx BA$. For example, there are operators A and B such that $AB = 0$ but $BA \neq 0$.

Theorem 269 can be very considerably strengthened:

272. For <u>any</u> two operators A and B

$$\det AB = \det A \cdot \det B\ .$$

(Formally, this is the well-known <u>multiplication theorem for determinants.</u>)

In the case $A \nabla B$ the last theorem may be considered as a simple consequence of (267) and the following <u>addition theorem</u> for the operator function exp:

273. $$A \nabla B \Rightarrow e^A e^B = e^{A+B}$$

The multiplication theorem becomes obvious also in the case that at least one of the operators A and B is <u>not</u> regular. A general proof is obtained as a result of the chain of lemmas (274) - (277).

274. $$\det(\alpha A) = \alpha^n \det A$$

275. The function $\mathcal{D}(\zeta;A,B) \underset{\text{def}}{=} \det(A(B-\zeta I))$ is a polynomial of degree $\leq n$ in ζ.

276. If B has a simple spectrum, then the polynomial $\mathcal{D}(\zeta;A,B)$ is divisible by the characteristic polynomial $\mathcal{D}(\zeta;B)$ and

$$\mathcal{D}(\zeta;A,B) = \mathcal{D}(\zeta;B) \det A\ .$$

Hence,

277. If the spectrum of B is simple, then

$$\det AB = \det A \det B.$$

Theorem (272) now follows from (62).

In connection with the multiplication theorem we note that

278. $\det A^n = (\det A)^n \quad (A \in \mathbb{R};\ n=0,\pm 1,\ \ldots).$

Definition 54. A functional $\delta(A)$, $(A \in \mathbb{M}(E))$, is *multiplicative* if

$$\delta(AB) = \delta(A)\delta(B) .$$

A trivial example is $\delta(A) \equiv 0$ (or 1) for every $A \in \mathbb{M}(E)$. In terms of D54, the multiplication theorem means that det A is a multiplicative functional on the space $\mathbb{M}(E)$. The question: "What kind of multiplicative functionals are there on $\mathbb{M}(E)$?" will be answered at the end of this section. Let us now introduce the multiplicative counterpart of the commutator.

Definition 55. The *group commutator* of two regular operators A and B is the operator $ABA^{-1}B^{-1}$ denoted by $\{A,B\}$.

279. Two regular operators A and B commute iff $\{A,B\} = I$.

Group commutators and commutators are directly connected by

280.
$$\frac{d}{d\zeta}\{e^A, e^B\}\bigg|_{\zeta=0} = [A,B].$$

In analogy with (204)

281.
$$\det\{A,B\} = 1 .$$

Definition 56. An operator, whose determinant is equal to unity, is called *properly unimodular*; it is called *unimodular* if the modulus (or absolute value) of its determinant is equal to unity. In analogy with (210) one would expect an affirmative answer to the question: Is it possible to represent every properly unimodular operator in the form of a group commutator?

282. The equation

$$\{X,B\} = C,$$

where B and C are given regular operators, can be solved for X, iff $CB \approx B$.

283. If C is a properly unimodular operator, then there exists a regular operator B with simple spectrum such that $CB \approx B$.

So it is in fact true that

284. Every properly unimodular operator C is representable

in the form

$$C = \{A,B\} ,$$

where B can be chosen from the class of operators with simple spectrum.

Applying this result we shall describe the set of all multiplicative functionals in $\mathbb{M}(E)$.

285. If $\delta(A)$ is a nonvanishing multiplicative functional, then $\delta(C) = 1$ for all properly unimodular operators C in the representation (284).

286. If $\delta(A)$ is a continuous nonvanishing multiplicative functional acting on a scalar operator A, then

$$\delta(\alpha I) = |\alpha|^{\gamma} \exp im \arg \alpha,$$

where γ and m are constants, and either $\mathrm{Re}\,\gamma > 0$ and m is integer or $\gamma = m = 0$.

287. The general form of a continuous nonvanishing multiplicative functional on $\mathbb{M}(E)$ is

$$\delta(A) = |\det A|^{\omega} \exp ik \arg (\det A) ,$$

where ω and k are constants, and either $\mathrm{Re}\,\omega > 0$ and k is integer or $\omega = k = 0$.

Finally we must emphasize:

288. There are no multiplicative <u>linear</u> functionals different from zero in $\mathbb{M}(E)$, if $\dim E > 1$.

This result is closely connected with (25) and follows from it easily, if one heeds this remark:

289. If $\delta(A)$ is a linear multiplicative functional, then $\mathrm{Ker}\,\delta$ is a two-sided ideal in the algebra $\mathbb{M}(E)$.

annihilator D9
annihilator for vector D11
annihilating polynomial (16)
associated vector (91)
automorphism D2

base of subalgebra D8
basis of diagonal representation D17
basis of triangular representation D22
biorthogonal system of functionals (236)
block-diagonal matrix D24

Cayley-Hamilton Theorem (74)
center of algebra D7
characteristic polynomial D28a
characteristic subspace D15a
characteristic value D14
classification theorem (104)
commutative operator D4
commutator D5
comparable invariant subspaces (95)
complementary projector D48
conjugate operator D3
connectedness of set of operators (246)
continuity of inverse (247)
cyclic subspace D30

decomposition of identity D51
decomposition of the resolvent (227)
∇ defined D4
$\nabla\nabla$ defined D43
determinant operator D53
direct sum of operators D23
Dunford decomposition D42

eigenvalue D14
eigenvector D15

factor operator D21
Frobenius basis D29
Frobenius' canonical form D29
Frobenius' Inequality (97)
function of operator D41

generating vector D29
generator of algebra D8
group commutator D55

hereditary (52)
Hermite interpolation problem (140)
Hermitian conjugate operator D3
Hilbert's equation (129)

idempotent operator (43)
identity operator D2
index of eigenvalue D26
invariant subspace D20
inverse matrix D17
inverse operator D2
invertible operator D2
irreducible operator D32

Jacobi's Identity (215)
Jordan basis D33
Jordan block (91)
Jordan canonical form D33
Jordan cell (91)
Jordan structure D35
Jordan's Theorem (91), (93)

Lagrange-Sylvester interpolation (234)
left ideal D13
Lie algebra D45
linear operator D1
logarithm of operator D39

matrix of operator D16
minimum polynomial D10
minimum polynomial of vector D12
monic polynomial (9)
monosemeion D25
multiple eigenvalue D28
multiplication theorem for determinants (272)
multiplicative functional D54
multiplicity of eigenvalue D28a
mutually orthogonal projectors D49

Neumann series (157)
nilpotent operator D40
nilpotent part of operator D42
nonderogatory operator D28b
nonsingular matrix D17

openness of set of operators (245)
operator D1
operator calculus (Riesz and Dunford) (134)
operator function (133)
operator of scalar type D18
operator of simple structure D28b
operator with simple spectrum D19
order of eigenvalue D26
order of operator D10
orthogonal system of projectors D50

partial operator D21
projection upon L parallel to M D47
projector D46
proper value D14

reducible operator D32
reducing a matrix D18
reducing invariant subspace D32
regular matrix D17
regular operator D2
resolvent of operator D37a
Riesz' Formulae (229)
right ideal D13
root of identity (44)
root of operator D38
root projector D52
root subspace D27

scalar operator D6
scalar part of operator D42
scalar-type operator D18
Schur's lemma (1)
similar operators D34
similar operators in different spaces D36
simple eigenvalue D28a
spectral decomposition (64)
spectral mapping theorem (144)
spectral radius D37b
spectral resolution D24 (64)
spectral theory (13)
spectrum of operator D15b
subalgebra D7
subspace of eigenvectors D15a

trace of operator D44
trivial ideal D13
truncated root subspaces (74)
two-sided ideal D13

unicellular operator D31
unimodular operator D56

Volterra operator D40

CHAPTER III

BILINEAR AND HERMITIAN-BILINEAR FUNCTIONALS.

THE UNITARY SPACE.

OPERATORS ACTING IN A UNITARY SPACE.

1. Bilinear and Quadratic Functionals.

In this section we shall study bilinear functionals $B(x,y)$ defined on $E\times E$ (cf. I,D71), or, briefly, *bilinear functionals on* E. We are interested in the special properties arising from the fact that both x and y belong to the same space. This makes $B(x,y)$ a Fredholm functional (I,D74).

Definition 1. Let $\Delta=\{e_k\}_1^n$ be some basis of E. The matrix of B relative to the pair Δ,Δ will be called the *matrix of the bilinear functional* $B(x,y)$ in the basis Δ. Thus it is the coefficient matrix in the expansion

$$B(x,y) = \Sigma_{j,k=1}^{n} \beta_{jk}\zeta_j\eta_k \quad (x = \Sigma_{j=1}^{n}\zeta_j e_j,\ y = \Sigma_{k=1}^{n}\eta_k e_k).$$

The second member defines a bilinear functional on the arithmetical space C^n, i.e. a *bilinear form*.

In the transition from basis Δ to basis Δ' the matrix of B transforms according to

$$B_1 = T'BT,$$

where B and B_1 are the matrices of the functional in the bases Δ and Δ_1 respectively, T is the matrix of Δ_1 relative to Δ, and T' is the transpose of T: with $T = (\tau_{jk})_{j,k=1}^{n}$ and $T' = (\tau'_{jk})_{j,k=1}^{n}$, we have τ'_{jk} = (j, k=1,2,...,n).

We introduce the quadratic functionals, closely connected with bilinear functionals on E, by

Definition 2. A functional Q(x) is said to be *quadratic* if there is a bilinear functional B(x,y) such that

$$Q(x) = B(x,x) \qquad (*)$$

1. The quadratic functionals in E form a linear space in regard to the usual definition of addition and multiplication by scalars.

We shall denote that space by $\mathbb{G}(E)$. Relation (*) defines then an epimorphism j: $\mathbb{B}(E) \to \mathbb{G}(E)$, which we want to examine. ($\mathbb{B}(E)$ is short for $\mathbb{B}(E,E)$, the space of bilinear functionals.)

Definition 3. The bilinear functional B(x,y) is called *antisymmetric* if B(y,x) = -B(x,y).

2. The kernel of the homomorphism j coincides with the subset of antisymmetric functionals in B(E).

Expressed differently, the bilinear functional B(x,y) is antisymmetric iff B(x,x) = 0 for all $x \in E$.

3. A bilinear functional is antisymmetric iff its matrix B is *antisymmetric* in some basis:

$$B' = -B .$$

This implies in particular $\beta_{kk} = 0$ (k = 1,2,...,n).

4. The dimension of the subspace $\mathbb{B}^{-}(E)$ of antisymmetric functionals is n(n-1)/2.

Formulated equivalently, def j = n(n-1)/2 (notation in I,D47).

5. The dimension of the space $\mathbb{G}(E)$ of quadratic functionals is n(n+1)/2.

Definition 4. The bilinear functional B(x,y) is called *symmetric* if B(y,x) = B(x,y).

6. A bilinear functional is symmetric iff its matrix B is

symmetric in some basis:

$$B' = B .$$

7. The dimension of the subspace $\mathbb{B}^+(E)$ of symmetric bilinear functionals is $n(n+1)/2$.

8. $$\mathbb{B}(E) = \mathbb{B}^+(E) \oplus \mathbb{B}^-(E) .$$

Thus every bilinear functional has a unique decomposition into a sum of a symmetric and an antisymmetric functional. This decomposition can be expressed in operator language.

9. The operator T acting on $\mathbb{B}(E)$ and defined by $(TB)(x,y) = B(y,x)$ is an <u>involution</u>, i.e. $T^2 = I$. The subspaces $\mathbb{B}^+(E)$ and $\mathbb{B}^-(E)$ are the characteristic subspaces of T, belonging to the eigenvalues +1, -1 respectively (cf. II,(44)).

Let us now introduce the restriction $j^+ = j|\mathbb{B}^+(E)$.

10. The homomorphism j^+ is an isomorphism between the spaces $\mathbb{B}^+(E)$ and $\mathbb{G}(E)$.

Thus to every quadratic functional Q belongs a unique symmetric bilinear functional B_Q that generates it by (*).

Definition 5. The functional B_Q is known as the *polar* of the functional Q.

The polar B_Q of a given quadratic functional Q can be constructed by a simple device. Use

11. $$Q(x+y) = Q(x) + Q(y) + 2B_Q(x,y) .$$

This identity, a generalization of an elementary identity, implies

12. $$B_Q(x,y) = \tfrac{1}{4}\{Q(x+y) - Q(x-y)\}.$$

Definition 6. The matrix $(\beta_{jk})_{j,k=1}^{n}$ of the polar B_Q will be called the *matrix of the quadratic functional* Q in the basis Δ. This matrix is symmetric, and we may evidently write

$$Q(x) = \Sigma_{j,k=1}^{n} \beta_{jk}\zeta_j\zeta_k \quad (x = \Sigma_{j=1}^{n} \zeta_j e_j).$$

The second member defines now a quadratic functional on the arithmetical space C^n, i.e. a *quadratic form*.

The bijective relation between quadratic functionals and their polars makes it possible to transfer "automatically" the concepts generator, support, kernel, defect, and rank to quadratic functionals (cf. I, Sect. 8). In doing so, the distinction between left and right concepts vanishes (the polar is symmetric!). For an antisymmetric bilinear functional the distinction becomes unessential, since

13. Left and right generators (I,D72) of an antisymmetric bilinear functional differ only by a factor -1.

For symmetric or antisymmetric bilinear functionals the relations of orthogonality expressed by I,(221) become

14. $S_B = (K_B)^{\perp}$ (here S_B is the carrier and K_B the kernel).

Every bilinear functional B generates B-orthogonality between vectors of the spaces E_1 and E_2 according to the general scheme laid out in I,D85-D87. In the present case, B-orthogonality is a relation between vectors of the same space, hence the left and right B-orthogonal complements ${}^{(\perp)}L$ and $L^{(\perp)}$ of a subspace L are again subspaces of E.

15. Any B-orthogonal complement of $L \subset E$ is its complement in the sense of I,D31, iff the restriction of B to L, defined by

$$B_{(L)}(x,y) = B(x,y) \quad \text{for } x,y \in L ,$$

is regular, i.e., iff $\operatorname{def} B_{(L)} = 0$ (cf. I,D88).

16. If one of the two B-orthogonal complements of L is its complement, then so is the other.

Definition 7. A subspace L is said to be B-*regular* if the functional $B_{(L)}$ is regular; a vector system is B-*regular* if its linear hull is B-regular.

17. A vector e is B-regular iff $B(e,e) \neq 0$.

We now turn to the problem of *reducing* a bilinear functional *to canonical form.* A general result in this direction was obtained in I,Sect.8. It can be made more precise, if the functional is symmetric or antisymmetric.

18. Let $\Gamma = \{g_k\}_1^r$ be some system of linear functionals. Then the bilinear functional $B = \Sigma_{k=1}^r g_k \otimes g_k$ is symmetric, and its support S_B is sp Γ (notation in I,D72,D76; $g_k \otimes g_k \equiv \{g_k(x)g_k(y) | x,y$ ranging over E independently).

This already implies rk $B \leq r$. If system Γ is linearly independent, then rk $B = r$.

It turns out that proposition (18) can be inverted:

19. Let B be a symmetric bilinear functional and rk B = $r > 0$. Then there is a basis $\{f_k\}_1^r$ of the support S_B such that $B = \Sigma_{k=1}^r f_k \otimes f_k$.

This result is fundamental for the construction of the canonical form of a <u>symmetric</u> bilinear functional. It may be formulated equivalently:

20. Let B and r be as in (19). Then there is a vector system $\{e_k\}_1^r$, linearly independent modulo K_B (I,D38), such that

$$B(x,y) = \Sigma_{k=1}^r B(x,e_k)B(y,e_k) .$$

The method of proof will be clear, if we observe that

21. The system $\{e_k\}_1^r$ is necessarily B-orthogonal, i.e.,

$$B(e_j,e_k) = \delta_{jk} \quad (j,k = 1,2,\ldots,n) .$$

The following lemma leads then directly to (20):

22. Let e_1 be a B-regular vector, $B(e_1,e_1) = 1$, and let $L = \text{sp}\{e_1\}$. If $x = \zeta e_1 + u$ and $y = \eta e_1 + v$ with $u,v \in L^{(\perp)}$, then

$$B(x,y) = B(x,e_1)B(y,e_1) + B(u,v) .$$

From (19) ensues the following theorem on the <u>canonical form of a quadratic functional</u>:

23. Let Q(x) be a quadratic functional and rk $Q = r > 0$. Then there is a linearly independent system $\{f_k\}_1^r$ of linear functionals such that

$$Q(x) = \Sigma_{k=1}^r [f_k(x)]^2 \quad (f_k \in E') .$$

Thus every quadratic functional $Q \neq 0$ may be represented as a sum of r (=rk Q) squares of linear functionals. Could it be done with fewer squares? The answer is:

24. In any representation of Q as a sum of squares,

$$Q(x) = \Sigma_{k=1}^{r}[g_k(x)]^2 \quad (g_k \in E')$$

$r \geqq \text{rk } Q$. Equality holds iff the system $\{g_k\}_1^r$ is linearly independent.

For antisymmetric bilinear functionals we use an analogous scheme.

25. Let $r > 0$ be an even integer $2m$ and $\Gamma = \{f_k\}_1^r$ be some linearly independent system of linear functionals. Then the bilinear functional

$$B = \Sigma_{k=1}^{m}(f_{2k-1} \otimes f_{2k} - f_{2k} \otimes f_{2k-1})$$

is antisymmetric and its support S_B is identical with the linear hull of the system Γ.

This already implies rk $B = r$. Is it possible to construct an antisymmetric bilinear functional of odd rank? We shall encounter the negative answer on the way to theorem (30), which describes the canonical form of an antisymmetric bilinear functional.

In the following three lemmas B is an antisymmetric bilinear functional.

26. The system $\{e_1, e_2\}$ is B-regular iff $B(e_1, e_2) \neq 0$.

27. A B-regular system $\{e_1, e_2\}$ is linearly independent.

28. Assume the system $\{e_1, e_2\}$ is B-regular and $B(e_1, e_2) = 1$; write $L \underset{\text{def}}{=} \text{sp}\{e_1, e_2\}$. If $x = \zeta_1 e_1 + \zeta_2 e_2 + u$ and $y = \eta_1 e_1 + \eta_2 e_2 + v$ with $u, v \in L^{(\perp)}$, then

$$B(x,y) = B(x,e_1)B(y,e_2) - B(x,e_2)B(y,e_1) + B(u,v) .$$

Now we are able to formulate the two fundamental theorems on antisymmetric bilinear functionals.

29. The rank of any antisymmetric bilinear functional is even.

30. Let B be an antisymmetric bilinear functional and $\operatorname{rk} B = 2m > 0$. Then there is a vector system $\{e_k\}_1^r$, linearly independent modulo K_B, such that

$$B(x,y) = \Sigma_{k=1}^{m} B(x,e_{2k-1})B(y,e_{2k}) - B(x,e_{2k})B(y,e_{2k-1}) .$$

Note the following corollary to (29):

31. If dim E is odd, then all antisymmetric bilinear functionals are nonregular.

Nonregular bilinear functionals on E are also called *degenerate*. The same terminology applies to Hermitian-bilinear functionals.

2. Hermitian-bilinear and Quadratic Functionals. Law of Inertia.

In this section we wish to transfer the theory developed in Section 1 to Hermitian-bilinear functionals H(x,y) (I,D84), but we shall stop only at those points where the Hermitian character becomes in some way manifest.

Definition 8. Let $\Delta = \{e_k\}_1^n$ be some basis of the space E. The matrix $(\gamma_{jk})_{j,k=1}^{n}$ of the coefficients in the expansion

$$H(x,y) = \Sigma_{j,k=1}^{n}\gamma_{jk}\zeta_j\bar{\eta}_k \quad \text{with } x = \Sigma_{j=1}^{n}\zeta_j e_j \text{ and } y = \Sigma_{k=1}^{n}\eta_k e_k$$

is called the *matrix of the Hermitian-bilinear functional* H in the basis Δ. The expansion itself is known as a *Hermitian-bilinear form* on C^n.

Definition 9. In the transition from Δ to Δ_1 the matrix of a Hermitian-bilinear functional transforms according to

$$G_1 = T^* G T .$$

Here G and G_1 are the matrices of the functional in the bases Δ and Δ_1, respectively, T is the matrix of Δ_1 relative to Δ, and T^* is the <u>conjugate</u> <u>transpose</u> of T, defined as follows:

if $T = (\tau_{jk})_{j,k=1}^{n}$ and $T^* = (\tau_{jk}^*)_{j,k=1}^{n}$ then

$$\tau_{jk}^* = \overline{\tau}_{kj} \quad (j,k = 1,2,\ldots,n) \quad .$$

Definition 10. The functional K(x) is called *Hermitian-quadratic*, if there is a Hermitian-bilinear functional H(x,y) such that

$$K(x) = H(x,x) \quad (x \in E) \quad . \qquad (*)$$

Relation (*) defines an epimorphism h from the space of Hermitian-bilinear functionals to the space of Hermitian quadratic functionals. But, unlike the situation in (2), we have now:

32. The homomorphism h is an isomorphism.

Definition 11. By virtue of (32) we may identify the matrix $(\gamma_{jk})_{j,k=1}^{n}$ of the Hermitian-bilinear functional $H_K(x,y)$, generating K(x) by (*), with the *matrix of the Hermitian-quadratic functional* K(x) in the basis Δ. The expansion of K(x) is then

$$\Sigma_{j,k=1}^{n}\gamma_{jk}\zeta_j\overline{\zeta}_k \quad \text{with } x = \Sigma_{j=1}^{n}\zeta_j e_j \ ,$$

which is a *Hermitian-quadratic form* on C^n. (Note that $H_K(x,y)$, the polar of K(x), is unique by (32).

33. The space of Hermitian-quadratic functionals on E has the dimension n^2.

The role of (11) is now played by

34. $$K(x+y) = K(x) + K(y) + H_K(x,y) + H_K(y,x) \ .$$

Whence

35. $$H_K(x,y) = \tfrac{1}{4}\Sigma_{m=0}^{3} i^m K(x + i^m y) \ .$$

Definition 12. A Hermitian-bilinear functional is said to be *symmetric* if $H(y,x) = \overline{H(x,y)}$, and *antisymmetric* if $H(y,x) = -\overline{H(x,y)}$. (The previous definition D4 loses its sense, since $H(x,y) = H(y,x) \Rightarrow H(x,y) \equiv 0$, if H is Hermitian-bilinear.)

Symmetry and antisymmetry are very closely connected for Hermitian-bilinear functionals because

36. A Hermitian-bilinear functional H is antisymmetric iff iH is symmetric.

Thus the two classes of functionals do not differ essentially.

37. A Hermitian-bilinear functional H(x,y) is symmetric iff the corresponding Hermitian-quadratic functional K(x) assumes only real values.

Definition 13. Such a Hermitian-quadratic functional (which has a Hermitian-symmetric polar) will be called *real*.

38. If K is a real Hermitian-quadratic functional, then

$$\operatorname{Re} H_K(x,y) = \tfrac{1}{4}[K(x+y) - K(x-y)] .$$

Observe also the following matrix criterion of symmetry: (cf. (6))

39. A Hermitian-bilinear functional is symmetric iff its matrix G is *Hermitian* in some basis, i.e., if it is its own conjugate transpose in that basis:

$$G^* = G .$$

A real Hermitian-quadratic functional on C^n is known as a real Hermitian-quadratic <u>form</u>. Written explicitly, it is $\Sigma_{j,k=1}^{n}\gamma_{jk}\zeta_j\bar{\zeta}_k$ with a Hermitian matrix $(\gamma_{jk})_{j,k=1}^{n}$. For the reduction of a symmetric Hermitian-bilinear functional to canonical form we shall use the same method as in Section 1.

Let H be a Hermitian-bilinear functional. We define the H-*regular* subspace and the H-*regular* vector system exactly as in D7. In doing so, theorems (15)-(17) remain valid.

40. Assume the vector e_1 as H-regular, $H(e_1,e_1) = \varepsilon$ with $\varepsilon = \pm 1$, and let $L = \operatorname{sp}\{e_1\}$. If $x = \zeta e_1 + u$ and $y = \eta e_1 + v$ with $u,v \in L^{(1)}$, then

$$H(x,y) = \varepsilon H(x,e_1)\overline{H(y,e_1)} + H(u,v) .$$

41. Let H be a symmetric Hermitian-bilinear functional and rk H = r > 0. Then there is a vector system $\{e_k\}_1^r$, linearly independent modulo K_H, such that

$$H(x,y) = \Sigma_{k=1}^{r}\varepsilon_k H(x,e_k)\ \overline{H(y,e_k)}$$

with $\varepsilon_k = \pm 1$ (k = 1,2,...,r).

Conversely (cf.(18)):

42. Let $\Gamma = \{g_k\}_1^r$ be some system of linear functionals and $\{\alpha_k\}_1^r$ some system of non-vanishing real numbers. Then the Hermitian-bilinear functional

$$H(x,y) = \Sigma_{k=1}^{r}\ \alpha_k g_k(x)\ \overline{g_k(y)}$$

is symmetric and its support is sp Γ.

This already implies rk H ≦ r. If system Γ is linearly independent, then rk H = r.

The canonical form of a real Hermitian-quadratic functional derives now directly from theorem (41):

43. Let K(x) be a real Hermitian-quadratic functional with rk K = r > 0. Then there is a linearly independent system $\{f_k\}_1^r$ of linear functionals such that (cf.(23))

$$K(x) = \Sigma_{k=1}^{r}\ \varepsilon_k|f_k(x)|^2 .$$

The counterpart to (24) is:

44. If the real Hermitian-quadratic functional K(x) is represented in the form

$$K(x) = \Sigma_{k=1}^{s}\ \alpha_k|g_k(x)|^2 , \qquad (**)$$

where the g_k are linear functionals and the α_k are non-vanishing real numbers, then s ≧ rk K. Equality is here possible iff the system $\{g_k\}_1^s$ is linearly independent.

Thus the number of terms in the representation (**) is uniquely determined by the functional K, provided the functionals g_k are linearly independent. But, in fact, even the number of terms with $\alpha_k > 0$ (and $\alpha_k < 0$) is uniquely determined. This remarkable result carries the name *Sylvester's law of inertia*. To formulate it more completely we need some auxiliary concepts.

Definition 14. Let K(x) be a real Hermitian-quadratic functional. A subspace L is called K-positive if

$$K(x) > 0 \quad \text{for all } x \in L - \{0\} .$$

The largest dimension occurring among all K-positive subspaces is known as the *positive index of inertia* of the functional K and denoted by $\text{ind}_+ K$. K-negative subspaces and the *negative index of inertia*, $\text{ind}_- K$, are defined correspondingly.

Definition 15. A subspace L is called K-*nonnegative* if

$$K(x) \geqq 0 \quad \text{for all } x \in L .$$

K-*nonpositive* subspaces are defined correspondingly.

45. If the real Hermitian-quadratic function K(x) is represented in the form

$$K(x) = \Sigma_{k=1}^{s} \alpha_k |g_k(x)|^2$$

with linearly independent linear functionals $\{g_k\}_1^s$, then the number of positive α_k equals ind_+K and the number of negative coefficients α_k equals ind_-K.

46. $$\text{ind}_+K + \text{ind}_-K = \text{rk } K .$$

Note also the following result, connected with the indices of inertia. Introducing first

Definition 16. A subspace L is called K-neutral, if

$$K(x) = 0 \quad \text{for all } x \in L .$$

We have:

47. The largest dimension of a K-neutral subspace is

$$\min(\text{ind}_+K, \text{ind}_-K) + \text{def } K .$$

In conclusion we give a classification of the real Hermitian-quadratic functionals:

K(x) is positive, if $ind_+K=n$. (Then rk K=n, ind_-K=def K=0.)

K(x) is nonnegative if $ind_-K = 0$.

K(x) is negative, if $ind_-K=n$. (Then rk K=n, ind_+K=def K=0.)

K(x) is nonpositive, if $ind_+K = 0$.

Positive and negative functionals are known as <u>definite</u>. Functionals with $ind_+K \neq 0$ <u>and</u> $ind_-K \neq 0$ are known as <u>indefinite</u>.

The law of inertia makes it possible to determine the type of a functional if it is given in the representation (45).

3. Unitary Space

Definition 17. *Unitary space* is a complex linear space in which a scalar product has been defined. *Scalar product* is the name given to some fixed symmetric Hermitian-bilinear functional, which generates a positive Hermitian-quadratic functional. This functional is denoted by (x,y). This definition implies the following properties of (x,y):

1. $(x_1 + x_2, y) = (x_1,y) + (x_2,y)$ (distributivity);
2. $(\alpha x,y) = \alpha(x,y)$ (homogeneity);
3. $(y,x) = \overline{(x,y)}$ (symmetry);
4. $(x,x) > 0$ iff $x \neq 0$ (positivity) .

Note that $(x,\alpha y) = \bar{\alpha}(x,y)$ is a formal consequence of 2 and 3.

Definition 18. Vectors x and y of a unitary space are said to be *mutually orthogonal* (symbol: $x \perp y$), if

$$(x,y) = 0 .$$

In other words, they are H-orthogonal with regard to the Hermitian-bilinear functional H which is the scalar product. Definition 18 provides the simplest possible theory of orthogonality for a unitary space; the concept of orthogonality of a vector and a linear functional, based on

I,D66, does the same for a general linear space. This is a formal consequence of the regularity I,D88 of the scalar-product functional considered in the more general framework of I, Sect. 10. The following fundamental theorem on the general form of a linear functional in unitary space thoroughly illuminates the situation.

48. Every linear functional f, defined on the unitary space E, can be represented in the form

$$f(x) = (x,y) ,$$

where y is some well-defined vector, which we denote here by y_f to emphasize its relation to the functional f (*Theorem of F. Riesz*).

49. The relation $y = y_f$, indicating the mapping $f \to y$ or $E' \to E$, is a Hermitian isomorphism (I,D82,(243)).

Definition 19. This mapping is the *canonical isomorphism* between the spaces E' and E; it appears only after the general linear space E has been converted into a unitary space by introducing a scalar product.

For the rest of this chapter we shall consider the space E as unitary and denote the orthogonal complement[§] of $L \subset E$ by $L^{\perp}$. The same notation was used for the orthogonal complement in I,D67. The two possible interpretations of the term "orthogonal complement" are equivalent in the following sense:

50. Let M be a subspace of E' and $N \subset E$ its image under the canonical Hermitian isomorphism $E' \to E$. Then $M^{\perp} = N^{\perp}$. Here $M^{\perp}$ refers to orthogonality of vector and linear functional by I,D66, while $N^{\perp}$ refers to orthogonality in E by D18.

We now enumerate the standard theorems about orthogonal completion.

51. $\dim L^{\perp} = \operatorname{codim} L$

52. $L \oplus L^{\perp} = E$

[§]Left and right orthogonal complements coincide because of the Hermitian symmetry of the scalar product.

53. $(L^{\perp})^{\perp} = L$

54. $L_1 \subset L_2 \Rightarrow L^{\perp}_1 \supset L^{\perp}_2$

55. $(L_1 + L_2)^{\perp} = L^{\perp}_1 \cap L^{\perp}_2$; $(L_1 \cap L_2)^{\perp} = L^{\perp}_1 + L^{\perp}_2$

In the unitary case duality generates orthogonality "without leaving the space." (The canonical Hermitian isomorphism compensates the transition to the conjugate space.)

To continue the presentation we need some new concepts.

Definition 20. A system of subspaces $\{L_k\}_1^m$ is called *orthogonal* if $L_j \perp L_k$ $(j \neq k)$. The system $\{L, L^{\perp}\}$ is an example.

56. If the system $\{L_k\}_1^m$ is orthogonal, then the sum $\Sigma_{k=1}^{m} L_k$ is direct.

Definition 21. The sum (I,D28) of an orthogonal system of subspaces will be called an *orthogonal sum* and denoted by

$$L_1 \boxplus L_2 \boxplus \ldots \boxplus L_m \quad \text{or} \quad \underset{k=1}{\overset{m}{\boxplus}} L_k$$

57. Let $L = \boxplus_{k=1}^{m} L_k$ and let $x, y \in L$ have the expansions $x = \Sigma_{k=1}^{m} x_k$ and $y = \Sigma_{k=1}^{m} y_k$ with $x_k, y_k \in L_k$. Then

$$(x,y) = \Sigma_{k=1}^{m} (x_k, y_k) .$$

In particular,

$$(x,x) = \Sigma_{k=1}^{m} (x_k, x_k) \quad \text{if } x \in \Sigma_{k=1}^{m} L_k .$$

This relation is usually called *Parseval's Equation;* it is a characteristic property of orthogonal systems:

58. If $\{L_k\}_1^m$ is a system of subspaces, and Parseval's equation holds for any $x = \Sigma_{k=1}^{m} x_k$ with $x_k \in L_k$, then the system is orthogonal.

Any subspace L generates the decomposition $E = L \boxplus L^{\perp}$ by (52). So any x can be uniquely represented in the form

$$x = u + v \quad \text{with } u \in L,\ u \in L^{\perp}\ .$$

Here the term u is called the orthogonal projection of the vector x on the subspace L. Clearly, u = Px, where P is the projector associated with the pair of subspaces $\{L, L^{\perp}\}$ (cf. II,(218)). Since there is only one $L^{\perp}$ to a given L, the operator P depends only on L; we indicate this by

$$P = P[L].$$

Definition 22. The projector P[L] is known as an *orthoprojector*; it projects onto the subspace L.

59. P is an orthoprojector iff $\operatorname{Ker} P = (\operatorname{Im} P)^{\perp}$.

Let us elaborate on the relation P = P[L] between a subspace and the operator it determines. Abbreviating $P[L^{\perp}]$ by $P^{\perp}$, we have

60. $P^{\perp} + \overline{P[L]}$ (notation in II,D48).

61. $L \perp M \Leftrightarrow P[L]P[M] = 0$.

Hence

62. $P[L]P[M] = 0 \Leftrightarrow P[M]P[L] = 0$.

Theorem (61) says that mutually orthogonal subspaces determine mutually orthogonal orthoprojectors and conversely.

63. If $\{L_k\}_1^m$ is an orthogonal system of subspaces, then

$$P[\boxplus_{k=1}^{m} L_k] = \Sigma_{k=1}^{m} P[L_k] \quad \text{(cf. II,(221))}.$$

Conversely,

64. If $P[\Sigma_{k=1}^{m} L_k] = \Sigma_{k=1}^{m} P[L_k]$, then $\{L_k\}_1^m$ is an orthogonal system (cf. II,(239)).

65. An orthogonal system of subspaces $\{L_k\}_1^m$ is basic (I,D34) iff $\Sigma_{k=1}^{m} P[L_k] = I$, that is, iff the corresponding orthoprojectors form a decomposition of the identity.

Definition 23. A decomposition of the identity, whose components are orthoprojectors, is called an *orthogonal decomposition* of the identity.

66. The formulas $P_k = P[L_k]$ $(k = 1,2,\ldots,m)$ establish a bijective relation between basic orthogonal subspaces $\{L_k\}_1^m$ and the components of an orthogonal decomposition of the identity, $\{P_k\}_1^m$.

Definition 24. Let $\{L_k\}_1^m$ be some orthogonal system of subspaces $\subset E$ (not necessarily basic). A mapping $p: E \to L_1 \times L_2 \times \ldots \times L_m$ defined by

$$px = (x_1, x_2, \ldots, x_m) \quad \text{where } x_k = P[L_k]x \ (k = 1,2,\ldots,m) ,$$

will be called a *formal decomposition* of E into the Cartesian product $L_1 \times L_2 \times \ldots \times L_m$. Correspondingly, we shall call px the formal decomposition of the vector $x \in E$.

67. The formal decomposition is a homomorphism; in fact, it is an epimorphism. Therefore a monomorphism j exists which is the right inverse of p (cf. (56) and I,(121)).

68. The homomorphism j has the form $j\{x_k\}_1^m = \Sigma_{k=1}^m x_k$. (*)

69. $pj = I_{L_1 \times L_2 \times \ldots \times L_m}$, $jp = P[\boxplus_{k=1}^m L_k]$.

Therefore

70. $x \in \boxplus_{k=1}^m L_k \Leftrightarrow x = \Sigma_{k=1}^m P[L_k]x$.

71. $\text{Ker } p = (\boxplus_{k=1}^m L_k)^\perp$.

In this way we obtain

72. $\text{Im } j \boxplus \text{Ker } p = E$.

73. The formal decomposition with respect to the system $\{L_k\}_1^m$ is a monomorphism (and hence an isomorphism), iff the system is basic.

This proposition can be considered as a criterion for the basic character of some system $\{L_k\}_1^m$, formulated in terms of the formal decomposition: If $\{L_k\}_1^m$ is basic, then the homomorphism j, being the right inverse of the formal decomposition p, becomes the inverse $j = p^{-1}$; formula (*) then becomes the decomposition of a vector with respect to the given system:

$$x = \Sigma_{k=1}^{m} P[L_k]x \ .$$

An important aspect of the formal decomposition becomes manifest in connection with Parseval's equation.

74. If $px = (x_1, x_2, \ldots, x_m)$ is a formal decomposition of $x \in E$, then

$$(x,x) - \Sigma_{k=1}^{m} (x_k, x_k) = (r,r) \ ,$$

where $r = x - jpx = x - \Sigma_{k=1}^{m} x_k$.

75. This gives us *Bessel's Inequality*

$$\Sigma_{k=1}^{m} (x_k, x_k) \leqq (x,x) \ ,$$

which becomes an equation (viz. Parseval's equation), iff $x \in \boxplus_{k=1}^{m} L_k$.

76. An orthogonal system of subspaces is basic iff Parseval's equation holds for all $x \in E$.

Let us now consider the parallel theory for systems of vectors (instead of subspaces). It could be reduced to the previous theory, but it can also be developed independently.

Definition 25. A vector system is called *orthogonal* if

$$e_j \perp e_k, \text{ for } j \neq k \ ;$$

it is *orthonormal* if in addition

$$(e_k, e_k) = 1 \qquad (k = 1,2,\ldots,m) \ .$$

Thus $(e_j, e_k) = \delta_{jk}$ for orthonormal systems.

Any subsystem of an orthogonal (orthonormal) vector system is of course orthogonal (orthonormal).

77. An orthogonal vector system is linearly independent, if it does not contain the zero vector.

Hence an orthonormal vector system is always linearly independent. In what follows we presuppose that none of our orthogonal systems $\{e_k\}_1^m$ contains 0.

78. Let $\{e_k\}_1^m$ be an orthogonal system of vectors. If $x = \Sigma_{k=1}^m c_k e_k$, then

$$c_k = (x,e_k)/(e_k,e_k) \quad (k = 1,2,\ldots,m)$$

(<u>Euler-Fourier</u> <u>Formulae</u>).

These formulae become particularly simple for an orthonormal system, viz. $c_k = (x,e_k)$ $(k = 1,2,\ldots,m)$.

Note that the c_k in (78) are well-defined for all $x \in E$.

Definition 26. The quantities $c_k(x) \underset{\text{def}}{=} (x,e_k)/(e_k,e_k)$, are called the *Fourier coefficients* of the vector x with respect to the orthogonal system $\{e_k\}_1^m$.

79. The operator P acting in E according to

$$Px = \Sigma_{k=1}^m c_k(x)e_k \quad (x \in E)$$

is the orthoprojector onto $sp\{e_k\}_1^m$.

80. Bessel's inequality,

$$\Sigma_{k=1}^m |c_k(x)|^2(e_k,e_k) \leqq (x,x) \ ,$$

holds. It turns into Parseval's equation,

$$\Sigma_{k=1}^m |c_k(x)|^2(e_k,e_k) = (x,x) \ ,$$

iff $x \in sp\{e_k\}_1^m$.

The simple form of these relations for an orthonormal system is obvious.

81. The orthogonal vector system $\{e_k\}_1^m$ is a basis iff Parseval's equation holds for all $x \in E$.

Parseval's equation plays a double role; it indicates completeness <u>and</u> orthogonality: (for completeness cf. I,D7 and the remark preceding I,D17):

82. Let $\Gamma = \{u_k\}_1^m$ be some vector system not containing 0. Parseval's equation $\Sigma_{k=1}^m |(x,u_k)|^2/(u_k,u_k) = (x,x)$ holds for

all $x \in \text{sp}\,\Gamma$, iff Γ is an orthogonal system.

Note that by (77) an orthogonal system is a basis iff it is complete. We can therefore formulate

Definition 27. An orthogonal (orthonormal) basis is a *complete orthogonal (orthonormal) vector system.*

An orthonormal basis could be defined as coincident with its conjugate [up to the canonical Hermitian isomorphismus (D19)].

83. The orthogonal vector system $\{e_k\}_1^m$ is complete iff the zero vector is the only vector that is orthogonal to all vectors e_k $(k = 1,2,\dots,m)$ (cf.I,(173)).

In other words, the criterion of completeness requires that any vector, whose Fourier coefficients with respect to the given $\{e_k\}_1^m$ vanish, equals 0.

Theorem (82) in conjunction with theorem (43) on the canonical form of a real Hermitian-quadratic functional leads to the fundamental result:

84. Any unitary space has an orthonormal basis.

One can even construct a basis that satisfies some special additional conditions such as the following:

85. Let $\{L_k\}_1^m$ be some basic orthogonal system of subspaces and let $\dim L_k = n_k$ $(k = 1,2,\dots,m)$. There is an orthonormal basis $\{e_k\}_1^n$ such that the subsystems

$$\{e_k\}_1^{n_1},\ \{e_k\}_{n_1+1}^{n_1+n_2},\ \dots,\ \{e_k\}_{n_1+\dots+n_{m-1}+1}^{n}$$

are bases of the subspaces $\{L_k\}_1^m$ respectively.

86. Any orthogonal (orthonormal) system $\{e_k\}_1^m$ can be completed to an orthogonal (orthonormal) basis $\{e_k\}_1^n$.

87. Let $\{L_k\}_o^n$ be some maximal chain of subspaces. There is an orthonormal basis $\{e_k\}_1^n$ such that each of the subsystems $\{e_k\}_1^m$ is a basis for L_m $(m = 1,2,\dots,n)$ (cf.I,(51)).

In other words,

88. For any basis $\{u_k\}_1^n$ of E there is an orthonormal basis $\{e_k\}_1^n$ of E such that $\text{sp}\{u_k\}_1^m = \text{sp}\{e_k\}_1^m$ for all $m = 1,2,\dots,n$.

This coincidence of the linear hulls means that one can pass from one basis to the other by means of triangular matrices:

$$e_k = \Sigma_{j=1}^{k} \alpha_{jk} u_j , \quad u_k = \Sigma_{j=1}^{k} \beta_{jk} e_j .$$

The actual construction of an orthonormal basis $\{e_k\}_1^n$, "triangular-equivalent" with a given basis $\{u_k\}_1^n$, can be carried out by the *Sonine-Schmidt orthogonalization*, also known as *Gram-Schmidt orthogonalization*, which is a simple recurrence algorithm:

89. Let $\{u_k\}_1^{m+1}$ be any linearly independent system and $\{e_k\}_1^m$ an orthonormal system such that $sp\{e_k\}_1^m = sp\{u_k\}_1^m$ holds for m vectors. Put

$$e'_{m+1} = u_{m+1} - \Sigma_{k=1}^{m} (u_{m+1}, e_k) e_k, \text{ and } e'_1 = u_1 \text{ for}$$

$m = 0$. Since $e'_{m+1} \neq 0$, we can write

$$e_{m+1} = e'_{m+1}/(e'_{m+1}, e'_{m+1})^{\frac{1}{2}}$$

Then $\{e_k\}_1^{m+1}$ is an orthonormal system, whose linear hull coincides with the linear hull of $\{u_k\}_1^{m+1}$.

The orthogonalization process leads again to theorem (86). Note also that

90. In an orthonormal basis the scalar product assumes the canonical form:

$$(x,y) = \Sigma_{k=1}^{n} \zeta_k \bar{\eta}_k \quad (x = \Sigma_{k=1}^{n} \zeta_k e_k, \; y = \Sigma_{k=1}^{n} \eta_k e_k) .$$

Thus the matrix of the scalar product (considered as a symmetric Hermitian-bilinear functional) is the unit matrix, and this is its characteristic property:

91. If the scalar product has the canonical form in some basis, then the basis is orthonormal.

In a similar way, an orthogonal basis is characterized by the fact that the scalar product has a diagonal matrix. In general, the matrix of the scalar product in the basis $\{v_k\}_1^n$ is

$$(\gamma_{jk})_{j,k=1}^{n} = ((v_j,v_k))_{j,k=1}^{n} .$$

Definition 28. For any vector system $\{w_k\}_1^m$, the matrix

$$(\gamma_{jk})_{j,k=1}^{m} = ((w_j,w_k))_{j,k=1}^{m}$$

is called the *Gram Matrix* of that vector system. It is a Hermitian matrix for any $\{w_k\}_1^m$.
Gram's matrix allows us to extend Bessel's inequality and Parseval's equation (which is also called the *completeness relation* because of (81)) to arbitrary linearly independent systems. We start with a preparatory remark.

92. A vector system is linearly independent, iff its Gram matrix is nonsingular.

93. Let $(\gamma_{jk})_{j,k=1}^{m}$ be the Gram matrix of the linearly independent system $\Gamma = (w_k{}_1^m$, and $(\gamma_{jk}^{(-1)})_{j,k=1}^{m}$ the inverse matrix. Then

$$\Sigma_{j,k=1}^{m} \gamma_{jk}^{(-1)} (x,w_j)\overline{(x,w_k)} \leqq (x,x)$$

for all $x \in E$, and the equality sign holds iff $x \in \mathrm{sp}\,\Gamma$.

94. The linearly independent system $\{w_k\}_1^m$ is complete iff

$$\Sigma_{j,k=1}^{m} \gamma_{jk}^{(-1)} (x,w_j)\overline{(x,w_k)} = (x,x) \quad (\text{all } x \in E).$$

Definition 29. In unitary space the concept of biorthogonal vector systems (I,D68) has a natural place. The systems $\Gamma = \{w_k\}_1^m$, $\Gamma^* = \{w_k{}^*\}_1^m$ are called *biorthogonal*, if $(w_j,w_k{}^*) = \delta_{jk}$ $(j,k = 1,2,\dots,m)$. A particular example is an orthonormal system, which is always biorthogonal to itself. The general theory of biorthogonal systems is obtained from I,(187)-(192) through the application of the canonical Hermitian isomorphism and thus does not need an independent construction. We only mention the following generalization of the Euler-Fourier formulas:

95. If $x = \Sigma_{k=1}^{m} \zeta_k w_k$, then $\zeta_k = (x, w_k^*)$ (cf. I,(168)),

implying

96. $(x,x) = \Sigma_{k=1}^{m} (x,w_k^*)(x,w_k)$ $(x \in \mathrm{sp}\,\Gamma)$,

This is Parseval's equation for biorthogonal systems. Bessel's inequality does not carry over but Parseval's equation continues to serve as a criterion of completeness:

97. The vector system $\{w_k\}_1^m$ is complete iff

$$\Sigma_{k=1}^{m} (x,w_k^*)\overline{(x,w_k)} = (x,x) \qquad (x \in E) .$$

Finally we want to clarify the role of Gram's matrix in the theory of biorthogonal systems:

98. The orthogonal projection of the vector w_j onto sp Γ^* equals $\Sigma_{k=1}^{m} \gamma_{jk}^{(-1)} w_k^*$, where $(\gamma_{jk})_{j,k=1}^{m}$ is Gram's matrix of the system Γ.

In particular,

99. The expansion of the vectors of any basis $\Delta = \{e_j\}_1^n$ in the biorthogonal basis $\Delta^* = \{e_k^*\}_1^n$ has the form $e_j = \Sigma_{k=1}^{n} \gamma_{jk}^{(-1)} e_k^*$, where $(\gamma_{jk})_{j,k=1}^{n}$ is the Gram matrix of Δ^*.

Hence

100. The Gram matrices of the biorthogonal bases Δ and Δ^* are mutually inverse.

The expansion (99) for the e_j may therefore be written in the form $e_k^* = \Sigma_{j=1}^{n} \gamma_{kj} e_j$.

4. The Adjoint Operator. Orthogonal Reduction by a Subspace.

Definition 30. Let A be a linear operator acting in the unitary space E and consider the *Hermitian-conjugate operator* A*, acting in the Hermitian-conjugate space E^* (cf. I,D80, also II,D3). $A^* : E^* \to E^*$ is defined by

$$(A^*g)(x) = g(Ax) \qquad (x \in E;\ g \in E^*).$$

(This is completely analogous to the definition of the conjugate homomorphism in I,D69.)

The canonical complex conjugation (I,D82) "translates" g from E^* to E', and the canonical Hermitian isomorphism (D19) "translates" it from E' to E. The mapping $E^* \to E$ so obtained will be denoted by T. Thus T: $g \to y$, $y \in E$.

101. T is an isomorphism.

Definition 31. The mapping T is called the *canonical isomorphism* between the spaces E^* and E.

It is easy to characterize T directly by an identity that follows from F. Riesz' theorem:

102. $g(x) = (Tg,x)$ $\qquad (x \in E,\ g \in E^*)$.

Conversely:

103. Let $g \in E^*$ be some Hermitian-linear functional. If $g(x) = (y,x)$ identically for all $x \in E$ and some fixed vector y, then $y = Tg$.

Correspondingly, we can reformulate D30 without reference to g. Using F. Riesz' theorem again we obtain:

104. $(Ax,y) = (x,TA^*T^{-1}y)$ $\qquad (x,y \in E)$.

Definition 32. The operator TA^*T^{-1}, which is similar (II,D34) to A^*, acts in the space E. Thus it ought to be distinguished from A^* by calling it, e.g., the internal Hermitian-conjugate to A and denoting it by another symbol. However, one traditionally uses the old symbol A^*, and in the Russian text A^* is called briefly the "conjugate of A". We shall use the same notation, but we shall call A^* the *adjoint (operator)* of A.

Definition 33. The identity (104) now assumes the form

$$(Ax,y) = (x,A^*y) \qquad (x,y \in E).$$

This can serve as the definition of the adjoint operator A^* (in Russian, "the new conjugate operator acting within E").

105. If the operators A and B acting in E satisfy

$$(Ax,y) = (x,By) \qquad (x,y \in E) ,$$

identically, then $B = A^*$.

The functional

$$H(x,y;A) \underset{\text{def}}{=} (Ax,y) \qquad (x,y \in E) \qquad (*)$$

is a Hermitian-bilinear functional (I,D84).

106. The mapping of the space of operators into the space of Hermitian-bilinear functionals, effected by (*), is an isomorphism.

This assertion can be made more precise:

107. In any orthonormal basis $\{e_k\}_1^n$, the matrix of the functional H(x,y;A) coincides with the transposed matrix of the operator A.

This means that the matrix elements of the operator A are

$$\alpha_{jk} = (Ae_k,e_j) \ .$$

108. If **A** is the matrix of A in an orthonormal basis, then the matrix of A^* in the same basis is $\mathbf{A}^*$ (D9).

Let us now list the most important properties of the concept "adjointness":

109. $A^{**} = A$. The equality sign is here literally correct rather than correct [c.i.].

110. $(A+B)^* = A^* + B^*$

111. $(\alpha A)^* = \bar{\alpha} A^*$

112. $(AB)^* = B^*A^*$

113. If A is regular, then so is A^* and $(A^*)^{-1} = (A^{-1})^*$, and

in addition

114. $A \nabla B \Leftrightarrow A^* \nabla B^*$

The Fredholm Theory is simpler in unitary space, because the adjoint operator now acts in the original space. The second and the third Fredholm theorems can now be formulated without "leaving" E. Omitting a formal account we list the principal relations:

115. $\operatorname{Im} A^* = (\operatorname{Ker} A)^\perp$, $\operatorname{Ker} A^* = (\operatorname{Im} A)^\perp$.

116. $\operatorname{rk} A^* = \operatorname{rk} A$, $\operatorname{def} A^* = \operatorname{def} A$.

Using the Fredholm theory, we can rederive

$$\sigma(A^*) = \overline{\sigma(A)}$$

(II,(119)), and also establish the following propositions:

117. If A is an operator of scalar type (II,D18) and $\{e_k\}_1^n$ its characteristic basis, then the biorthogonal system $\{e^*\}_1^n$ is the characteristic basis for A^*, and the eigenvalues associated with e_k and e_k^* are conjugate complex.

118. Those characteristic subspaces of A and A^* that are associated with eigenvalues λ and μ are mutually orthogonal, if $\lambda \neq \bar{\mu}$.

Thus $(W_k(A))^\perp = \Sigma^{\oplus}_{j\neq k} W_j(A^*)$, provided the two spectra have been numbered correspondingly.

Let us now study the connection between the invariant subspaces of A and A^*.

119. If the subspace L is invariant for A, then $L^\perp$ is invariant for A^*, and the reductions are connected by $(A|L)^* = P[L]A^* L^\perp$. Here P[L] is the orthoprojector projecting on L (D22).

Definition 34. We shall say that the subspace L *orthogonally reduces* the operator A, if both L and $L^\perp$ are invariant with regard to A.

120. L orthogonally reduces A iff it is invariant with regard to A and A^*. In that case $(A|L)^* = A^*|L$.

Definition 35 introduces the orthogonal sum of operators. Let $E = \boxplus_{k=1}^{m} L_k$ be the decomposition of E into an orthogonal sum of (nontrivial) invariant subspaces L_k for the operator. Then A is said to be the *orthogonal sum* of its parts:

$$A = \boxplus_{k=1}^{m} A_k \qquad \text{(cf. II,D23).}$$

Here it is clear that every subspace L_k orthogonally reduces the operator A.

Definition 36. The operator A is called *orthogonally irreducible*, if it cannot be decomposed into a nontrivial orthogonal sum (i.e., if it does not possess a nontrivial orthogonally reducing subspace).

121. If $A = \boxplus_{k=1}^{m} A_k$, then $A^* = \boxplus_{k=1}^{m} A_k^*$

The set of orthogonally reducing subspaces of A has a definite algebraic structure (in general this is not true for reducing subspaces):

122. Sum and intersection of any set of orthogonally reducing subspaces are again orthogonally reducing subspaces (cf. II,(51)).

5. Spectral Theory of Self-Adjoint Operators. The Algebra of Orthoprojectors.

Definition 37. The operator S acting in the unitary space E is called *self-adjoint* if

$$S^* = S, \quad \text{that is}$$

$$(Sx,y) = (x,Sy) \quad \text{for all } x,y \in E .$$

The *class of self-adjoint operators* in E will be denoted by $\mathbb{S}(E)$.

123. $A \in \mathbb{S}(E)$ iff the Hermitian-bilinear functional $H(x,y;A)$ is symmetric.

124. An operator is self-adjoint iff its matrix in some orthonormal basis is Hermitian (defined in (39)).

Here are some examples of self-adjoint operators.

125. Every orthoprojector (D22) is a self-adjoint operator.

Conversely,

126. Every self-adjoint projector is an orthoprojector.

127. The operators A^*A and AA^* are self-adjoint for any operator A.

128. The operators $S = (A + A^*)/2$ and $T = (A - A^*)/2i$ are self-adjoint for any operator A.

They are called the real and imaginary parts of A. Clearly,

$$A = S + iT . \qquad (*)$$

Conversely,

129. If $A = S_1 + iT_1$ and $S_1, T_1 \in \mathbb{S}(E)$, then $S_1 = S$ and $T_1 = T$.

Definition 38. Formula (*) is called the *Cartesian representation* of the operator A. It is the analog of the usual representation of complex numbers for operators.

130. If L is an invariant subspace of the self-adjoint operator S, then the restriction $S|L$ is a self-adjoint operator in L.

131. $S_1, S_2 \in \mathbb{S}(E) \Rightarrow (S_1 + S_2) \in \mathbb{S}(E)$. For real α, $\alpha S \in \mathbb{S}(E)$, if $S \in \mathbb{S}(E)$.

Note that for $S \in \mathbb{S}(E)$ and $S \neq 0$, $\alpha S \in \mathbb{S}(E)$ only if α is real. Thus (131) means that $\mathbb{S}(E)$ is a real linear space in the sense of Chapter VI.

Together with (130) this implies a remarkable property of the spectrum of self-adjoint operators:

132. The spectrum of a self-adjoint operator lies on the real axis.

In addition we have:

133. The characteristic subspaces of a self-adjoint operator form an orthogonal system.

Is this system a basic system (I,D34)? This question, which is central in the theory of self-adjoint operators, can be answered by virtue of lemma

134. For a self-adjoint operator every invariant subspace is an orthogonally reducing subspace.

Now it can be easily established that

135. The characteristic subspaces of a self-adjoint operator form a basic system.

Thus $S \in \mathfrak{S}(E)$ implies that S is an operator of scalar type (II,D18), and (84) implies in addition:

136. Every self-adjoint operator has an orthonormal characteristic basis.

We can formulate this result in terms of the decomposition of the identity:

137. The decomposition of the identity by a self-adjoint operator is orthogonal.

Thus we have established the <u>spectral</u> <u>theorem</u> in the theory of <u>self-adjoint</u> <u>operators</u>:

138. The spectral resolution of a self-adjoint operator has the form

$$S = \Sigma_{k=1}^{m} \lambda_k P_k \ .$$

where $\{P_k\}_1^m$ is an <u>orthogonal</u> decomposition of the identity, and $\{\lambda_k\}_1^m$ is a set of real numbers (cf. II,(227)).

This representation is essentially unique:

139. If a self-adjoint operator admits the representation $S = \Sigma_{k=1}^{m} \lambda_k P_k$ with $\lambda_j \neq \lambda_k$ for $j \neq k$, where the P_k are projectors and $\Sigma_{k-1}^{m} P_k = I$, then $\{P_k\}_1^m$ is the decomposition of the identity by the operator S and $\{\lambda_k\}_1^m$ is its spectrum.

Every operator of the form $S = \Sigma_{k=1}^{m} \lambda_k P_k$, where $\{P_k\}_1^m$ is an orthogonal decomposition of the identity and the $\{\lambda_k\}_1^m$ are real numbers, is clearly self-adjoint. This holds in particular for an operator with orthonormal characteristic basis and real spectrum. Thus, exactly those operators are self-adjoint that possess a real spectrum and an orthonormal characteristic basis.

Let us now quickly point out which class of operators is

obtained, if only the requirements "real spectrum and scalar type" are retained.

140. An operator of scalar type has a real spectrum iff it is similar (II,D34) to a self-adjoint operator.

Here it is convenient to introduce

Definition 39. An operator that is similar to a self-adjoint operator is said to be *symmetrizable.*

The foregoing results give a considerably stronger version of theorem (42) on the canonical form of a symmetric Hermitian-bilinear functional.

141. For any symmetric Hermitian-bilinear functional $H(x,y)$ in a unitary space there exists an orthonormal basis $\{e_k\}_1^n$ such that

$$H(x,y) = \Sigma_{k=1}^{n} \lambda_k \zeta_k \bar{\eta}_k \qquad (x = \Sigma_{k=1}^{n} \zeta_k e_k,\ y = \Sigma_{k=1}^{n} \eta_k e_k).$$

Whence it follows that

142. For any real Hermitian-quadratic functional $K(x)$ in a unitary space there exists an orthonormal basis $\{e_k\}_1^n$ such that

$$K(x) = \Sigma_{k=1}^{n} \lambda_k |\zeta_k|^2 \qquad (x = \Sigma_{k=1}^{n} \zeta_k e_k) .$$

Definition 40. This basis is known as a *system of principal axes* of the functional K.

Note also:

143. In the canonical representations (141) and (142) the coefficients $\{\lambda_k\}_1^n$ are unique except for a change of order (cf.(45)).

Theorem (142) has an important interpretation:

144. If $K(x)$ and $K_o(x)$ are real Hermitian-quadratic functionals in a unitary space and if $K_o(x)$ is positive, then there exists a basis $\{e_k\}_1^n$ such that

$$K(x) = \Sigma_{k=1}^{n} \lambda_k |\zeta_k|^2,\quad K_o(x) = \Sigma_{k=1}^{n} |\zeta_k|^2 \qquad (x = \Sigma_{k=1}^{n} \zeta_k e_k).$$

This is the theorem on the *simultaneous reduction* of a pair of functionals to canonical form.

Let us now return to self-adjoint operators and start with an elementary property of the resolvent (II,D37) of $S \in \mathbf{S}$.

145. $$[R_\lambda(S)]^* = R_{\bar{\lambda}}(S) .$$

A particular case of (145) is

146. For any real $\lambda \notin \sigma(S)$, $R_\lambda(S)$ is a self-adjoint operator.

A deeper property reveals itself, if we consider scalar-valued functions $(R_\lambda x,x)$ of the variable λ, where x is some fixed vector $\in E$, and R_λ stands for $R_\lambda(S)$.

147. The function $(R_\lambda x,x)$ is real on the real axis and satisfies in the upper halfplane the inequality

$$\mathrm{Im}(R_\lambda x,x) > 0 \qquad \text{(cf. II,(129))}.$$

Thus $R_\lambda(x,x)$ belongs to a class of functions $f(\lambda)$, which are holomorphic on $\mathrm{Im}\,\lambda > 0$ and satisfy there $\mathrm{Im}\, f(\lambda) \geqq 0$. It is not difficult to show that the most general form of a rational $f(\lambda)$ which approaches zero as $\lambda \to \infty$ is

$$f(\lambda) = \Sigma_{k=1}^{m} \frac{c_k}{\lambda_k - \lambda} ,$$

where the poles λ_k are real and the residues c_k positive. This remark could serve as another starting point for the spectral theory of self-adjoint operator.

For self-adjoint operators the operational calculus of II,Sect.5 allows considerable broadening. Instead of the class Φ_A (recall the remarks after II,(134)), we can take Φ_S as the class of all $\varphi(\lambda)$ defined only for $\lambda \in \sigma(S)$. The definition of $\varphi(A)$ by a line integral is replaced by

$$\varphi(S) \underset{\text{def}}{=} \Sigma_{k=1}^{m} \varphi(\lambda_k) P_k ,$$

use being made of the spectral resolution (138). (This procedure works also for the wider class of operators of scalar type.)

For the adjoint of $\varphi(S)$ we have

148. $[\varphi(S)]^* = \overline{\varphi}(S)$,

and hence

149. The operator $\varphi(S)$ is self-adjoint iff $\varphi(\lambda)$ is real on the spectrum $\sigma(S)$.

The theory of the operator equation $\psi(x) = A$ becomes simpler for the class of self-adjoint operators.

150. Let S be a self-adjoint operator and ψ any function, defined on some point set M of the complex plane. The operator equation

$$\psi(x) = S$$

has a self-adjoint solution x, iff

$$\psi(\mu) = \lambda$$

has a <u>real</u> solution μ_λ for every $\lambda \in \sigma(S)$. The solution x is unique iff μ_λ is unique for every $\lambda \in \sigma(S)$.

A particular case of (150) is

151. Let p be a natural number. The equation

$$X^p = S \qquad (S \text{ is self-adjoint})$$

always has a unique self-adjoint solution X, if p is odd. For even p it has a self-adjoint solution iff the spectrum of S is nonnegative. In the subclass of self-adjoint operators with nonnegative spectrum this solution is unique.

Definition 41. A self-adjoint operator is said to be *nonnegative (positive)*, if its spectrum is nonnegative (positive).

Theorem (151) now says that $X^p = S$ has a unique self-adjoint solution X for any nonnegative S. It coincides with the arithmetical root, S^p, introduced in II,D38.

152. The equation $e^X = S$ with positive S has a unique self-adjoint solution X.

It coincides with the principal value, log S, introduced in II,D39.

Note here:

153. Every orthoprojector P is a nonnegative operator. P is positive only if it equals I.

154. The self-adjoint operator S is nonnegative iff

$$(Sx,x) \geqq 0 \quad (x \in E) .$$

(Note that (Sx,x) is real for any self-adjoint operator.) S is positive iff

$$(Sx,x) > 0 \quad (x \in E,\ x \neq 0) .$$

Thus the positivity (nonnegativity) of a self-adjoint operator is equivalent to the positivity (nonnegativity) of the corresponding Hermitian-quadratic functional.

155. For any operator A the operators A^*A and AA^* are nonnegative. They are positive iff A is regular.

156. $\mathrm{Ker}\, A^*A = \mathrm{Ker}\, A$, $\mathrm{Ker}\, AA^* = \mathrm{Ker}\, A^*$.

157. $\mathrm{Im}\, A^*A = \mathrm{Im}\, A^*$, $\mathrm{Im}\, AA^* = \mathrm{Im}\, A$.

It follows that $\mathrm{rk}\, A^*A = \mathrm{rk}\, A^* = \mathrm{rk}\, A = \mathrm{rk}\, AA^*$.

Note the following theorem which is a converse of (155).

158. To every nonnegative self-adjoint operator S there exists an operator A such that $S = A^*A$.

The two theorems that follow are connected with Gram's matrix (cf.(91)).

159. Let $\{u_k\}_1^n$ be a basis and $(\gamma) = (\gamma_{jk})_{k=1}^n$ its Gram matrix. Then (γ) is the matrix of a positive operator in any orthonormal basis $\{e_j\}_1^n$.

Conversely,

160. If (γ) is the matrix of a positive operator in some orthonormal basis, then there is a basis such that (γ) is its Gram matrix.

In conclusion, we stop briefly to describe some of the algebraic properties of orthoprojectors (cf. II,Sect.8).

161. If the sum of some orthoprojectors is a projector, then it is an orthoprojector.

162. The sum $\Sigma_{k=1}^m P_k$ of orthoprojectors is an orthoprojector

iff the system $\{P_k\}_1^m$ is orthogonal (II,D50).

163. If the difference of two orthoprojectors is a projector, then it is an orthoprojector.

As a special case, $\overline{P}$, the complement I - P of the orthoprojector P, is itself an orthoprojector.

164. Let P_1 and P_2 be orthoprojectors and set $P = P_1 - P_2$. P is an orthoprojector iff $\overline{P}_1$ and P_2 are orthogonal.

Turning to the product PQ, we first note the lemma:

165. With S and T $\in \mathbf{S}(E)$, the product ST $\in$ S(E), iff S∇T.

Hence

166. The product PQ of two orthoprojectors is an orthoprojector iff P∇Q.

167. Let $\{P_k\}_1^m$ be a system of orthoprojectors. If $P = P_1P_2 \ldots P_m$ does not change under cyclic permutations of the factors nor under a reversal of their order, then P is an orthoprojector.

This lemma is helpful in the following problem, in which the orthoprojector projecting onto the intersection of m subspaces $L_1L_2 \ldots L_m$ is calculated in terms of the orthoprojectors $P_k = P[L_k]$ (notation in D22).

For q > m, set $P_q = P_k$, where k is the remainder after division of q by m. Then

168. $$P[\bigcap_{k=1}^{m} L_k] = \prod_{q=1}^{\infty} P_q$$ (*Formula of Kaczmarz - von Neumann*).

6. Spectral Theory of Unitary Operators. Cayley's Transformation. Polar Representation.

Definition 42. The operator U acting in a unitary space E is called *unitary* if it is regular and if

$$U^* = U^{-1}$$

We shall denote the class of unitary operators by U(E).

169. A is unitary iff it preserves the scalar product:

$$(Ax,Ay) = (x,y) \qquad (x,y \in E).$$

The concept of a unitary matrix is introduced by

Definition 43. The matrix A is called *unitary* if $A^* = A^{-1}$ (notation in D9).

170. The operator A is unitary iff its matrix is unitary in some orthonormal basis.

This means for the matrix elements of A:

$$\Sigma_{j=1}^{n} \alpha_{jk}\overline{\alpha}_{js} = \delta_{ks} \qquad (k,s = 1,2,\ldots,n).$$

171. An operator is unitary iff it maps some orthonormal basis into an orthonormal basis.

172. The product of unitary operators is a unitary operator.

173. An operator which is inverse (and, hence, adjoint) to a unitary operator is unitary.

This means that the unitary operators in (the unitary space) E form a group, which is a subgroup of the group of automorphisms of E. For $n > 1$, this subgroup is non-Abelian. It is called the n-dimensional *unitary group*. -- It is an important topological property of the unitary group that

174. The unitary group is compact.

The spectral theories of unitary and self-adjoint operators are not only analogous, but there is a very direct connection between them, as we shall see below.

175. If L is an invariant subspace of the unitary operator U, then the restriction $U|L$ is a unitary operator in L.

176. If $U \in \mathbb{U}(E)$ and $|\alpha| = 1$, then $\alpha U \in \mathbb{U}(E)$, and this holds only if $|\alpha| = 1$.

177. The spectrum of a unitary operator lies on the unit circle.

Definition 44. A spectrum that lies on the unit circle is called a *unitary spectrum*.

178. The characteristic subspaces of a unitary operator form an orthogonal system.

179. Each invariant subspace of a unitary operator is orthogonally reducing (D34).

180. The characteristic subspaces of a unitary operator form a basic system (I,D34).

Hence $U \in \mathbb{U}(E) \Rightarrow U$ is of scalar type.

181. Every unitary operator has an orthonormal characteristic basis.

We now come to the spectral theorem for unitary operators.

182. The spectral decomposition of a unitary operator is

$$U = \Sigma_{k=1}^{m} (\exp i\vartheta_k) P_k ,$$

where $\{P_k\}_1^m$ is the orthogonal decomposition of I effected by U and the $\{\vartheta_k\}_1^m$ are real numbers.

183. If a unitary operator is written in the form

$$U = \Sigma_{k=1}^{m} \lambda_k P_k ,$$

where $\lambda_j \neq \lambda_k$ $(j \neq k)$, if the P_k are projectors and $\Sigma_{k=1}^{m} P_k = I$, then $\{P_k\}_1^m$ is the decomposition of I effected by U and $\{\lambda_r\}_1^m$ is its spectrum.

184. Every operator of the form $U = \Sigma_{k=1}^{m} (\exp i\vartheta_k) P_k$, where $\{P_k\}_1^m$ is an orthogonal decomposition of I and the $\{\vartheta_k\}_1^m$ are real numbers, is a unitary operator.

It follows in particular that an operator is unitary, if it has a unitary spectrum and an orthonormal characteristic basis. Thus unitary operators are exactly those operators (in a unitary space) that possess a unitary spectrum and an orthonormal characteristic basis.

185. An operator of scalar type has a unitary spectrum, iff it is similar (II,D34) to a unitary operator.

One of the consequences of this theorem is:

186. If A satisfies $A^p = I$ for integer $p \geq 1$, then A is similar to a unitary operator.

This theorem permits a far-reaching generalization.

187. Let $\Gamma = \{A_k\}_1^p$ be a finite group of regular operators.

There is an automorphism T such that all operators

$$U_k = TA_kT^{-1} \qquad (k = 1,2,\ldots,p)$$

are unitary.

This is a consequence of (188) - (190). Introduce in E another scalar product $(x,y)_\Gamma$, which is generated by the group Γ, viz.

$$(x,y)_\Gamma = (1/p)\Sigma_{k=1}^{p} (A_kx, A_ky) .$$

and show:

188. The functional $(x,y)_\Gamma$ has all the properties of a scalar product.

189. All operators belonging to the group Γ are unitary relative to the scalar product $(x,y)_\Gamma$.

190. If another scalar product $[x,y]$ is introduced in the unitary space E, then an automorphism T exists such that for any operator A, unitary relative to $[x,y]$, the operator $U = TAT^{-1}$ is unitary relative to the "native" scalar product in E.

The direct connection between unitary and self-adjoint operators announced above can be realized by at least two methods, one of which is based on the so-called Cayley transformation. The applicability of this transformation is rooted in the spectral mapping theorem (II, (144)). In order to pass from a self-adjoint to a unitary operator, we must transform the real into a unitary spectrum. This can be done by the linear fractional transformation

$$\zeta = (\lambda - \overline{\omega})/(\lambda - \omega)$$

where the imaginary part of the parameter ω is not zero.

Definition 45. The operator

$$U_\omega = (S - \overline{\omega}I)(S - \omega I)^{-1} , \text{ Im } \omega \neq 0,$$

is called the *Cayley transformation* (or Cayley transform) of the self-adjoint operator S.

191. The operator U_ω is unitary.

192. $1 \notin \sigma(U_\omega)$.

193. S can be expressed in terms of its Cayley transform U_ω:

$$S = (\omega U_\omega - \bar{\omega} I)(U_\omega - I)^{-1} .$$

Definition 46. We shall also use the term Cayley transformation in reverse.

$$S_\omega = (\omega U - \bar{\omega} I)(U - I)^{-1}, \quad (\text{Im } \omega \neq 0),$$

is the Cayley transform of the unitary operator U. S_ω exists if $\sigma(U)$ does not contain unity.

194. The operator S_ω is self-adjoint.

195. U can be expressed in terms of its Cayley transform S_ω:

$$U = (S_\omega - \bar{\omega} I)(S_\omega - \omega I)^{-1}$$

In this way a bijective correspondence has been established between self-adjoint operators and unitary operators whose spectrum does not contain the point 1.

The other method of passing from $\mathbb{S}$ to $\mathbb{U}$ uses the mapping by the exponential function $\zeta = e^{i\lambda}$ (which also maps the real axis of the λ-plane onto the unit circle of the ζ-plane).

196. If S is self-adjoint, then $U = e^{iS}$ is unitary.

197. For any unitary U there is a self-adjoint S such that

$$U = e^{iS} \qquad (*)$$

This equation is the analog of the polar form of a point on the unit circle. We shall develop this analogy further.

Definition 47. For any operator A acting in unitary space, we introduce its *left and right operator modules* by

$R_A^{(\ell)} = (AA^*)^{1/2}$, $R_A^{(r)} = (A^*A)^{1/2}$ (cf. (155) and II, D38).

We shall omit the subscript, if it is implied by the context.

If A is regular, then $R^{(\ell)}$ and $R^{(r)}$ are regular.

198. For any regular A there is a unitary operator U such that $A = R^{(\ell)}U$.

Together with the proper limiting process (cf. (174)) this implies that

199. <u>Any</u> operator A acting in unitary space may be represented in the form

$$A = R^{(\ell)}U \qquad (U \text{ is unitary}) .$$

Similarly,

200. <u>Any</u> operator A acting in unitary space may be represented in the form

$$A = UR^{(r)} \qquad (U \text{ is unitary}) .$$

Definition 48. The foregoing representations are known as the *polar representations* of the operator A. The unitary operators occurring in (199) and (200) are called the right and left *phase factors* of A, respectively. Writing the phase factors in the form (*), we obtain complete analogy with the polar form of a complex number. Are these representations unique? To be definite, we restrict ourselves to a representation of the form

$$A = RU, \qquad (**)$$

where R is nonnegative and U is unitary. Then we have

201. For regular A, (**) is unique: $R = R^{(\ell)}$, $U = R^{(\ell)})^{-1}A$.

If A has no inverse, one can only assert the uniqueness of the <u>module</u>:

202. The representation (**) implies $R = R_A^{(\ell)}$ for any A.

203. The general form of the <u>phase factor</u> U in the representation (**) is

$$U = (I - T)U_o .$$

Here U_o is one of the phase factors of A, and T runs through the set of operators satisfying

$$TT^* = T + T^* , \quad \operatorname{Im} T \subset \operatorname{Ker} A^* .$$

Definition 49. The operators A and B are called

unitarily similar, if there is a unitary operator U such that

$$B = UAU^{-1} \qquad (= UAU^{*}) .$$

Unitary similarity is an equivalence relation.

204. A and B are unitarily similar, iff for any orthonormal basis Δ there exists an orthonormal basis Δ_1 such that the matrix of B in the basis Δ_1 and the matrix of A in the basis Δ are equal (cf. II,(98)).

Using the polar representation we can easily establish:

205. The operators AA^* and A^*A are unitarily similar, whatever the operator A (cf. II,(271)).

206. If two self-adjoint (or unitary) operators are similar, then they are unitarily similar.

7. Spectral Theory of Normal Operators

Definition 50. The operator N acting in unitary space is called *normal* if

$$N^* \nabla N .$$

Self-adjoint and unitary operators are obvious examples.

207. An operator is normal iff its real and imaginary parts commute.

208. An operator is normal iff its module and its phase factor commute.

For the proof the following lemma is helpful.

209. If N is a normal operator, then Ker N^* = Ker N.

Normal operators are the largest class of operators of scalar type which produce an orthogonal decomposition of the identity. The construction of the spectral theory of normal operators follows the same scheme that worked in the self-adjoint and unitary cases.

210. Each characteristic subspace of a normal operator N is orthogonally reducing (D34).

The same is even true for the invariant subspaces of N (cf. (134) and (179)), but the proof is not quite as simple. However, (210) suffices to obtain the spectral theorem for the class N, and the generalization to arbitrary invariant subspaces will follow from the spectral theorem.

211. For a normal operator N, the characteristic subspace associated with the eigenvalue λ coincides with the characteristic subspace associated with the eigenvalue $\bar{\lambda}$ of N^*.

212. The system of characteristic subspaces of a normal operator is orthogonal.

213. The system of characteristic subspaces of a normal operator is basic.

Thus a normal operator is an operator of scalar type (II,D18).

214. Any normal operator N has an orthonormal characteristic basis.

Each such basis is also a characteristic basis for N^* because of (211).

Now we can obtain the <u>spectral theorem for normal operators:</u>

215. The spectral resolution of a normal operator has the form

$$N = \Sigma_{k=1}^{m} \lambda_k P_k ,$$

where $\{P_k\}_1^m$ is an <u>orthogonal</u> resolution of the identity and $\{\lambda_k\}_1^m$ are scalars.

216. If the normal operator N is represented in the form

$$N = \Sigma_{k=1} \lambda_k P_k ,$$

where $\lambda_j \neq \lambda_k$ $(j \neq k)$, the P_k are projectors, and $\Sigma_{k=1}^{m} P_k = I$, then $\{P_k\}_1^m$ is the resolution of I effected by N, and $\{\lambda_k\}_1^m$ is $\sigma(N)$.

217. Every operator of the form $N = \Sigma_{k=1}^{m} \lambda_k P_k$, where $\{P_k\}_1^m$ is an orthogonal resolution of I and the $\{\lambda_k\}_1^m$ are some scalars, is normal.

If, in particular, the operator has an orthonormal characteristic basis, then it is normal. Normal are therefore exactly those operators that have an orthonormal characteristic basis. There are no restrictions of any kind on the spectrum of a normal operator. Self-adjoint operators are _exactly_ those normal operators that have a real spectrum; and unitary operators are _exactly_ those normal operators that have a unitary spectrum.

218. An operator is of scalar type iff it is similar to a normal operator.

What follows is a generalization of (206):

219. Any two similar normal operators are unitarily similar.

The _theorem on simultaneous reduction_ that follows offers another access to the spectral theorem for normal operators.

220. If two self-adjoint operators S_1 and S_2 commute, then they possess a common orthonormal characteristic basis (cf. II,(193) and III,(144)).

This theorem and its converse follow from the spectral theorem for normal operators by virtue of (207). But (220) can be strengthened (cf. II,(194)):

221. For any set $\{S_\nu\}$ of pairwise commuting self-adjoint operators there exists an orthonormal characteristic basis that is common to all S_ν.

222. We obtain the form of (221) for normal operators, if in (221) S_ν is replaced by N_ν and self-adjoint by normal.

The connection between normal and self-adjoint operators also becomes manifest when we consider functions of operators:

223. Any function of a self-adjoint operator is a normal operator.

Conversely,

224. For any normal N there is a self-adjoint S and a function φ (cf. text preceding (148)) such that $N = \varphi(S)$.

In some cases it is possible to "split off" the normal part from an operator that is not normal. An example is

225. Suppose the spectrum $\sigma(A)$ of a general operator A is split according to

$$\sigma(A) = \sigma^{(1)} \cup \sigma^{(2)} \qquad (\sigma^{(1)} \cap \sigma^{(2)} = \emptyset)$$

in such a way that the characteristic subspaces associated with the points $\lambda \in \sigma^{(1)}$ are invariant for the operator A^*. Then $A = N \boxplus K$, $\sigma^{(1)} = \sigma(N)$ and $\sigma^{(2)} = \sigma(K)$, and N is normal.

8. The Extremal Properties of the Eigenvalues of a Self-adjoint Operator.

Let S be a self-adjoint operator and let the eigenvalues

$$\lambda_1 \geq \lambda_2 \geq \dots \geq \lambda_n$$

be counted with regard to their multiplicity: each eigenvalue occurs as often as its multiplicity indicates. (This mode of numeration of the eigenvalues of self-adjoint operators will recur through the whole book.) The eigenvectors will be counted accordingly: $e_1, e_2, \dots, e_n$. The self-adjoint S generates a Hermitian-quadratic functional $K_S(x) = (Sx,x)$, see D10. In the *Theory of E. Fischer and R. Courant*, the eigenvalues of S are characterized as the extreme values of the functional $K_S(x)$, if $x \in \{(x,x) = 1\}$ (unit sphere).

226. $$\lambda_1 = \max_{(x,x)=1} (Sx,x) , \qquad \lambda_n = \min_{(x,x)=1} (Sx,x) ;$$

the maximum and the minimum are attained for $x = e_1$ and e_n, respectively.

Definition 51. The vector $x \in E$ will be called *normed* if $(x,x) = 1$.

A complete description of the extreme vectors e_1, e_n is now possible:

227. The set of normed vectors x, satisfying

$$(Sx,x) = \lambda_1 , \quad \text{respectively} \quad (Sx,x) = \lambda_n$$

coincides with the set of normed eigenvectors of S that are associated with the eigenvalue λ_1, respectively λ_n.

The intermediate eigenvalues also satisfy certain extremal conditions:

228. $$\lambda_k = \max_{\substack{(x,x)=1 \\ (x,e_j)=0 \text{ for } 1 \leq j \leq k-1}} (Sx,x) \qquad (k = 2,3,\ldots,n)$$

The maximum is attained for $x = e_k$.
Or, in analogous formulation:

$$\lambda_k = \min_{\substack{(x,x)=1 \\ (x,e_j)=0 \text{ for } k+1 \leq j \leq n}} (Sx,x) \qquad (k = 1,2,\ldots,n-1)$$

The minimum is attained for $x = e_k$.

This description of the intermediate eigenvalues suffers from the disadvantage that it is "hung up by the characteristic basis of S". The removal of this disadvantage leads to a deeper (independent) formulation of the extremal conditions, known as the *minimax characterization* (230).

229. Suppose L_k is any subspace of codimension k-1. Then

$$\min_{\substack{(x,x)=1 \\ x \in L_k}} (Sx,x) \leq \lambda_{n-k+1} , \qquad \max_{\substack{(x,x)=1 \\ x \in L_k}} (Sx,x) \geq \lambda_k .$$

The precise characterization of the intermediate eigenvalues is

230. $$\lambda_{n-k+1} = \max_{L_k} \min_{\substack{(x,x)=1 \\ x \in L_k}} (Sx,x), \quad \lambda_k = \min_{L_k} \max_{\substack{(x,x)=1 \\ x \in L_k}} (Sx,x) .$$

Consider now some applications of the Fischer-Courant theory.

231. Let P be an orthoprojector that projects onto some (n-1) - dimensional subspace L_{n-1}. Let $\check{S} = PS|L_{n-1}$ be a self-adjoint operator acting in L_{n-1}, and let $\{\check{\lambda}\}_1^{n-1}$ be its eigenvalues. Then the eigenvalues of S and $\check{S}$ intertwine:

$$\lambda_1 \geq \check{\lambda}_1 \geq \lambda_2 \geq \check{\lambda}_2 \ldots \lambda_{n-1} \geq \check{\lambda}_{n-1} \geq \lambda_n .$$

232. Let $\{L_k\}_0^n$ be some maximal chain of subspaces (I,D26), let P_k be an orthoprojector projecting onto L_k, and put

$$S_k \underset{\text{def}}{=} P_k\, S|L_k \qquad (k = 1,2,\ldots,n) .$$

Then the eigenvalues of any two neighboring members of the chain $\{S_k\}_1^n$ interlace.

From this proposition the determinantal criterion for the positiveness of a self-adjoint operator (D41) can be easily derived. It is usually quoted as the *Criterion of Sylvester and Jacobi*. With $d_k \underset{\text{def}}{=} \det S_k$ the criterion reads:

233. The operator S is positive iff all d_k are positive.

This is a special case of a more general result:

234. If $d_k \neq 0$ for all k, then the number of negative eigenvalues of the operator S equals the number of sign-changes in the sequence 1, d_k, d_2, ..., d_n.

Suppose now that $\mathcal{D}_k(\lambda)$ is the characteristic polynomial (II,D28) of the operator S_k. Denote the number of sign-changes in the sequence

$$1,\ \mathcal{D}(\lambda),\ \mathcal{D}_2(\lambda),\ \ldots,\ \mathcal{D}_n(\lambda)$$

for those λ-values, for which all $\mathcal{D}_k(\lambda) \neq 0$, by $V_s(\lambda)$.

235. The number of eigenvalues of S in the (open) interval (α,β) is equal to $V_s(\beta) - V_s(\alpha)$ (provided that $V_s(\alpha)$ and $V_s(\beta)$ are defined).

236. The function $V_s(\lambda)$ gives the number of eigenvalues of S in the interval $(-\infty,\lambda)$.

Continuing the development of the Fischer-Courant theory, we notice first of all that (231) is precise in the following sense:

237. If some numerical sequence $\{\hat{\lambda}_k\}_1^{n-1}$ satisfies the inequalities (231), then there is an orthoprojector P of rank n-1 such that $\{\hat{\lambda}\}_1^{n-1}$ is the spectrum of $\hat{S} = P(S|\text{Im } P)$.

The results of (231) and (237) are contained in the more

general assertion:

238. Let the increasing numerical sequence $\{\hat{\lambda}_k\}_1^r$, $0 < r < n$, be given. An orthoprojector P of rank r, such that $\hat{S} = PS|\mathrm{Im}\,P$ has the spectrum $\{\hat{\lambda}_k\}_1^r$, exists iff $\lambda_{k+(n-r)} \leq \hat{\lambda}_k \leq \lambda_k$ for $k = 1,2,\ldots,r$.

239. For a fixed integer r, $0 < r < n$, let the vector sequence $\{x_k\}_1^r$ run through all orthonormal vector systems in E. Then

$$\lambda_1 + \lambda_2 + \ldots + \lambda_r = \max_{\{x_k\}_1^r} \Sigma_{k=1}^r (Sx_k, x_k)$$

and, analogously,

$$\lambda_n + \lambda_{n-1} + \ldots + \lambda_{n-(r-1)} = \min_{\{x_k\}_1^r} \Sigma_{k=1}^r (Sx_k, x_k).$$

The extrema are attained for $\{x_k\}_1^r = \{e_k\}_1^r$ and for $\{x_k\}_1^r = \{e_k\}_{n-(r-1)}^n$, respectively, where $\{e_k\}_1^n$ is the orthonormalized characteristic basis associated with S. For $r = n$, the quantity $\Sigma_{k=1}^n (Sx_k, x_k)$ in the second members is constant and equals tr S.

Theorem (239) is a generalization of the basic extremal property (228). Analogous generalizations of the minimax characterization (230) are the following two *Theorems of Wielandt*.

240. Suppose $m_1 < m_2 < \ldots < m_r$ $(m_r \leq n)$ are natural numbers, and $L^{(1)} \subset L^{(2)} \subset \ldots \subset L^{(r)}$ is some fixed chain of subspaces with dimensions m_j; let $\{x_j\}_1^r$ be any orthonormal vector system satisfying $x_j \in L^{(j)}$, $j = 1,2,\ldots,r$. Then

$$\lambda_{m_1} + \lambda_{m_2} + \ldots + \lambda_{m_r} = \max_{\{L_j\}_1^r} \min_{\{x_j\}_1^r} \Sigma_{j=1}^r (Sx_j, x_j)$$

241. Suppose again $m_1 < m_2 < \ldots < m_r$ $(m_r \leq n)$ are natural numbers, but $M^{(1)} \subset M^{(2)} \subset \ldots M^{(r)}$ some fixed chain of subspaces with dimensions $m_j - 1$; let $\{x_j\}_1^r$ be any ortho-

normal vector system satisfying $x_j \perp M^{(j)}$, $j = 1,2,\ldots,r$. Then

$$\lambda_{m_1} + \lambda_{m_2} + \ldots + \lambda_{m_r} = \min_{\{M^{(j)}\}_1^r} \max_{\{x_j\}_1^r} \Sigma_{j=1}^{r} (Sx_j, x_j) .$$

We finally note the *Inequalities of H. Weyl*, which also follow from the minimax characterization.
Let $\lambda_1^+(S)$, $\lambda_2^+(S)$, ... be the positive eigenvalues of the operator S, counted in decreasing order; let $\lambda_1^-(S)$, $\lambda_2^-(S)$, ... be the negative and vanishing eigenvalues of S, counted in decreasing order of their absolute values; let finally $\mu_1(S) \geqq \mu_2(S) \geqq \ldots \geqq \mu_n(S)$ be the absolute values of the eigenvalues of S. Then Weyl's inequalities read:

242.
$$\lambda^+_{j+k+1}(S_1 + S_2) \leqq \lambda^+_{j+1}(S_1) + \lambda^+_{k+1}(S_2) ,$$
$$\lambda^-_{j+k+1}(S_1 + S_2) \geqq \lambda^-_{j+1}(S_1) + \lambda^-_{k+1}(S_2) ,$$
$$\mu_{j+k+1}(S_1 + S_2) \leqq \mu_{j+1}(S_1) + \mu_{k+1}(S_2) ,$$

valid for any pair S_1, S_2 of self-adjoint operators.

9. Schur's Theorem. The Singular Numbers (s-Numbers) of Operators.

Unless the operator A acting in a unitary space is normal, it does not possess an orthonormal characteristic basis, even though it may be an operator of scalar type. Thus the question for the "unitary" canonical form of such an operator arises. *Schur's Theorem*, which follows here, is a significant contribution to the answer.

Definition 52. An orthonormal basis of triangular representation for the matrix of A is called a *Schur basis* of A.

243. Every operator has a Schur basis.

The proof of (243) can be carried out by I,D26 and (88). More specifically, choose a maximal chain of invariant subspaces

$$0 = L_o \subset L_1 \subset \dots \subset L_{n-1} \subset L_n = E$$

as in II,(58) and a basis $\{x_i\}$ of E such that $x_i \in L_i$. Now apply (88) to this basis.

244. An operator A is normal iff all its Schur bases are characteristic.

It is, of course, sufficient that (only) one Schur basis be characteristic. The necessity ensues from the following criterion mentioned in Section 7 after (210):

245. The operator A is normal iff all its invariant subspaces are orthogonally reducing.

In the case of a nilpotent (= Volterra) operator A (II,D38) Schur's theorem allows it to be expressed by its imaginary part (128). Suppose A is nilpotent, $\{e_j\}_1^n$ its Schur basis, and Q_k the orthoprojector onto $sp\{e_j\}_1^k$, and $Q_o \underset{def}{=} 0$, then the following *Formula of M. S. Brodskiĭ* holds:

246. $$A = 2i\Sigma_{k=1}^{n-1} Q_k \frac{A-A^*}{2i} (Q_{k+1}-Q_k)$$

Definition 53. The eigenvalues of the operator module of the operator A are known as the *singular numbers* or *s-numbers* of A and are denoted by $s_k(A)$ or, briefly, s_k. Thus

$$s_k(A) \underset{def}{=} \lambda_k((A^*A)^{\frac{1}{2}}).$$

The singular numbers are unitarily invariant:

248. If U is a unitary operator, then

$$s_k(UA) = s_k(AU) = s_k(A) \qquad (k = 1,2,\dots,n) .$$

249. $$s_k(A) = s_k(A^*) \qquad (k = 1,2,\dots,n) .$$

Hence $\lambda_k((A^*A)^{\frac{1}{2}})$.

250. If $A = RU$ is a polar decomposition of A, if $\{e_k\}_1^n$ is a characteristic basis of the self-adjoint operator R, and if $e'_k \underset{\text{def}}{=} U^{-1}e_k$, then

$$Ax = \Sigma_{k=1}^n s_k(A)(x_k, e'_k)e_k$$

holds for any $x \in E$ (<u>Schmidt's decomposition</u>).

251. For a normal operator N

$$s_k(N) = |\lambda_k(N)| \qquad (k = 1,2,\ldots,n)$$

if the eigenvalues are counted suitably.

Conversely,

252. If for some operator A

$$s_k(A) = |\lambda_k(A)| \qquad (k = 1,2,\ldots,n) ,$$

then A is a normal operator.

A stronger assertion is

253. If for some operator A and some orthonormal basis $\{e_k\}_1^n$

$$|(Ae_k, e_k)| = s_k(A) \qquad (k = 1,2,\ldots,n) ,$$

then A is a normal operator.

The proof can be based on Schmidt's decomposition.

254. For an orthonormal basis

$$\Sigma_{k=1}^n |(Ae_k, e_k)| \leq \Sigma s_k(A) .$$

This, of course, implies

$$|\Sigma_{k=1}^n (Ae_k, e_k)| \leq \Sigma_{k=1}^n s_k(A) \qquad (*)$$

255. Equality in (*) holds only for a normal operator of the form $A = e^{i\omega}B$, where B is a nonnegative operator and $\omega = \arg(\operatorname{tr} A)$.

256. Equality in (254) holds only for an operator of the form $A = US$, where S is a nonnegative self-adjoint operator and

Sx vanishes on the linear hull of those e_k for which $(Ae_k, e_k) = 0$, while $Ue_k = e^{i\omega}e_k$ with $\omega_k = \arg (Ae_k, e_k)$.

From (254) follows

257. $$\Sigma_{k=1}^{n} |\lambda_k(A)| \leq \Sigma_{k=1}^{n} s_k(A) ,$$

and hence

258. $$\max_{1 \leq k \leq n} |\lambda_k(A)| \leq \max s_k(A)$$

Schur's Inequality, which follows here, is in some sense intermediate between (257) and (258):

259. $$\Sigma_{k=1}^{n} |\lambda_k(A)|^2 \leq \Sigma_{k=1}^{n} s_k^2(A) .$$

260. Equality in each of the relations (257) - (259) holds only for normal operators.

Note here the *Mirsky Inequality*, an interesting consequence of Schur's inequality:

261. $$\max_{j,k} |\lambda_j(A) - \lambda_k(A)| \leq \sqrt{2(\operatorname{tr} AA^* - |\operatorname{tr} A|^2/n)} .$$

To prove it, one can use the formula

262. $$\Sigma_{j,k=1}^{n} |\lambda_j(A) - \lambda_k(A)|^2 = n\Sigma_{k=1}^{n} |\lambda_k(A)|^2 - |\operatorname{tr} A|^2$$

We finally mention some inequalities for the s-numbers of sums and products of operators.

263. $$s_{j+k+1}(A+B) \leq s_{j+1}(A) + s_{k+1}(B) \qquad (j+k \leq n-1) .$$

264. $$\Sigma_{k=1}^{n} s_k(A+B) \leq \Sigma_{k=1}^{n} s_k(A) + \Sigma_{k=1}^{n} s_k(B) .$$

To prove (264), use

265. $\Sigma_{k=1}^{n} s_k(A) = \sup |\operatorname{tr} AU|$, taken over all $U \in U(E)$.

266. $$s_{j+k+1}(AB) \leq s_{j+1}(A)\, s_{k+1}(B) \qquad (j+k \leq n-1) .$$

10. The Hausdorff Set of an Operator.

Definition 54. The *Hausdorff set* or the *numerical range* of an operator A is the set $H(A)$ of complex numbers (Ax,x), where $x \in \{x | (x,x) = 1\}$ (= unit circle).

267. For a normal operator N, $H(N)$ coincides with the convex hull of the spectrum $\sigma(N)$, i.e., with the smallest convex set containing $\sigma(N)$. (More on convex sets in VII, Sect. 2; see in particular D14 and D19).

Special cases are:

268. For a self-adjoint operator S, $H(S)$ coincides with the segment $[\lambda_n, \lambda_1]$ of the real axis.

269. For a unitary operator U, $H(U)$ coincides with the polygon whose corners are the set $\sigma(U)$.

270. For a scalar operator αI, $H(\alpha I) = \{\alpha\}$.

Conversely:

271. If $H(A)$ consists only of a single point, then A is a scalar operator.

We shall investigate the geometrical properties of $H(A)$. Our first observation is an elementary consequence of Schur's theorem.

272. For two-dimensional E, the Hausdorff set of any operator A is an ellipse, and its foci are the set $\sigma(A)$. This ellipse degenerates into a segment (or a single point) iff A is a normal operator.

Now we note:

273. If L is any subspace of E (dim E may now be > 2) and P is an orthoprojector onto L, and if $\check{A} \underset{\text{def}}{=} PA|L$, then $H(\check{A}) \subset H(A)$.

From this result taken together with (272) ensues the remarkable *Theorem of Hausdorff*:

274. The Hausdorff set of any operator is convex,

and

275. $$\sigma(A) \subset H(A) .$$

It follows that the convex hull of $\sigma(A)$ is always contained in $H(A)$.

Consider now the boundary of the set $H(A)$. It is a closed convex curve in the complex plane (disregarding the degenerate case of a segment or a point). The following lemma is basic for the later propositions in this section.

276. If λ is an angular point of the boundary (i.e., a point with two different tangents), and if x is that vector for which $(Ax,x) = \lambda$ and, of course, $(x,x) = 1$, then the orthogonal complement of x is invariant for A and A^*.

Consequently:

277. All corners on the boundary of the set $H(A)$ are contained in $\sigma(A)$, and the characteristic subspaces associated with those corners are orthogonally reducing subspaces (D34) for the operator A. ((120) is useful here.)

The boundary of $H(A)$ has therefore only a finite number of corners.

278. If $\{\lambda_j\}_1^m$ are the corners of $H(A)$, then from (225)

$$A = N \boxplus K \qquad \text{(notation in D35)}$$

where N is normal, $\sigma(N) = \{\lambda_j\}_1^m$ and $\sigma(K) = \sigma(A) - \sigma(N)$. (This implies that for the minimum polynomial of A, $M(\lambda;A)$, the roots $\{\lambda_j\}_1^m$ are simple.) If $K = 0$, then A is normal and $\sigma(A) = \{\lambda_j\}_1^m$. This must happen if $m = n$. On the other hand, "A is normal" does not imply $K \equiv 0$, but only "K is normal".

279. If $H(A)$ is a (convex) polygon with corners $\{\lambda_j\}_1^m$ and N and K are as in (278), then $H(I) \subset H(N)$.

These results contain a partial converse of (267). A complete converse would be false, as is shown by (282) below.

280. If $A = \boxplus_{k=1}^{m} A_k$, then $H(A)$ is the convex hull of $\bigcup_{k=1}^{m} H(A_k)$.

Consequently:

281. If N is a normal operator in the subspace L, and if

K is any operator in the subspace $L^{\perp}$ and K satisfies $\mathcal{H}(I) \subset \mathcal{H}(N)$, then the Hausdorff set associated with $A \underset{def}{=} N \boxplus K$ coincides with $\mathcal{H}(N)$.

Thus we obtain the result:

282. $\mathcal{H}(A)$ and the convex hull of $\sigma(A)$ coincide iff a is either normal or has the form $N \boxplus K$, where N is normal and $\mathcal{H}(I) \subset \mathcal{H}(N)$.

Definition 55. We shall call a finite point set M in the plane c - *independent*, if none of the points of M belongs to the convex hull of all the other points of M. (This means that there is a convex polygon which has M as the set of its corners.)

283. If $\sigma(A)$ is c-independent and if the convex hull of $\sigma(A)$ equals $\mathcal{H}(A)$, then A is normal.

284. If A has a unitary spectrum (D45) and if the convex hull of $\sigma(A)$ equals $\mathcal{H}(A)$, then A is unitary.

285. If $\mathcal{H}(A)$ is a segment of the real axis, then A is self-adjoint.

We show now one of the applications of the Hausdorff set.

286. Let tr A = 0. There is an orthonormal basis $\{e_k\}_1^n$ such that $(Ae_k, e_k) = 0$ for all k (cf. II,(209)).

Starting from here and using the method developed in II, Sect. 6, we can establish the following proposition:

287. If S is a self-adjoint operator and tr S = 0, then there exists a self-adjoint operator S_1 with simple spectrum (II,D19) and a self-adjoint operator S_2 such that $S = (1/i)[S_1, S_2]$ (notation in II,D5).

From (287) ensues, among other things,

288. If S is self-adjoint and tr S = 0, then there is an operator A such that $S = (1/i)[A, A^*]$.

adjoint operator (A^*) D32
antisymmetric bilinear functional D3
antisymmetric Hermitian bilinear functional D12
antisymmetric matrix (3)

Bessel's Inequality (75)
bilinear form D1
biorthogonal vector system D29
B-regular subspace D7
Brodskiĭ's Formula (246)

canonical form of quadratic functional (23)
canonical form of real Hermitian quadratic functional (43)
canonical isomorphism between E' and E D19
canonical isomorphism between E^* and E D31
Cartesian representation of operator (128)
Cayley transformation D45
Cayley transformation, converse D46
c-independent point set D55
completeness relation (57)
complete orthogonal (orthonormal) vector system D27
conjugate transpose of a matrix D9

degeneracy of antisymmetric bilinear functional (31)

Euler-Fourier Formulas (78)

Fischer-Courant theory (227)ff.
formal decomposition of space E (vector x) D24
Fourier coefficients D26

Gram Matrix D28
Gram-Schmidt orthogonalization (89)

Hausdorff set D54
Hausdorff's Theorem (274)
Hermitian-bilinear form D8
Hermitian-bilinear functional D8
Hermitian-conjugate operator D30
Hermitian matrix ($G=G^*$) (39)
Hermitian quadratic form D11
Hermitian-quadratic functional D10
H-regular subspace (40)

intertwining eigenvalues (231)
involution (9)

Kaczmarz-von Neumann F. (168)
K-neutral subspace D16
K-nonpositive (nonnegative) subspace D15

left operator module D47

matrix of adjoint operator (A^*) (108)
matrix of bilinear functional D1
matrix of Hermitian-bilinear functional D8
matrix of Hermitian-quadratic functional D11
matrix of quadratic functional D6
minimax characterization of eigenvalues (230)
Mirsky's Inequality (261)
mutually orthogonal vectors ($x \perp y$) D18

negative index of inertia D14
nonnegative self-adjoint operator D41
normal operator D50
number of eigenvalues in $(-\infty, \lambda)$ (236)
numerical range of operator D54

operational calculus for self-adjoint operators (14)ff.

orthogonal (orthonormal) basis D27
orthogonal decomposition of identity D23
orthogonally irreducible operator D36
orthogonally reducing subspace D34
orthogonal sum (⊞) D21
orthogonal sum of operators D35
orthogonal system of spaces D20
orthogonal vector system D25
orthoprojector D22

Parseval's Equation (57)
Parseval's equation for biorthogonal systems (96)
phase factor of operator D48
polar of functional D5
polar representation of operator D48
positive index of inertia D14
positive self-adjoint operator D41

quadratic form D6
quadratic functional D2

real Hermitian-quadratic form (39)
real Hermitian quadratic functional D13
F. Riesz' Theorem (48)
right operator module D47

scalar product D17
Schmidt's decomposition (25)
Schur basis D52
Schur's Inequality (259)
Schur's Theorem (243)
self-adjoint operator D37
simultaneous reduction of normal operators (222)
simultaneous reduction of quadratic functionals to canonical form (144)
simultaneous reduction of self-adjoint operators (221)
singular numbers of operator D53
s-numbers of operator D53
Sonine-Schmidt orthogonalization (89)
spectral theorem for normal operator (215)
spectral theorem for self-adjoint operators (138)
spectral theorem for unitary operators (182)
splitting off the normal part of an operator (225)
Sylvester-Jacobi Criterion (233)
Sylvester's law of inertia (45)
symmetric bilinear functional D4
symmetric Hermitian-bilinear functional D12
symmetrizable operator D39
system of principal axes D40

transpose of matrix D1

unitarily similar operators D49
unitary group (173)
unitary matrix ($A^*=A^{-1}$) D43
unitary operator D42
unitary space D17
unitary spectrum D44

Volterra operator (246)

Wielandt's Theorems (240-241)
Weyl's Inequalities (242)

CHAPTER IV

NORMED SPACES; FUNCTIONALS AND OPERATORS IN NORMED SPACES

1. Norm, Metric, and Topology.

Definition 1. A functional $\nu(x)$ defined on the linear space E is called a *norm* if it possesses the following properties:

(1) $\nu(x) > 0 \quad (x \neq 0)$ (*positivity*),

(2) $\nu(\alpha x) = |\alpha| \nu(x)$ (*absolute homogeneity*),

(3) $\nu(x_1+x_2) \leqq \nu(x_1) + \nu(x_2)$ (*triangle inequality*).

The norm is the natural generalization of the elementary concept "length of a vector".

We note the immediate consequences of D1:

1. $\nu(0) = 0$.
2. $\nu(-x) = \nu(x)$ (symmetry).
3. $|\nu(x_1) - \nu(x_2)| \leqq \nu(x_1 \pm x_2) \leqq \nu(x_1) + \nu(x_2)$.

Definition 2. A space is *normed* if some particular norm functional has been assigned to all $x \in E$. The preferred notation is then $\|x\|$. A normed space is often called a *Minkowski Space* if dim E is finite. (The infinite-dimensional analog of a Minkowski space is called a *Banach Space* if it is complete.) A subspace of a normed space is clearly a normed space.

There are infinitely many ways to introduce a norm in a given E, since

4. If $\nu(x)$ is a norm, $\alpha\nu(x)$ is also a norm, when α is some positive scalar.
5. Let $\Delta = \{e_k\}_1^n$ be some basis in E. Then the functional

$$\nu_c(x;\Delta) = \max_{1\leq k\leq n} |\xi_k| \qquad (x = \Sigma_{k=1}^n \xi_k e_k)$$

is a norm.

Definition 3. ν_c is the so-called *c-norm* or *uniform norm* relative to the given basis Δ.

6. The functional

$$\nu_\ell(x;\Delta) = \Sigma_{k=1}^n |\xi_k|$$

is a norm.

Definition 4. ν_ℓ is the *ℓ-norm* relative to Δ. It can be generalized as follows:

7. Let p be some fixed number ≥ 1. The functional

$$\nu_{\ell^p}(x;\Delta) = (\Sigma_{k=1}^n |\xi|^p)^{1/p}$$

is a norm.

Definition 5. ν_{ℓ^p} is the *ℓ^p-norm* relative to Δ. It becomes the c-norm for $p\to\infty$; for $p=1$ it is the ℓ-norm.

Definition 6. The arithmetical space, furnished with the ℓ^p-norm relative to some basis, is known as the *ℓ^p-space*; for $p=1$ and $p=\infty$ one speaks of *ℓ- and c-spaces*, respectively.

The triangle inequality in the ℓ^p-space is known from analysis as *Minkowski's Inequality*:

$$(\Sigma_{k=1}^n|\xi_k + \eta_k|^p)^{1/p} \leq (\Sigma_{k=1}^n|\xi_k|^p)^{1/p} + (\Sigma_{k=1}^n|\eta_k|^p)^{1/p} .$$

The ℓ^2-norm is a particular case of great importance. It arises naturally in unitary space:

8. In unitary space the functional

$$\|x\| = \sqrt{(x,x)} \qquad (*)$$

is the ℓ^2-norm relative to any fixed orthonormal basis. In this way the unitary space obtains its natural norm.

Definition 7. The norm (*) in unitary space E is known as *Euclidean norm*, and E considered as a normed space is the (complex) *Euclidean space*. Any ℓ^2-space is thus Euclidean. (The infinite-dimensional analog of a Euclidean space is said to be a Hilbert space if it is complete.)

The triangle inequality for the Euclidean norm is closely connected with the important *Inequality of H. A. Schwarz:*

9. In Euclidean space

$$|(x,y)| \leq \|x\|\cdot\|y\| \quad .$$

Here equality holds iff x and y are linearly dependent.

Schwarz' inequality is the geometrical version of what in analysis goes under the name of Cauchy's inequality, viz.

$$|\Sigma_{k=1}^{n} \xi_k\eta_k| \leq \sqrt{\Sigma_{k=1}^{n} |\xi_k|^2}\cdot\sqrt{\Sigma_{k=1}^{n} |\eta_k|^2} \quad .$$

The connection between Schwarz' inequality and the triangle inequality in n-dimensional Euclidean space is obtained by the proper generalization of the elementary law of cosines:

10. $$\|x + y\|^2 = \|x\|^2 + \|y\|^2 + 2\mathrm{Re}(x,y) \ .$$

This "law of cosines" also yields the *parallelogram law* in the same form as in elementary plane geometry:

11. $$\|x + y\|^2 + \|x - y\|^2 = 2(\|x\|^2 + \|y\|^2) \ .$$

Relation (11) is also characteristic for the Euclidean norm:

12. If a norm in the space E satisfies (11), then E is Euclidean.

From (12) we can derive the proposition:

13. If every two-dimensional subspace of the space E is Euclidean, then so is E.

Definition 8. In connection with (12) we can introduce

a (numerical) *measure for the deviation* of the space E *from being Euclidean:*

$$\mu(E) \underset{\text{def}}{=} \sup_{x,y\in E} \frac{\|x+y\|^2+\|x-y\|^2}{2(\|x\|^2+\|y\|^2)}$$

14. In every normed space the exact inequality $1 \leq \mu(E) \leq 2$ holds.

15. E is Euclidean iff $\mu(E) = 1$.

Every normed linear space has the natural metric

$$d(x,y) = \|x-y\| \quad . \qquad (**)$$

It follows from the axioms of norm that

(1) $d(x,y) > 0$ for $x \neq y$, $d(x,x)=0$ (*positivity*),
(2) $d(y,x) = d(x,y)$ (*symmetry*)
(3) $d(x,y) \leq d(x,z) + d(z,y)$ (*triangle inequality*),
(4) $d(\alpha x,\alpha y) = |\alpha| d(x,y)$ (*absolute homogeneity*).

Here we shall also stipulate

(5) $d(x+z, y+z) = d(x,y)$ (*translational invariance*).

Axioms 4. and 5. need not hold in the most general case of a metric.

16. If the functional $d(x,y)$ defined on a linear space satisfies relations (1) - (5), then it generates a unique norm by (**).

Definition 9. The *metric* (**) here introduced originates in the norm, and it is indeed possible to consider every normed linear space as a *metric space*, hence as a topological space. The topology established in E already in I,D89 is known as the weak topology§.

17. Any norm-generated topology (= norm topology) in E coincides with the weak topology.

---

§The topologies determined by the various choices of the functionals $\{f_k\}_1^m$ are identical, i.e., their families of open sets are identical (cf. the pertinent remarks in the Dictionary of General Terms).

This fundamental result ensues from theorems (18) - (23) below.

18. The weak topology is generated by the c-norm, regardless what basis Δ is used in (5).

Definition 10. Let $\nu_1(x)$ and $\nu_2(x)$ be two norms in E. Norm ν_1 is said to be *dominated* by norm ν_2, if there is a constant C such that

$$\nu_1(x) \leqq C\nu_2(x) \quad \text{for all } x \in E .$$

19. Every norm is dominated by the c-norm.

Hence it follows that

20. Every norm is a continuous functional in the weak topology.

Definition 11. Two *norms* dominated by each other are called *equivalent.*

21. Equivalent norms generate identical topologies.

Using (18) and elementary theory of continuous functions, we find that

22. Every norm is equivalent to the c-norm.

Hence

23. Any two norms on E are equivalent.

Now theorem (17) follows directly from (23), (21), and (18).

Definition 12. *Convergence in norm* (or *norm convergence*) is introduced in normed space by interpreting the meaning of

$$\lim_{k\to\infty} x_k = x \quad \text{as} \quad \lim_{k\to\infty} \|x_k - x\| = 0 .$$

This definition agrees with the definition of convergence in a general metric space. By I,(305) norm convergence in E coincides with coordinate-wise convergence.

Also the class of fundamental sequences is described by means of the norm concept: $\{x_k\}_1^\infty$ is a fundamental sequence iff

$$\lim_{j,k\to\infty} \|x_j - x_k\| = 0 ,$$

giving that term a definite sense in metric space.
In order to formulate some propositions about series in a normed space we put down

Definition 13. The vector series $\Sigma_{k=1}^{\infty} x_k$ in a normed space is absolutely convergent, if $\Sigma_{k=1}^{\infty} \|x_k\| < \infty$.

4. A vector series is absolutely convergent if it con-
erges unconditionally. (I, D100).

Absolute convergence thus does not depend on the choice of the norm, as follows easily from (21).--Note also the following lemma:

5. If $\Sigma_{k=1}^{\infty} x_k$ converges absolutely, then $\|\Sigma_{k=1}^{\infty} x_k\| \leq \Sigma_{k=1}^{\infty} \|x_k\|$.

For vector power series it is important to note the modified Cauchy-Hadamard theorem (I,(334)):

6. Let $\Sigma_{k=1}^{\infty} a_k(\lambda-\lambda_o)^k$ ($a_k \in E$, λ and λ_o complex numbers)
e a vector power series, and put

$$\rho = \frac{1}{\overline{\lim_{k\to\infty}} \|a_k\|^{1/k}} .$$

he number ρ is independent of the choice of the norm and
epresents the radius of convergence of the power series.

2. Seminorms and Induced Norms

Definition 14. The functional $p(x)$ on E is called a *seminorm* if it satisfies the triangle inequality, $p(x_1 + x_2) \leq p(x_1) + p(x_2)$, and if it is absolutely homogeneous, $p(\alpha x) = |\alpha|p(x)$. Note the following elementary properties of this concept:

7. $p(0) = 0$.

8. $p(-x) = p(x)$.

9. $p(x) \geq 0$.

Thus a seminorm is a norm iff $p(x) \neq 0$ for all $x \neq 0$.

30. The set $\operatorname{Ker} p = \{x | p(x) = 0\}$ is a subspace.

31. If L is some complement of Ker p, then the functional $p|L$ is a norm on L.

32. The congruence $x \equiv y \pmod{\operatorname{Ker} p}$ implies $p(x) = p(y)$. Consequently we can define a functional $\hat{p}([x])$ on the factor space E/Ker p by $\hat{p}([x]) \underset{\text{def}}{=} p(x)$.

33. The functional $\hat{p}$ is a norm on E/Ker p.

Definition 15. As in D10 we shall say that p_1 is *dominated* by p_2 if there is a constant C such that

$$p_1(x) \leq Cp_2(x) \quad \text{for all } x \in E.$$

34. A seminorm is dominated by any norm.

Let p be a seminorm on a normed space. Then, by (34), there is a C such that

$$p(x) \leq C\|x\| \quad \text{for all } x \in E.$$

Definition 16. The *smallest C admissible* in this inequality is called the *norm of the functional* p and is denoted by $\|p\|$. Clearly,

$$\|p\| = \sup_{x \in E} \frac{p(x)}{\|x\|} ,$$

which we can also write in the form

$$\|p\| = \sup_{\|x\|=1} p(x) .$$

Definition 17. If $\|x\| = 1$ we shall call the vector x *normed*, and we shall speak of a *normed vector system* $\{x_k\}_1^m$ if all x_k are normed. The set of all normed $x \in E$ is the *unit sphere* of E.

35. The seminorm p_1 is dominated by the seminorm p_2 iff $\operatorname{Ker} p_2 \subset \operatorname{Ker} p_1$.

Definition 18. As in D11, two *seminorms* dominated by each other are called *equivalent*.

36. Seminorms p_1 and p_2 are equivalent iff $\operatorname{Ker} p_1 = \operatorname{Ker} p_2$.

In this way we have established a bijective relation between the classes of equivalent seminorms and the subspaces of E. The class of norms corresponds to the subspace {0} (since Ker of any norm = {0}, and the class of the (unique) zero-seminorm ($p(x) = 0$, all $x \in E$) corresponds to E. Using the seminorm we can introduce induced norms in factor spaces, spaces of functionals, and of homomorphisms.

Definition 19. Let L be a subspace of the normed space E. By the *distance of* x *from* L we mean

$$d(x;L) = \inf_{y \in L} \|x - y\| .$$

37. The functional $d(.;L)$ is a seminorm on E, and Ker $d(.;L)$ = L; we denote it by $\|.\|_d$ for brevity.

By (33) a norm is induced on the factor space E/L; denote it by $\|.\|_{E/L}$ to distinguish it from $\|.\|_d$ on E.

38. This induced norm is equal to

$$\|[x]\|_{E/L} = \inf_{x \in [x]} \|x\|_d .$$

On the left, [x] is the symbol for the class as an element of E/L; on the right it indicates the set of elements of E in the class [x].

39. Let E and E_1 be normed spaces and $h \in \mathrm{Hom}(E,E_1)$. Then the functional

$$p_h(x) \underset{\mathrm{def}}{=} \|hx\|_1 \qquad (x \in E \text{ and } \|.\|_1 \text{ is the norm on } E_1)$$

is a seminorm on E, and Ker p_h = Ker h.

Definition 20. The norm of the functional p_h serves as the *norm of the homomorphism* h and is denoted by $\|h\|$. Thus, by D16,

$$\|h\| = \sup_x \frac{\|hx\|_1}{\|x\|} = \sup_{\|x\|=1} \|hx\|_1 .$$

Definition 21. In the special case $E = E_1$ we so obtain the *norm of the linear operator* A operating in the normed space E:

$$\|A\| = \sup_x \frac{\|Ax\|}{\|x\|} = \sup_{\|x\|=1} \|Ax\|.$$

Definition 22. From D21 we obtain in particular the *norm of a linear functional* defined on the normed space E,

$$\|f\| = \sup_x \frac{|f(x)|}{\|x\|} = \sup_{\|x\|=1} |f(x)|.$$

40. The functional $\|h\|$ defined on the space $Hom(E,E_1)$ has all the properties of a norm.

Indeed, it is the norm induced by $\|.\|_E$ and $\|.\|_{E_1}$ in the space $Hom(E,E_1)$ of which $\mathbb{M}(E)$ and E' are special cases In the space of Hermitian homomorphisms the induced norm can be defined in exactly the same way, and this covers, of course, the norm on E^* as a special case.

For homomorphisms, the induced norm has certain special properties that originate in Hom(E,E) being an algebra (I,D55) and also in I,D54.

41. $\|h_2h_1\| \leq \|h_2\|.\|h_1\|$ ($\|.\|$ is a *ring norm*).

42. $\|I\| = 1$ ($\|.\|$ *preserves unity*).

Definition 23. Also the induced *norm of a bilinear functional* is constructed after the model of D20:

$$\|B\| = \sup_{x,y} \frac{|B(x,y)|}{\|x\|\|y\|} = \sup_{\|x\|=\|y\|=1} |B(x,y)|,$$

and likewise the *norm of a Hermitian-bilinear functional*.

Definition 24. Finally, the induced *norm of a quadratic functional* is

$$\|Q\| = \sup_x \frac{|Q(x)|}{\|x\|^2} = \sup_{\|x\|=1} |Q(x)| ,$$

and likewise the *norm of a Hermitian-quadratic functional*.

43. Let B_Q denote the polar (III,D5) of a quadratic functional Q: We have

$$\|Q\| \leq \|B_Q\| \leq \mu(E)\|Q\| . \quad (\mu \text{ in D8}).$$

Therefore:

44. In Euclidean space $\|B_Q\| = \|Q\|$.

An analogous result holds for Hermitian-quadratic functionals, if we redefine $\mu(E)$ of D8 as follows:

$$\mu^{(c)}(E) = \sup_{x,y} \frac{\|x+y\|^2+\|x-y\|^2+\|x+iy\|^2+\|x-iy\|^2}{4(\|x\|^2+\|y\|^2)} .$$

45. As in (14), the exact inequality $1 \leqq \mu^{(c)}(E) \leqq 2$ holds, and E is Euclidean iff $\mu^{(c)}(E) = 1$.

46. Let H_K be the polar (III,D10) of a Hermitian-quadratic functional K: here

$$\|K\| \leqq \|H_K\| \leqq 2\mu^{(c)}(E)\|K\| .$$

47. For a real Hermitian-quadratic functional (III,D13),

$$\|K\| \leqq \|H_K\| \leqq \mu^{(c)}(E)\|K\| .$$

Thus,

48. For a real Hermitian-quadratic functional in Euclidean space

$$\|H_K\| = \|K\| .$$

Also note:

49. The norm of a bilinear (or Hermitian-bilinear) functional is equal to the norm of each of its generators (I,D72).

Hence the norm of left and right generators are equal.

3. Absolutely Convex Sets and Seminorms. Extended Seminorms.

Definition 25. The set $W \subset E$ is said to be *absolutely convex* if for any pair $x,y \in W$ and any pair of scalars satisfying $|\alpha|+|\beta| = 1$

$$\alpha x + \beta y = W .$$

50. An absolutely convex set W is symmetric about 0:

$$x \in W \Leftrightarrow -x \in W .$$

51. Every nonvoid absolutely convex set W contains the zero vector.

52. For every such W, $x \in W \Rightarrow \alpha x \in W$, provided $|\alpha| \leq 1$ (W is *balanced*).

Definition 26. An absolutely convex set which contains a neighborhood of 0 is called an *absolutely convex body*.

53. If p is a seminorm, then the *unit ball* $p(x) \leq 1$ is a closed absolutely convex body.

54. If the unit balls associated with the seminorms p_1 and p_2 coincide, then $p_1(x) = p_2(x)$ for all $x \in E$.

This means that the unit ball of a seminorm uniquely determines the seminorm with which it is associated. Conversely, let W be some closed, absolutely convex body, consider for some fixed $x \in E$ those positive numbers α for which $\alpha^{-1}x \in W$, and denote $\inf \alpha$ by $p_W(x)$. Then

55. The functional $p_W(x)$ is a seminorm, and W is its unit ball.

Definition 27. The seminorm $p_W(x)$, defined on E with regard to W, is known as the *Minkowski Functional* of the body W.

This establishes a bijection between the set of seminorms on E and the set of closed absolutely convex bodies in E.

56. Let the seminorm p_W be defined on the normed space E. Iff the associated unit ball W is bounded in the norm-topology of E, then p_W is a norm.

Note here the connection with "boundedness" in I,D93:

57. A set M in a normed space is bounded iff it is contained in some ball $\|x\| \leq r$.

There is a useful generalization of our theory, which considers functionals in the range $[0,+\infty]$.

Definition 28. A seminorm, considered in the extended range $[0,+\infty]$, but with D14 in force, is now called an *extended seminorm* or *écart*.

58. Let p be an extended seminorm and set

Dom p $\underset{\text{def}}{=} \{x \mid p(x) < \infty\}$. Then Dom p is either empty or a subspace.

If Dom P $\neq \emptyset$, then the functional p|Dom p is a seminorm, defined on Dom p.

59. If Dom p $\neq \emptyset$, then $p(0) = 0$ (and conversely).

60. The unit ball $p(x) \leq 1$ for an extended seminorm p is a closed absolutely convex set; it is nonvoid iff Dom p is nonvoid, and it is a closed absolutely convex body, iff p is a seminorm.

61. If the unit balls associated with the extended seminorms p_1 and p_2 coincide, then $p_1(x) = p_2(x)$ for all $x \in E$.

Now let W be some absolutely convex set and introduce the Minkowski functional $p_W(x)$ exactly as in the case of a body. Then

62. $p_W(x)$ is an extended seminorm, and the associated unit ball is equal to W.

This establishes a bijection between the set of extended seminorms on E and the family of closed absolutely convex sets in E. Dominance and equivalence transfer without change to extended seminorms.

63. Let p_1 and p_2 be extended seminorms. p_1 is dominated by p_2 iff

$$\text{Ker } p_2 \subset \text{Ker } p_1 \quad \text{and} \quad \text{Dom } p_2 \subset \text{Dom } p_1 .$$

64. With p_1, p_2 as in (63), p_1 is equivalent to p_2 iff

$$\text{Ker } p_1 = \text{Ker } p_2 \quad \text{and} \quad \text{Dom } p_1 = \text{Dom } p_2 .$$

In order to formulate criteria (63) and (64) in terms of unit balls we define: set W_1 *absorbs* set W_2, if there is a $q > 0$ such that $qW_2 \subset W_1$.

Let now p_1 and p_2 be extended seminorms, and W_1, W_2 be absolutely convex sets. Then

65. p_1 is dominated by p_2 iff W_1 absorbs W_2; p_1 and p_2 are equivalent iff W_1 and W_2 absorb each other.

4. Hahn-Banach Theory.

The title of this section refers to the extension problem of homomorphisms given on subspaces of a normed space E.

66. Let $L \neq \{0\}$ be a subspace of E; let E_1 be another normed space, and let $h \in \mathrm{Hom}(E,E_1)$. Then $\|h|L\| \leqq \|h\|$.

In other words, extending a homomorphism either increases its norm or leaves it unchanged.

The basic problem is to demonstrate the possibility of extending a homomorphism while preserving its norm.

67. If $P \neq 0$ is a projector, then $\|P\| \geqq 1$.

Definition 29. For a given subspace $L \neq \{0\}$ let $\Pi[L]$ be the set of all projectors $P[L]$ upon L§ and define

$$\omega(L,E) = \inf_{P \in \Pi[L]} \|P\| .$$

$\omega(L,E)$, a numerical function of $L \subset E$ for fixed E, is called the *projection constant* of L. Clearly, $\omega(L,E) \geqq 1$; $E = \ell$ is an example for $\omega(L,E) > 1$, if $n = \dim E > 2$ and $\dim L = n-1$. (See Sp. Lit. to IV, Nr. 31a.)

68. There is a projector $P[L] \in \Pi[L]$ such that $\omega(L,E) =$ $P[L]$.

Consequently, $\omega(L,E) = \min_{P \in \Pi[L]} \|P\|$.

69. Every homomorphism $h \in \mathrm{Hom}(L,E_1)$, $L \neq \{0\}$, can be extended to $\tilde{h} \in \mathrm{Hom}(E,E_1)$ under preservation of its norm, iff $\omega(L,E) = 1$.

Applying this general result to a Euclidean E, we find:

70. For an orthoprojector $P \neq 0$ in Euclidean space, $\|P\| = 1$.

Hence,

---

§The notation $P[L]$ is the same as in III,D22, but the normed space in which P operates, is here not necessarily unitary, neither is P an orthoprojector.

71. $\omega(L,E) = 1$ is satisfied for all subspaces $L \neq \{0\}$ of a Euclidean space. (This condition is characteristic for Euclidean E; cf. VII,(81).)

72. If E is a Euclidean space, then for each of its subspaces $L \neq \{0\}$ and for each homomorphism $h \in \mathrm{Hom}(L,E_1)$ there is an extension $\tilde{h} \in \mathrm{Hom}(E,E_1)$ which preserves the norm of h.

However, theorem (70) can be inverted:

73. If P is a projector in Euclidean space and $\|P\| = 1$ or $P = 0$, then P is an orthoprojector.

Definition 30. This result is the motivation for designating a projector P with $\|P\| = 1$ as an *orthoprojector in any normed space*; $\omega(L,E) = 1$ then asserts that there exists an orthoprojector upon L.

Definition 31. We now introduce *orthogonality in a normed space* by stipulating: vector y is orthogonal to vector x if

$$\|x + \alpha y\| \geqq \|x\| \qquad (\text{all } \alpha) \qquad (*)$$

Notice that this "orthogonality" concept is in general not symmetric, but in Euclidean space it is identical with our previous formulation:

74. In Euclidean space, relation (*) holds for all α, iff $(x,y) = 0$.

Definition 32. We shall call a subspace M of a normed space *orthogonal to subspace* L, if (by extension of the previous terminology) every $y \in M$ is orthogonal (D31) to every $x \in L$. In the remainder of this chapter "orthogonal" is used in the sense of D31, unless something else is specified.

75. If subspace M is orthogonal to subspace L, then M and L are mutually independent (I,D29a).

Yet an arbitrary subspace L need not have an orthogonal complement in the sense of D32, and even if that complement exists, it need not be unique. However,

76. The subspace L possesses an orthogonal complement, iff

$\omega(L,E) = 1$.

This proposition is closely related to the following one:

77. The projector P is an orthoprojector iff Ker P is orthogonal to Im P.

But since (*) is not symmetric, the complementary projector $\overline{P}$ (II,D48) need not be an orthoprojector now.

Let T_x denote the set of those vectors $\in$ E which are orthogonal (D31) to $x \in E$. T_x need not be a subspace; nevertheless:

78. $y \in T_x \Rightarrow \alpha y \in T_x$ (all α).

In particular $0 \in T_x$.

79. $x \neq 0 \Rightarrow x \notin T_x$.

80. $x = 0 \Rightarrow T_x = E$.

81. $T_{\alpha x} = T_x$ $(\alpha \neq 0)$.

There are always "enough" vectors in T_x, as we shall see soon.

Definition 33. Subspace M is said to be *orthogonal to the vector* x, if $M \subset T_x$, or expressed differently, if M is orthogonal to sp x (cf. D32). It is evident that every subspace of M is also orthogonal to x.

82. Every subspace orthogonal to x is contained in some maximal subspace orthogonal to x.

If $x \neq 0$ and M is the maximal subspace orthogonal to x, then $M \neq E$, or codim $M \geq 1$, as implied by (79).

In fact:

83. If $x \neq 0$ and M is the maximal subspace orthogonal to x, then codim M = 1 (*Theorem of Ascoli and Mazur*).

In other words, the unique subspace M with dim M = n-1 deserves the designation "orthogonal complement of the one-dimensional subspace generated by x".

The proof of this fundamental theorem is based on the two lemmas that follow.

84. Let the subspace N be orthogonal to x and let h_N be the homomorphism that contracts the space modulo N (I,D45). Then

$$\|h_N y\| \le \|y\| \quad \text{(all } y \in E) \quad \text{and} \quad \|h_N x\| = \|x\|.$$

It follows that $\|h_N\| = 1$.

85. If dim $E > 1$, then for any x there is a y orthogonal to x.

Consider now the most important consequences of the Ascoli-Mazur theorem.

86. If $x \neq 0$ and T_x is a subspace, then codim $T_x = 1$.

87. For a one-dimensional L, $\omega(L,E) = 1$.

Hence any homomorphism given on a one-dimensional subspace can be extended to the entire space E under preservation of its norm. As a special case,

88. Any linear functional given on a subspace of E can be so extended (*Hahn-Banach Theorem*). (A basis, in which the given functional has the simplest form, and D27 will be useful.

This is the central theorem of the theory under consideration. We can use it to establish without difficulty the following *theorem due to Minkowski*.

89. For any given vector x a linear functional $f_x \neq 0$ exists such that

$$|f_x(x)| = \|f_x\| . \|x\| .$$

Definition 34. f_x is known as the *supporting functional* at the point x (viz. f_x supports the sphere $\{y \mid \|y\| = \|x\|\}$); f_x is determined only up to a factor $\lambda \neq 0$.-- Supporting functionals are associated with orthogonal subspaces as follows:

90. A linear functional $f \neq 0$ is supporting at the point x, iff Ker f is orthogonal to x.

91. The supporting functional at $x \neq 0$ is uniquely determined (up to some scalar factor), iff T_x is a subspace.

Independently of the Hahn-Banach theorem it is true that

92. Every linear functional $f \neq 0$ is supporting at some $x \neq 0$.

This proposition permits us to construct the analog of

the Sonine-Schmidt orthogonalization in a normed space:

93. To any basis $\{u_k\}_1^n$ a normed basis $\{e_k\}_1^n$ can be constructed such that $sp\{e_k\}_1^{j-1}$ $(j = 2,3,\ldots,n)$ is orthogonal to e_j and coincides with $sp\{u_k\}_1^{j-1}$.

Definition 35. A basis $\{e_k\}_1^n$ with the properties enounced in (93) is called *semi-orthogonal.* Theorem (93) asserts that such a basis exists in any normed space.

Definition 36. A basis $\{e_k\}_1^n$ is called an *orthogonal* or *Auerbach basis*, if for each $j = 1,2,\ldots,n$ the subspace $sp\{e_k\}_{k\neq j}$ is orthogonal to the vector e_j. An Auerbach basis exists also in any normed space, but its construction requires more complicated tools (see V,(106,107)).

94. The canonical basis in ℓ^p $(1 \leq p \leq \infty)$ is a normed Auerbach basis.

The Hahn-Banach theory transfers without essential change to a space endowed with a seminorm. Let us first introduce

Definition 37. The linear functional f is said to be *dominated* by the extended seminorm p, if the seminorm $|f|$[§] is dominated by p. We may then put

$$\|f\|_p \underset{\text{def}}{=} \sup_x \frac{|f(x)|}{p(x)} .$$

96. If p is an extended seminorm on E, and g is a linear functional defined on a subspace and dominated there by p, then an extension $\tilde{g} \in E'$ exists, which is dominated by p and satisfies $\|\tilde{g}\|_p = \|g\|_p$.

5. Isometry, Universality, and Embedding.

Definition 38. A normed space E is called *isometric* to a normed space E_1, if E and E_1 are isomorphic under an

---

[§] $|f(x)|$ is clearly a seminorm on E.

isomorphism $j \in \mathrm{Hom}(E,E_1)$ that preserves the norm:

$$\|jx\| = \|x\| \qquad (\text{all } x \in E).$$

The isomorphism j is then called an *isometry* between the spaces E and E_1. It is clearly an equivalence relation.

97. Isometry is a distance-preserving relation, viz.

$$d(jx,jy) = d(x,y) \qquad (x,y \in E).$$

Conversely, every distance-preserving surjection $E \to E_1$ is an isometry.

Definition 39. A distance-preserving Hermitian isomorphism between E and E_1 is called a *Hermitian isometry*.

98. The canonical complex conjugation $E' \to E^*$ (I,D82) is a Hermitian isometry.

But this does not mean that E' and E^* are isometric.

However:

99. The canonical isomorphisms $E \to E''$, $E \to E^{**}$ are isometries.

This fundamental theorem follows easily from Minkowski's theorem (89). With the aid of (99) and (49) one obtains

100. $\|h^*\| = \|h'\| = \|h\|$ (notation in I,(249)).

101. All Euclidean spaces of the same dimension are isometric.

102. If $p \neq p_1$ $(1 \leq p, p_1 \leq \infty)$, then the spaces ℓ^p and ℓ^{p_1} are not isometric, with the exception of the two-dimensional spaces ℓ and c (i.e. ℓ^1 and ℓ^∞) and the one-dimensional spaces for all p.

In particular:

103. ℓ^p for $p \neq 2$ is non-Euclidean (the one-dimensional case excepted).

Definition 40. For any $p \in (1,\infty)$ the *dual index* $q = p/(p-1)$ satisfies the symmetric relation

$$\frac{1}{p} + \frac{1}{q} = 1$$

104. The conjugate space to ℓ^p is isometric to ℓ^q.

Special cases are: the conjugate space to ℓ is isometric to c, and the conjugate space to c is isometric to ℓ.

Theorem (104) expresses in geometrical form *Hölder's Inequality*, well-known from analysis,

$$|\Sigma_{k=1}^{n} \xi_k \eta_k| \leqq (\Sigma_{k=1}^{n} |\xi_k|^p)^{1/p} (\Sigma_{k=1}^{n} |\eta_k|^q)^{1/q} ,$$

which becomes Cauchy's inequality (9) for $p = 2$. Given two isomorphic but not necessarily isometric spaces E_1 and E_2, we can construct a numerical characterization of their "anisometry".

Definition 41. $\delta(E_1,E_2) \underset{\text{def}}{=} \inf_j \|j\| \|j^{-1}\|$ is said to be the *measure of anisometry* of the normed spaces E_1 and E_2. (Of course, $\dim E_1 = \dim E_2$, and j runs through the set of isomorphisms $(E_1 \to E_2)$.) We also define

$$\rho(E_1,E_2) \underset{\text{def}}{=} \log \delta(E_1,E_2)$$

and call it the *Banach-Mazur distance* between the isometric spaces E_1 and E_2.

105. The Banach-Mazur distance has the basic properties of a metric:

(1) $\rho(E_1,E_2) \geqq 0$; equality holds iff E_1 and E_2 are isometric;

(2) $\rho(E_1,E_2) = \rho(E_2,E_1)$;

(3) $\rho(E_1,E_3) \leqq \rho(E_1,E_2) + \rho(E_1,E_3)$.

106. $\rho(E_1',E_2') = \rho(E_1,E_2)$

107. Let Δ_1 and Δ_2 be canonical bases of ℓ^{p_1} and ℓ^{p_2} respectively (I,D18). The following relations hold for an isomorphism $j \in \mathrm{Hom}(\ell^{p_1},\ell^{p_2})$ that maps Δ_1 upon Δ_2:

$$\|j\| = 1, \quad \|j^{-1}\| = n^{(1/p_1)-(1/p_2)} \quad (1 \leqq p_1 \leqq p_2).$$

Therefore

108. $$\delta(\ell^{p_1},\ell^{p_2}) \leqq n^{(1/p_1)-(1/p_2)}, \quad (1 \leqq p_1 \leqq p_2),$$

and in particular

109. $$\delta(\ell^p,c) \leqq n^{1/p}.$$

On the other hand

110. $\delta(\ell^p, \ell^2) \leq n^{(1/2)-(1/p)}$ $(2 \leq p \leq \infty)$.

This implies by (106)

111. $\delta(\ell^p, \ell^2) \leq n^{(1/p)-(1/2)}$,

and so

112. $\delta(\ell^p, \ell^2) = n^{|(1/2)-(1/p)|}$ $(1 \leq p \leq \infty)$.

Together with the triangle inequality for the Banach-Mazur distance, (112) leads to the Formula of *Gurarii*, *Kadec*, and *Macaev*:

113. $\delta(\ell^{p_1}, \ell^{p_2}) = n^{|(1/p_1)-(1/p_2)|}$, $(p_1-2)(p_2-2) \geq 0$.

Definition 42. A normed space E_o is called *universal* for some class $\mathbb{E}$ of normed spaces, if for any $E \in \mathbb{E}$ there is a subspace of E_o, which is isometric to E.
The following definition and propositions are preparatory to theorem (117) on the universality of the space c.
Definition 43. Let L be a subspace $\in$ E with codim L = 1, and $x_1 \in L$. The set

$$F(x_1, L) \underset{\text{def}}{=} \{x \mid x = y + \alpha x_1,\ y \in L,\ \|\alpha\| \leq 1\}$$

will be called a *hyperlayer*, and the intersection W of a finite number of hyperlayers a *symmetric polyhedron*; if m is the smallest number of hyperlayers that is needed to form a given symmetric polyhedron by intersection, we shall call it a symmetric *2m-hedron*.

114. A symmetric polyhedron is an absolutely convex body.

115. A symmetric polyhedron $W = \bigcap_{k=1}^{r} F(x_k, L_k)$ is bounded, iff $\bigcap_{k=1}^{r} L_k = 0$.

This is only possible if $r \geq n$.
Let W_m be a bounded symmetric 2m-hedron. $E(W_m)$ will denote a normed space whose norm is generated by considering W_m as the unit ball in E (cf.(56)).

116. Let W_m be a bounded symmetric 2m-hedron $W_m = \bigcap_{k=1}^{m} F(x_k, L_k)$, and $z = y_k + \alpha_k x_k$ ($y_k \in L_k$, $k = 1,2,\ldots,m$). Then the norm of $z \in E(W_m)$ is equal to

$$\|z\| = \max_{1 \leq k \leq m} |\alpha_k|.$$

A special case of (116) is $\|x_k\| = 1, k = 1,2,\ldots,m$.

117. Any space $E(W_m)$ is isometric to some subspace of c (dim c = m $\geq$ n).

Thus the space c is universal for the class of all spaces of the form $E(W_m)$, where $m \leq \dim c$.

The universality of c remains valid for the class of all normed spaces, if it is used in a weaker sense (see (120) below).

118. For an n-dimensional space E normed in any way and any given $\varepsilon > 0$, there is a number $m = m(n,\varepsilon)$ and a norm-generating W_m such that $\rho(E, E(W_m)) < \varepsilon$ (D41).

This can be proved if we approximate the unit ball of E by a symmetric polyhedron, using the following proposition:

119. Let $\{e_k\}_1^n$ be a normed semi-orthogonal basis. Then the distance

$$d_j = d(e_j, \mathrm{sp}\{e_k\}_{k \neq j}) \quad \text{(cf. D19)}$$

satisfies the inequality $d_j \geq 2^{1-n}$ ($j = 1,2,\ldots,n$).

Definition 44. The normed space E_o is said to be ε-*universal* for some class $\mathbb{E}$ of normed spaces, if for any $E \in \mathbb{E}$ there is a subspace $L_o \subset E_o$ such that $\rho(E, L_o) < \varepsilon$.

120. For any $\varepsilon > 0$ there is an $m = m(n,\varepsilon)$ such that the m-dimensional space c is ε-universal for the class of all n-dimensional normed spaces.

This can be regarded as an approximation-theoretical analog of the Banach-Mazur Theorem. (See e.g. General Textbooks, Nr. 34, Ch. III, § 26).

In conclusion we consider estimates of projection constants as they occur in the isometric embedding of one normed space into another.

121. If E contains a subspace L (dim L = m) which is isometric to ℓ^p, then $\omega(L,E) \leqq m^{1/p}$ $(1 \leqq p \leqq \infty)$.

In the proof one can use the Hahn-Banach theorem adapted to canonical coordinates, transferred from ℓ^p to L. For $p = \infty$, (121) gives:

122. If E contains a subspace L, which is isometric to c (hence dim L = dim c), then $\omega(L,E) = 1$.

In connection with (121), note that the existence of an Auerbach basis in any normed space E implies:

123. For any subspace $L \subset E$ (dim L = m)

$$\omega(L,E) \leqq m \ .$$

6. Best Approximations.

Definition 45. Recalling the functional d(x;L), introduced in D19 to characterize the distance of x from a subspace L, we now call the vector $y \in L$ a *best approximation* to the vector x by vectors from L, if

$$\|x - y\| = d(x;L) \ .$$

124. For any given x and subspace L there is a best approximation y. Thus we may write D19 as $d(x;L) = \min_{y \in L} \|x-y\|$.

Obviously, the best approximation need not be unique, but

125. If $x \in L$, then its best approximation by vectors from L is unique and = x.

126. The vector y is a best approximation to x by vectors from L, iff L is orthogonal to the difference vector x-y.

Thus, for example,

127. In Euclidean space, y is a best approximation to x by vectors from L, iff y = P[L]x, where P[L] is the orthoprojector upon L.

Hence the best approximation is unique in Euclidean space.

Definition 46. A normed space E is called *strictly normed* if the equality sign in the triangle inequality $\|x+y\| \leq \|x\| + \|y\|$ holds only for linearly dependent x and y (then $y = \alpha x$ with $\alpha > 0$, when x and y are $\neq 0$).

128. Euclidean spaces are strictly normed.

129. All ℓ^p spaces, $1 < p < \infty$, are strictly normed.

However,

130. The spaces ℓ and c are not strictly normed.

This example is typical in the following sense:

131. In a space that is not strictly normed there are linearly independent vectors e_1, e_2 such that

$$\|\alpha e_1 + \beta e_2\| = \alpha + \beta$$

for any pair of real $\alpha, \beta \neq 0$.

The usefulness of this concept is illustrated by the following proposition:

132. The best approximation of any given $x \in E$ by vectors from any given subspace $L \subset E$ is unique, iff the space E is strictly normed.

Definition 47. The preceding theorem determines for any $L \subset E$ an *operator* $A[L]$ *of best approximation*, which maps any vector x into its best approximation by vectors from L. This operator is in general nonlinear, although homogeneous since

133. $$A[L]\alpha x = \alpha A[L]x ,$$

and, like a projector, it is idempotent:

134. $$(A[L])^2 = A[L] .$$

Look at it as a *nonlinear orthoprojector* upon L; in a Euclidean E it actually coincides with the orthoprojector $P[L]$.

135. The operator $A[L]$ is continuous.

In view of (135) and (133), we can introduce a norm for $A[L]$:

$$\|A[L]\| = \sup_x \frac{\|A[L]x\|}{\|x\|} = \sup_{\|x\|=1} \|A[L]x\| .$$

136. $\|A[L]\| \geq 1$ (all $L \neq 0$).

Consider the complementary operator $\overline{A[L]} \underset{def}{=} I - A[L]$. It is also homogeneous and continuous, hence we can again introduce

$$\|\overline{A[L]}\| = \sup_x \frac{\|\overline{A[L]x}\|}{\|x\|} = \sup_{\|x\|=1} \|\overline{A[L]x}\|.$$

This norm coincides with the norm of the nonlinear functional $d(.;L)$, constructed in the same way as the norm of the functional p in D16. One obtains in fact

137. $$\|\overline{A[L]}\| = 1 \quad (\text{all } L \neq E)$$

Let us now consider the problem of best approximation in regard to a chain of subspaces. Here the space E is no longer assumed to be strictly normed.

Let $\{L_k\}_o^n$ be some maximal chain of subspaces (I,D26) and abbreviate $d(x;L_k) = d_k(x)$. Obviously $d_o(x) = \|x\|$, $d_n(x) = 0$.

138. $$d_o(x) \geq d_1(x) \geq \dots \geq d_n(x) \quad (\text{all } x \in E).$$

But the converse is also true:

139. For any system of numbers $\delta_o \geq \delta_1 \geq \dots \geq \delta_n$ ($\delta_n = 0$) there is a vector x such that $d_k(x) = \delta_k$, $k = 0,1,\dots,n$. This *theorem* is due to *S. N. Bernstein*.

We now look at the special case when the chain $\{L_k\}_o^n$ is generated by a basis $\{e_k\}_1^n$: $L_k = sp\{e_j\}_o^k$ $(k = 0,1,\dots,n)$.

Definition 48. The basis $\{e_k\}_1^n$ is said to be a *basis of best approximation* if for any $x = \Sigma_{j=1}^n \xi_j e_j$ the distances $d_k(x)$ defined above satisfy

$$d_k(x) = \|\Sigma_{j=k+1}^n \xi_j e_j\| .$$

Let R_k denote the projector upon the linear hull $sp\{e_j\}_{k+1}^n$ parallel to L_k. Clearly,

$$R_k x = \Sigma_{j=k+1}^{n} \xi_j e_j ,$$

and the best approximation to x by vectors from L_k is $\overline{R}_k x = \Sigma_{j=1}^{k} \xi_j e_j$.

140. The basis $\{e_j\}_1^n$ is a basis of best approximation, iff for any $x \in E$

$$\|R_o x\| \geqq \|R_1 x\| \geqq \ldots \geqq \|R_{n-1} x\| \qquad (\|R_o x\| \underset{def}{=} \|x\|).$$

Such a basis has very special properties (indeed so special that in a three-dimensional space it need not even exist.

141. The basis $\{e_j\}_1^n$ is a basis of best approximation, iff e_k is orthogonal to $sp\{e_j\}_{k+1}^n$ for $k = 1,2,\ldots,n-1$.

This construction is dual to the (always existing) semi-orthogonal basis of D35.

142. The canonical basis of ℓ^p $(1 \leqq p \leqq \infty)$ is a basis of best approximation.

7. The Aperture of Two Subspaces. The Metric Space of Subspaces.

Definition 49. The $L \neq 0$ be a subspace, denote by S_L the unit sphere of L

$$S_L = \{x \mid x \in L, \|x\| = 1\},$$

and set $S_0 = 0$. The *aperture of the subspaces* L and M is defined as

$$\theta(L,M) = \max\{\sup_{x \in S_M} d(x;L), \ \sup_{x \in S_L} d(x;M)\}.$$

The aperture satisfies

143. $$\theta(L,M) \leqq 1.$$

The aperture shares with the metric the properties

144. 1. $\theta(L,M) > 0$ if $L \neq M$, $\theta(L,L) = 0$;

2. $\theta(L,M) = \theta(M,L)$.

However, the triangle inequality need not hold. The definition of the aperture will be slightly modified in what follows, in order to obtain a metric on the set of subspaces.

In Euclidean space the aperture of D49 <u>is</u> a metric as will be shown now.

145. For any two subspaces L and M of a Euclidean space we have

$$d(x;L) = \|(P[M] - P[L])x\| \qquad (x \in M) ,$$

Therefore

146. $$\theta(L,M) \leq \|P[M] - P[L]\| .$$

But in fact

147. $$\theta(L,M) = \|P[M] - P[L]\| .$$

In proving (147) one can use the identity

148. $$\|(P[M]-P[L])x\|^2 = \|P[M](I- P[L])x\|^2 + \|(I-P[M])P[L]x\|^2.$$

(147) implies immediately:

149. In Euclidean space, θ obeys the triangle inequality

$$\theta(L,M) \leq \theta(L,N) + \theta(N,M) .$$

Another straightforward consequence of (147) is the self-duality of the aperture:

150. $$\theta(L^{\perp},M^{\perp}) = \theta(L,M) .$$

Now we can also clarify the conditions for equality in (143):

151. $\theta(L,M) = 1$ in Euclidean space, iff at least one of the subspaces L,M contains a non-vanishing vector that is orthogonal to the other subspace, i.e., $L \cap M^{\perp} \neq 0$ or $L^{\perp} \cap M \neq 0$.

An important consequence of (150) is the *Theorem of Sz.-Nagy:*

52. If in a Euclidean space $\theta(L,M) < 1$, then

$$\dim L = \dim M .$$

The converse is of course not true. That is, if L and M are subspaces of a Euclidean space and dim L = dim M, then $\theta(L,M) < 1$ need not hold.

The following proposition supplements (152):

153. Let P and Q be orthoprojectors upon L and M respectively and suppose again $\theta(L,M) < 1$. Then the operator $S = I+P(Q-P)P$ is self-adjoint and positive, and the operator $V = QS^{-\frac{1}{2}}P$ maps L onto M isometrically.

Sz.-Nagy's theorem can be generalized to finite-dimensional spaces with arbitrary norm, but such a generalization needs special topological methods[§].

Definition 50. The number

$$\tilde{\theta}(L,M) = \max\{\sup_{x\in S_M} d(x;S_L),\ \sup_{x\in S_M} d(x;S_M)\}$$

is called the *spherical aperture* of the subspaces L and M (both $\neq 0$); as usual, $d(x;F)$ denotes the distance of x from the set F which happens to be compact now.

We complete this definition by setting $\tilde{\theta}(0,M) = \tilde{\theta}(M,0) = 1$, if $M \neq 0$ and $\theta(0,0) = 0$.

This modification of the aperture concept is quantitatively not very important:

154. The spherical aperture satisfies $\theta(L,M) \leq \tilde{\theta}(L,M) \leq 2\theta(L,M)$.

Its advantage shows itself in that

155. The spherical aperture is always a metric.

Note that aperture and spherical aperture are different even in Euclidean space. There is still another modification of the aperture concept, given by

Definition 51. $\check{\theta}(L,M) = \max \sup_{x\in V_M} d(x;S_L),\ \sup_{x\in V_L} d(x;S_M),$

where V_L and V_M are the <u>unit balls</u> in the subspaces L

[§]See M.G. Kreĭn, M.A. Krasnosel'skiĭ, D.P. Mil'man, On deficiency indices of linear operators in Banach spaces and some geometrical questions. Sbornik Trudov Akad. Nauk. Ukr.S.S.R, 11(1948) 97-112. (Russian)

and M; it could be called the *globular aperture* $\check{\theta}$. This time,

156. $\check{\theta}(L,M) = \theta(L,M)$ in a Euclidean space,

and again:

157. The globular aperture is always a metric.

158. $\theta(L,M) \leq \check{\theta}(L,M) \leq C\theta(L,M)$, where C is an absolute constant.

The set of all subspaces of the normed space E, endowed with the metric of the spherical or globular apertures, is a metric space. We shall denote it by $\mathfrak{G}(E)$ and investigate its properties. Let us first consider the convergence of sequences. By (154) and (158) convergence may be formulated in terms of the aperture θ:

159. The sequence of subspaces $\{L_k\}_1^\infty$ converges to L, i.e., $\lim_{k\to\infty} L_k = L$, iff $\lim_{k\to\infty} \theta(L_k, L) = 0$.

160. If $\lim_{k\to\infty} L_k = L$, then L is the set of all vectors x that are limits of those convergent sequences $\{x_k\}_1^\infty$ whose elements $x_k \in L_k$ $(k = 1,2,\ldots)$.

Definition 52a. The vector set described in (160) is well-defined for any sequence of subspaces $\{L_k\}_1^\infty$ (of course, it may only contain the 0 vector). In the general case we shall call this set the *lower limit* of the sequence $\{L_k\}_1^\infty$ and denote it by $\underline{\lim}_{k\to\infty} L_k$. If $\lim_{k\to\infty} L_k$ exists, then it coincides with $\underline{\lim}_{k\to\infty} L_k$.

Definition 52b. The *upper limit* of a sequence $\{L_k\}_1^\infty$ shall mean the sum of subspaces (I,D28), which have the form $\underline{\lim}_{j\to\infty} L_{k_j}$. (These are the lower limits of all possible subsequences $\{L_{k_j}\}_{j=1}^\infty$ of $\{L_k\}_1^\infty$.) We denote it by $\overline{\lim}_{k\to\infty} L_k$ and have

161. $$\underline{\lim}_{k\to\infty} L_k \subset \overline{\lim}_{k\to\infty} L_k .$$

162. The sequence of subspaces $\{L_k\}_1^\infty$ converges iff $\underline{\lim}_{k\to\infty} L_k = \overline{\lim}_{k\to\infty} L_k$.

It is clear that all norms, which one can introduce in the basic space E, generate the same topology in $\mathfrak{G}(E)$. Thus in studying $\mathfrak{G}(E)$ as a topological space, we may as well assume that E is Euclidean, which makes the task much easier, and we shall maintain this assumption until and including problem (175). For instance, from Sz.-Nagy's theorem immediately ensues the following proposition:

163. Every $L \in \mathfrak{G}(E)$ has a neighborhood N of subspaces such that all subspaces $\in N$ have the same dimension.

164. The set $\mathfrak{G}_m(E)$ of all subspaces of $\mathfrak{G}(E)$ with fixed dimension m is at the same time open and closed in $\mathfrak{G}(E)$.

The topological space $\mathfrak{G}(E)$ is therefore not connected.

Definition 53. The set $\mathfrak{G}_m(E)$, which is also a topological space, is known as a *Grassmann manifold* of rank m.

Is the Grassmann manifold $\mathfrak{G}_m(E)$ a connected set? Or, in other words, is every "homogeneous part" of $\mathfrak{G}(E)$ connected? To answer this question we must carry the investigation further.

Let $\mathfrak{S}_m(E)$ denote the set of all orthonormal systems $\{u_k\}_1^m$ in the Euclidean space E; let $\Gamma = \{u_k\}_1^m$ and $\Gamma_1 = \{v_k\}_1^m$ be two such systems and put

$$\delta(\Gamma,\Gamma_1) = \{\Sigma_{k=1}^m \|u_k - v_k\|^2\}^{\frac{1}{2}} .$$

165. The functional δ is a metric on $\mathfrak{S}_m(E)$; hence $\mathfrak{S}_m(E)$ is now a metric space (and, of course, a topological space).

Definition 54. $\mathfrak{S}_m(E)$ with the metric δ is called a *Stiefel Manifold* of rank m.

Now consider the mapping $\mathsf{L}_m : \mathfrak{S}_m(E) \to \mathfrak{G}_m(E)$, defined by

$$\mathsf{L}_m(\Gamma) = \text{sp } \Gamma \qquad (\Gamma \in \mathfrak{S}_m(E)).$$

166. L_m is a continuous mapping of the (topological) space

$S_m(E)$ onto the (topological) space $\mathfrak{G}_m(E)$.

However, we can prove that

167. The space $\mathfrak{S}_m(E)$ is connected.

Therefore

168. The space $\mathfrak{G}_m$ is connected.

The mapping L_m is useful also in other questions. The following proof that the Grassmann manifold $\mathfrak{G}(E)$ is compact is an example:

169. $\mathfrak{S}_m(E)$ is compact (cf. III,(174)).

170. $\mathfrak{G}_m(E)$ is compact.

Consequently

171. $\mathfrak{G}(E)$ is compact.

This is a fundamental result. One of its immediate consequences is that $\mathfrak{G}(E)$ is complete§.

The mapping L_m can, in a certain sense, be inverted. Denote by $\tilde{\mathfrak{S}}_m(E)$ the set of all linearly independent systems of m vectors in E and introduce the mapping $T:\tilde{\mathfrak{S}}_m(E) \to \mathfrak{S}_m(E)$, determined by the process of orthonormalization.

172. Suppose $L_o \in \mathfrak{G}_m(E)$ and Γ_o is an orthonormal basis in L_o. Then the system $\Gamma_L \underset{def}{=} TP[L]\Gamma_o$ is an orthonormal basis for the subspace L that belongs to the neighborhood $\theta(L,L_o) < 1$ of L_o (cf.(153)).

The last topic in this section is the theory of the perturbation of bases. Note first that the metric δ can be applied to the set of all systems $\{u_k\}_1^m$ (not only of orthonormal systems). This permits the formulation of the following theorem on quadratically close bases.

173. If Δ is a normed orthogonal basis, then all points of the ball $\delta(\Gamma,\Delta) < 1$ are bases.

This is a "best" theorem, because

---

§"complete" is used here in the sense in which it occurs in the theory of metric spaces. Cf. Nr. 25 in the list of General Textbooks, pp. 20 and 22.

174. There is a system on the sphere $\delta(\Gamma,\Delta) = 1$, which is not a basis.

Related to (173) is the *Theorem of Wiener and Paley*.

175. If $\{e_k\}_1^n$ is a normed orthogonal basis, then every system $\{u_k\}_1^n$, which satisfies the inequality

$$\|\Sigma_{k=1}^n \alpha_k(u_k - e_k)\|^2 \leqq \vartheta\Sigma_{k=1}^n|\alpha_k|^2 \qquad (0 \leqq \vartheta < 1)$$

for all sets $\{\alpha_1,\alpha_2,\ldots,\alpha_n\}$ is a basis. For the coordinates of $x = \Sigma_{k=1}^n \xi_k u_k$ one has the bounds

$$(1 - \vartheta)\Sigma_{k=1}^n|\xi_k|^2 \leqq \|x\|^2 \leqq (1 + \vartheta)\Sigma_{k=1}^n|\xi_k|^2 .$$

This also is a best theorem.

For any basis $\{e_k\}_1^n$ and even for an arbitrarily normed space E the following *Theorem of Kreĭn, Mil'man, and Rutman* holds:

176. Let $\Delta = \{e_k\}_1^n$ be a basis in E and $\Delta' = \{f_{k_n}\}_1^n$ the conjugate basis in E' (I,D62). Then every system $\{u_k\}_1^n$ is a basis if it satisfies

$$\Sigma_{k=1}^n \|f_1\|.\|u_k - e_k\| < 1 .$$

A particular case is

177. If $\{e_{k_n}\}_1^n$ is a normed Auerbach basis (D36), then every system $\{u_k\}_1^n$ is a basis if it satisfies

$$\Sigma_{k=1} \|u_k - e_k\| < 1.$$

Theorem (177) is again a best theorem:

178. If $\{e_k\}_1^n$ is a normed Auerbach basis, then there is a linearly dependent system $\{u_k\}_1^n$ for which $\Sigma_{k=1}^n\|u_k - e_k\| = 1$.

8. Isometric Operators and Compressions. Ergodic Theory.

Definition 55. Recalling D38 we introduce here the

isometric operator in the normed space E, defining it as an isometric mapping E E.

179. The isometric operators form a multiplicative group.

Isometricity can be described in terms of the operator norm:

180. A regular operator is isometric iff $\|A\| = 1$ and $\|A^{-1}\| = 1$.

Using (105,(1)) we can prove that

181. For any regular operator A

$$\|A\|.\|A^{-1}\| \geqq 1 .$$

Here equality holds iff $A = \rho U$, where U is an isometric operator.

Thus the sufficient condition in (180) can be weakened:

182. If the regular operator A satisfies the conditions $\|A\| \leqq 1$ and $\|A^{-1}\| \leqq 1$, then it is isometric.

Definition 56. An operator A that satisfies $\|A\| \leqq 1$ is called a *compression*[§]. It is a *strict compression* if $\|A\| < 1$.

183. The compressions form also a multiplicative semigroup.

From (100) ensues

184. An operator conjugate (II,D3) to a compression (to an isometric operator) is a compression (an isometric operator).

The same is true for a Hermitian-conjugate operator. In Euclidean space the set of isometric operators can be characterized very lucidly:

185. An operator in a Euclidean space is isometric, iff it is unitary.

Also in a general normed space a considerable part of the spectral theory of unitary operators carries over

[§]We do not introduce the usual term "contraction" to avoid confusion with I,D45.

to isometric operators.

186. An isometric operator has a unitary spectrum (III,D45).

187. An isometric operator is of scalar type (II,D18).

In a more general fashion we can prove:

188. Let A be some regular operator. If the cyclic group $\{A^k\}_{-\infty}^{+\infty}$ generated by A is bounded, i.e., if $\sup_{-\infty<k<+\infty} \|A^k\|<\infty$, then A is an operator of scalar type with unitary spectrum.

The converse assertion is also true.

However, (188) is not more than a formal generalization of (186) and (187) for the following reason:

189. If $\sup_{-\infty<k<\infty} \|A^k\| < \infty$, then the functional $\sup_{-\infty<k<\infty} \|A^k x\|$ defined on E is a norm on E, and the operator A is isometric under that norm.

Let us now look at the orthogonality (D32) of the characteristic subspaces of an isometric operator.

190. Suppose U is an isometric operator, $\{P_j\}_1^m$ its decomposition of the identity, and $\{e^{i\vartheta_j}\}_1^m$ the corresponding eigenvalues. Then the projectors are given by

$$P_j = \lim_{r\to\infty} \frac{1}{2r+1} \Sigma_{k=-r}^{r} e^{-ik\vartheta_j} U^k .$$

This formula implies that

191. All projectors P_j, which appear in the decomposition of the identity effected by an isometric operator, are orthoprojectors: $\|P_j\| = 1$, $(j = 1,2,..,m)$.

It follows:

192. The characteristic subspaces of an isometric operator are mutually orthogonal. In addition, the orthogonal complement of each characteristic subspace is the direct sum of the remaining characteristic subspaces.

Definition 57. A *basis* Δ in E is called *absolute*, if the norm of a vector depends on the absolute values of the coordinates in that basis. In the space ℓ^p $(1 \leq p \leq \infty)$ the canonical basis is an example of an absolute basis. In Euclidean space, the class of absolute bases and the class of orthogonal bases are identical.

193. Every absolute basis is an Auerbach basis (D36).

The converse is false. Worse than that: there are normed spaces without absolute basis.

194. If an operator with unitary spectrum has a characteristic basis which is absolute, then it is isometric.

The converse is, of course, not true, but it becomes true if the condition of ergodicity is added.

Definition 58. An isometric operator is called *ergodic*, if its spectrum $\{e^{i\vartheta_j}\}$ is simple (II,D19) and the exponents ϑ_j are <u>rationally independent</u>. This means: $\Sigma_{j=1}^{n} \rho_j \vartheta_j = 0$ with rational coefficients ρ_j implies that all ρ_j vanish. This assertion is equivalent to the following one: for every selection of real numbers $\{\omega_j\}_1^n$ and for every $\varepsilon > 0$ there exist integers k and $q_1, q_2, \ldots, q_n$ such that

$$|k\vartheta_j + 2\pi q_j - \omega_j| < \varepsilon \qquad (j = 1,2,\ldots,n) .$$

195. The characteristic basis of an ergodic operator is absolute.

From our spectral theory now follows directly an *ergodic theorem:*

196. The limit

$$P = \lim_{r\to\infty} \frac{1}{r+1} \Sigma_{k=0}^{r} U^k$$

exists for every isometric operator U. The operator P is an orthoprojector which projects upon the subspace of the fixed points of U, given by Im P = $\{x|U_x = x\}$ (cf.(190)).

The ergodic theorem can be proved without recourse to spectral theory. If this is done, there appears the possibility of generalizing the ergodic theorem to compressions.

197. If A is a compression, then all arithmetical means

$A_r = (1/(r+1))\Sigma_{k=0}^{r} A^k$, $r = 0,1,2,\ldots$, are also compressions.

198. If P is the limit of some subsequence of the sequence $\{A_r\}_0^\infty$, then AP = PA = P.

Hence

199. $$P^2 = P,$$

and P is a projector. More than that:

200. P is an orthoprojector upon the subspace of fixed points of the operator A.

We must still establish that P is independent of the choice of the convergent subsequence in (198). For that the following remark suffices:

201. The limits of any two convergent subsequences of the sequence $\{A_r\}_0^\infty$ commute.

We have thus obtained an ergodic theorem for compressions:

202. For any compression A the limit

$$P = \lim_{r\to\infty} \frac{1}{r+1} \Sigma_{k=0}^{r} A^k$$

exists. P is an orthoprojector upon the subspace of fixed points of A.

The operator P can be completely described in spectral terms:

203. The subspace Ker P coincides with the sum of those root subspaces (I,D28) of the operator A, which belong to eigenvalues different from unity.

In proving (203), keep in mind, first, that

204. The spectrum of a compression lies on the closed unit disc $|\lambda| \leq 1$;

and, second, that

205. The orders (II,D26) of those eigenvalues of a compression which lie on the unit circle, are equal to unity.

Supplementing (205), we observe:

206. For a compression, the characteristic subspaces associated with eigenvalues "on the rim" are mutually orthogonal.

On the way to the generalized ergodic theorem one more

(formal) step can be made.

207. If the cyclic semigroup $\{A^k\}_0^\infty$ generated by the operator A is bounded (i.e., $\sup_{k\geq 0} \|A^k\| < \infty$), then the limit

$$P = \lim_{r\to\infty} \frac{1}{r+1} \Sigma_{k=0}^{r} A^k$$

exists and is a projector upon the subspace of fixed points of A. (A is now not assumed to be a compression.)

But, in analogy with (189), this generalization is again purely formal, since

208. If $\sup_{k\geq 0} \|A^k\| < \infty$, then $\sup_{k\geq 0} \|A^k x\|$ $(x \in E)$ is a norm on E, and A is a compression under that norm.

A further generalization of this ergodic theorem is impossible, because

209. $[\lim_{r\to\infty} (1/(r+1))\Sigma_{k=0}^{r} A^k \text{ exists}] \Rightarrow \sup_{k\geq 0} \|A^k\| < \infty.$

Finally we note an interesting modification of (208) in the case of Euclidean space:

210. If A operates in a Euclidean space E and satisfies $\sup_{k\geq 0} \|A^k\| < \infty$, then it becomes a compression when a certain Euclidean norm, $k \geq 0$ (D7), which in general does not coincide with the original norm, is introduced in E.

9. Norm and Spectral Radius of an Operator.

In II,(128) the spectral radius was calculated from the powers of the operator by a rather complicated limiting process. This problem simplifies considerably, if we base the calculation on the norm concept. We have first

211. $$\rho(A) \leq \|A\|$$

and shall need the following lemma, which is equivalent to *Fekete's Theorem* (Pólya and Szegö, Aufgaben und Lehrsätze, Vol. 1,I,Nr.98):

212. The sequence of nonnegative numbers $\{\alpha_\nu^{1/\nu}\}_1^\infty$ with

$\alpha_{m+n} \leqq \alpha_m \alpha_n$ converges, and $\lim\limits_{\nu\to\infty} \alpha_\nu^{1/\nu} = \inf\limits_{1\leqq\nu<\infty} \alpha_\nu^{1/\nu}$.

From II,(128), which is a consequence of II,(127) and I,(333), we obtain

213. $$\rho(A) = \overline{\lim_{k\to\infty}} (\|A^k\|)^{1/k} .$$

Now (212) can be applied to $\{\|A^k\|^{1/k}\}_1^\infty$:

214. $\lim\limits_{k\to\infty} (\|A^k\|)^{1/k}$ exists and is equal to $\inf\limits_{1\leqq k<\infty} \|A^k\|^{1/k}$.

Thus we obtain the *Formula of Gel'fand:*

215. $$\rho(A) = \lim_{k\to\infty} (\|A^k\|)^{1/k}$$

and the corollary

216. $$\rho(A) = \inf_{1\leqq k<\infty} (\|A^k\|)^{1/k} .$$

An interesting consequence of (215) is

217. If $A\nabla B$, then $\rho(A+B) \leqq \rho(A) + \rho(B)$ and $\rho(AB) \leqq \rho(A)\rho(B)$.

It is evident that $\rho(A) \geqq 0$ and $\rho(\alpha A) = |\alpha|\rho(A)$.

Returning to (211), we now want to point out conditions for the equality sign there.

218. If A is regular, $\|A\| = \rho(A)$, and $\|A^{-1}\| = 1/\rho(A)$, then $A = \rho U$, where $\rho = \rho(A) > 0$, and U is an isometric operator.

219. $\|A\| = \rho(A)$, iff $\|A^k\| = \|A\|^k$, $k = 2,3,\ldots$.

Consider now the Euclidean case:

220. If N is a normal operator (III,D51), then $\|N\| = \rho(N)$.

221. The operator A acting in Euclidean space satisfies $\|A\| = \rho(A)$, iff it is either normal or can be written as $A = N \boxplus R$ (notation in III,D35) where N is normal and $\|R\| \leqq \|N\|$ (cf.III,(282)).

We notice in passing that the extremal properties of the eigenvalues imply (cf.III,(227))

222. In Euclidean space, $\|A\| = \sqrt{\rho(A^*A)} = \sqrt{\rho(AA^*)}$.

223. Also in Euclidean space, $\|A\| = \max\limits_{\|x\|=1,\|y\|=1} |(Ax,y)|$.

Thus the norm of an operator coincides here with the norm of the Hermitian-bilinear functional which it generates. This is no longer true for the corresponding Hermitian-quadratic functional, since

224. $\max_{\|x\|=1} (Ax,x) \leqq \|A\|.$

In other words, the Hausdorff set of A (III,D55) is a <u>subset</u> of the disc $|\lambda| \leqq \|A\|$.

225. The relation $\max_{\|x\|=1} |(Ax,x)| = \|A\|$ is true, iff $\|A\| = \rho(A)$.

The companion relation to (224) is

226. $\|A\| \leqq 2 \max_{\|x\|=1} |(Ax,x)|,$

which cannot be improved, because

227. If $n > 1$, there is an operator A such that $\|A\| = 1$ and $\max_{\|x\|=1} |(Ax,x)| = \frac{1}{2}$.

Again returning to (211), we notice that the first member is independent of the choice of the norm. Should it be impossible to choose a norm for a given operator A so that (211) becomes an equality? We can first prove by means of (207) that

228. $\rho(A) = \inf_{\|\cdot\|} \|A\|$ (i.e., taken over all norms on E).

Using (209), we obtain an even more precise assertion:

229. If A is an operator in Euclidean space, then $\rho(A) = \inf_{T} \|TAT^{-1}\|$.

This means that we can restrict the set of norms in (228) to Euclidean norms, when determining the infimum.

230. The infimum in (228) is attained, iff all orders (II, 26) of eigenvalues, located on $|\lambda| = \rho(A)$, equal unity. Thus equality in (211) can be obtained for compressions (cf.(205)) by the appropriate choice of the norm, but not for a general operator A.

We now turn to the norm of the resolvent. Similarly to D19, we denote the distance of λ from a point set M in the complex plane by $d(\lambda;M) = \inf_{\mu \in M} |\lambda-\mu|$ and abbreviate $d(\lambda;\sigma(A))$ by $d(\lambda;A)$. Then we obtain from the Gel'fand formula

231. $d(\lambda;A) = 1/\lim_{k\to\infty} \|R_\lambda^k\|^{1/k}$ ($R_\lambda = R_\lambda(A)$ as in II,D37a), and also

232. $$d(\lambda;A) = \sup_{k\geq 1} 1/\|R_\lambda^k\|^{1/k} .$$

This implies an inequality, which is often useful:

233. $$\|R_\lambda\| \geq 1/d(\lambda;A) .$$

If A in this relation is a normal operator in Euclidean space, then

234. $$\|R_\lambda\| = 1/d(\lambda;A).$$

Conversely,

235. If A operates in Euclidean space and $\|R_\lambda\| = 1/d(\lambda;A)$, then A is normal.

In the special case of a self-adjoint operator in Euclidean space,

236. $$\|R_\lambda\| \leq 1/|Im\ \lambda|.$$

Conversely,

237. If A operates in Euclidean space and $\|R_\lambda\| \leq 1/|Im\ \lambda$ then A is self-adjoint.

In analogy with (236) and (237) we have:

238. If U is a unitary operator in Euclidean space, then $\|R_\lambda\| \leq 1/(1 - |\lambda||$; conversely, this inequality implies that U is a unitary operator.

The last result holds also in general normed spaces in the following form:

239. If U is an isometric operator in a normed space, then $R_\lambda \leq 1/(1 - |\lambda||$; conversely, this inequality implies that U is isometric.

10. Norms in the Space of Operators.

We can consider the space of operators $\mathbb{M}(E)$ as an independent n^2-dimensional complex linear space; hence we may invest it with any norm we like, independently of the norm on E. The following

Definition 59 introduces the needed terms. If the norm on $\mathbb{M}(E)$ is *induced* by the norm on E, we call it an *operator norm*. Any norm on $\mathbb{M}(E)$ that has the ring property

$$\|AB\| \leq \|A\|\|B\| \qquad (A,B \in \mathbb{M}(E))$$

will be called a *ring norm*. Finally, if $\|I\| = 1$, the norm will be called *unity-preserving*.

Every operator norm is clearly a unity-preserving ring norm.

240. If E is a unitary space, then so is $\mathbb{M}(E)$ for the scalar product

$$(A,B) \underset{\text{def}}{=} \text{tr } AB^* .$$

241. The norm corresponding to this scalar product,

$$\|A\| = \sqrt{\text{tr } AA^*} ,$$

is a ring norm, but for $n > 1$ it is not unity-preserving, hence not an operator norm.

Definition 60. The norm just described is known as the Hilbert norm (also as the *absolute norm*). It has occurred in III,(261).

242. Let Δ be some basis in E, and $(\alpha_{j,k})_{j,k=1}^{n}$ the matrix of an arbitrary operator in that basis. Putting

$$\|A\| = \frac{1}{n} \Sigma_{j,k=1}^{n} |\alpha_{j,k}|,$$

we obtain a unity-preserving norm on $\mathbb{M}(E)$, but it is not a ring norm, if $n > 1$.

243. Putting

$$\|A\| = C \max_{j,k} |\alpha_{jk}|, \qquad 0 < C < n, \quad C \neq 1, \qquad (*)$$

we obtain a norm which is neither a ring norm nor unity-preserving.

Incidentally, for $C \geqq n$ the norm (*) is a ring norm but not unity-preserving; for $C = 1$ it is unity-preserving but not a ring norm if $n > 1$.

In the following two propositions, Δ is the canonical basis and $(\alpha_{jk})_{j,k=1}^{n}$ the matrix of the operator A in that basis for the spaces ℓ and c respectively. One obtains:

244. The <u>operator</u> ℓ-<u>norm</u> is $\max_k \Sigma_{j=1}^{n} |\alpha_{jk}|$.

245. The <u>operator</u> c-<u>norm</u> is $\max_j \Sigma_{k=1}^{n} |\alpha_{jk}|$.

246. $\|I\| \geqq 1$ for every ring norm.

Definition 61. Suppose some <u>initial</u> norm now denoted by $\|.\|$ has been introduced in $\mathbb{M}(E)$. Then we define the *derived norm* $\|.\|'$ by

$$\|A\|' = \sup_{B\neq 0} (\|AB\|/\|B\|) .$$

247. The derived norm is always a unity-preserving ring norm.

248. If the initial norm $\|.\|$ is a ring norm, then $\|A\|' \leqq \|A\|$.

249. If the initial norm $\|.\|$ is unity-preserving or even if $\|I\| < 1$, then $\|A\|' \geqq \|A\|$.

250. A norm on E is a unity-preserving ring norm, iff it coincides with its derived norm.

Proposition (248) is a good motivation to introduce an order relation in the set of all norms on $\mathbb{M}(E)$. If $\|A\|_1 \leqq \|A\|_2$ for all $A \in \mathbb{M}(E)$, we can speak of $\|.\|_1$ being smaller than or equal to $\|.\|_2$ (or being dominated by $\|.\|_2$). It is of course possible that two norms cannot be compared. E.g., the derived norm cannot in general be compared with the initial norm, since the converse theorems to (248) and (249) hold:

251. If the derived norm $\|.\|'$ is smaller than or equal to the initial norm $\|.\|$, then the latter is a ring norm.

252. If the derived norm $\|.\|'$ is larger than or equal to the initial norm $\|.\|$, then $\|I\| \leqq 1$.

We shall now introduce a more general concept of "derived norm" by

Definition 62. Let $\mathbb{J} \neq 0$ be some left ideal of the algebra $\mathbb{M}(E)$ (II,D13). For every norm on $\mathbb{M}(E)$ we define a *derived norm relative to* $\mathbb{J}$

$$\|A\|'_{\mathbb{J}} \underset{\text{def}}{=} \sup_{\substack{B\neq 0\\ B\in J}} (\|AB\|/\|B\|)$$

253. The norm $\|A\|'_{\mathbb{J}}$ is also a unity-preserving ring norm.
254. The derived norm of a ring norm relative to any left ideal $\neq 0$ is dominated by the initial norm.

This is a more general form of (248). However, (249) cannot be generalized to ideals $\mathbb{J} \neq \mathbb{M}(E)$, since such ideals do not contain the identity operator.

We shall use the derived norms relative to the ideals in $\mathbb{M}(E)$ in order to characterize the class of operator norms on $\mathbb{M}(E)$. Recalling from II,(21) and (22), that there is a bijective relation between the left ideals in $\mathbb{M}(E)$ and the subspaces $L \subset E$, we first can "mark" a subspace L by a functional f ($f \in E'$, $f \neq 0$) so that $L \subset \text{Ker } f$. Then a left ideal $\mathbb{J}$ is the set of all operators $A \in \mathbb{M}(E)$ such that

$$\text{Ker } A \supset \text{Ker } f .$$

This ideal will be denoted by $\mathbb{J}(f)$.

255. The operator A belongs to the ideal $\mathbb{J}(f)$, iff for any $y \in E$ there exists an $x \in E$ such that $Ay = f(y)x$.

If we suppress here the vector variable y, we recognize the operator A represented as a tensor product $f\otimes x$. However, mixed tensor products as this one did not occur in I, Sect. 8; they will appear in V, Sect. 1.

Following the pattern of I,(226) and (227), we consider $f\otimes x$ (fixed $f \in E'$, fixed $x \in E$) as a bilinear functional $B(y,g)$ ($y \in E'' \approx E$ [c.i.], $g \in E'$). Then

256. $B(y,g) \underset{\text{def}}{=} f\otimes x = f\otimes Ax$, where the operator $A \in \mathbb{M}(E)$, and the matrix of A coincides with the matrix of B (I,D75). (As basis in E' the conjugate of the basis chosen in E (I,D62) could be taken, if an invariant representation is desired.)
257. If E is a normed space, then the following formula

holds for the induced norms on $\mathbb{B}$ and on E': $\|f\otimes x\| = \|f\|.\|x\|$.

The role of the left ideals $\mathbb{J}(f)$ is now clarified by the following proposition:

258. If for any norm on $\mathbb{M}(E)$ the derived norm relative to any left ideal $\mathbb{J}(f)$ is formed, then the norm so obtained is an operator norm.

Hence it follows:

259. If the derived norm relative to any left ideal $\mathbb{J}(f)$ of some norm on $\mathbb{M}(E)$ coincides with the initial norm, then the initial norm is an operator norm.

On first sight this criterion seems only sufficient, but it is in fact necessary, since

260. The derived norm relative to any $\mathbb{J}(f)$ of every operator norm coincides with the initial norm.

Thus we have obtained a complete characterization of the operator norms on $\mathbb{M}(E)$, but it does not permit us to say something definite about those unity-preserving ring norms that are not operator norms. Hence we must continue the investigation and find:

261. Every ring norm dominates some operator norm.

Definition 63. A ring norm will be called *minimal*, if there is no other ring norm which it dominates.

262. A minimal ring norm on $\mathbb{M}(E)$ is an operator norm.

This characteristic is also a necessary property of operator norms:

263. A ring norm is an operator norm, iff it is minimal.

The proof of necessity is based on the lemmas 264-266 which follow.

264. If the norms $\|.\|_1$ and $\|.\|_2$ are introduced in E and if the corresponding operator norms on $\mathbb{M}(E)$ satisfy $\|A\|_1 \leq \|A_2\|$ for all $A \in \mathbb{M}(E)$, then $\sup_{x\in E}(\|x\|_1/\|x\|_2) \leq \inf_{f\in E'}(\|f\|_2/\|f\|_1)$ $(x \neq 0,\ f \neq 0)$.

265. For any two norms on E

$$\sup_{x\in E}(\|x\|_1/\|x\|_2)=\sup_{f\in E'}(\|f\|_2/\|f\|_1),\ \inf_{x\in E}(\|x\|_1/\|x\|_2)=\inf_{f\in E'}(\|f\|_2/\|f\|_1)$$

So the (numerical) ranges of the two functionals $(\|x\|_1/\|x\|_2)$ and $(\|f\|_2/\|f\|_1)$ are the same.

266. If two operator norms on $\mathbb{M}(E)$ are comparable in the sense of the order relation introduced above, then the norms on E, by which they are generated, are proportional, and the initial operator norms coincide.

Criterion (263) implies in particular:

267. If $\|.\|_1$ and $\|.\|_2$ are two different operator norms on $\mathbb{M}(E)$, then the norm $\|.\| = \max(\|.\|_1, \|.\|_2)$ is a unity-preserving ring norm, but not an operator norm.

An example follows:

268. $$\|A\| = \max\left(\max_j \Sigma_{k=1}^{n} |\alpha_{jk}|, \ \max_k \Sigma_{j=1}^{n} |\alpha_{jk}|\right),$$

constructed from the elements of the matrix $(\alpha_{jk})_{j,k=1}^{n}$ of the operator A in some fixed basis, is not an operator norm for $n > 1$, but it is a ring norm and unity-preserving.

Note finally that it would make no sense to consider seminorms on $\mathbb{M}(E)$, and for the following reason:

269. If the seminorm $p \neq 0$ on $\mathbb{M}(E)$ has the ring property $p(AB) \leq p(A)p(B)$, then p is a norm (cf. II,(25)).

11. Inequalities between Norms of Operator Powers

Definition 64. The operator A in a normed space E is known as an *operator of class K*, if it satisfies the inequalities

$$\|A^k x\| \leq C_{m,k} \cdot \|x\|^{(m-k)/m} \cdot \|A^m x\|^{k/m}$$

for all $x \in E$, $m = 2,3,\ldots$, $k = 1,2,\ldots, m-1$; the $C_{m,k}$ are constants.

For an operator A of Class K, $C_{m,k}(A)$ shall denote the least possible value of the constant $C_{m,k}$ (for fixed m and k, of course).

270. If A satisfies the inequality $\|Ax\| \leq C\|x\|^{1/2}\|A^2x\|^{1/2}$ for

all $x \in E$, then A is of class K, and $C_{m,k}(A) \leq c^{k(m-k)}$.

Hence $C_{m,k}(A) \leq [C_{2,1}(A)]^{k(m-k)}$ is correct for any operator of class K.

271. A is an operator of class K, iff $\operatorname{Ker} A^2 = \operatorname{Ker} A$. This means that A is either regular or the order (II,D26) of the eigenvalue $\lambda = 0$ is equal to 1.

The last theorem makes clear that membership in the class K is independent of the choice of the norm on E. But the numbers $C_{m,k}(A)$ depend of course on the particular norm chosen.

272. If A is a regular operator, then

$$C_{m,k}(A) \leq \|A^k\|^{(m-k)/m} \|(A^{-1})^{m-k}\|^{k/m} .$$

273. If A is a regular operator, then $C_{m,k}(A^{-1}) = C_{m,m-k}(A)$.

274. Every operator of scalar type is an operator of class K, and $\sup_{m,k} C_{m,k}(A) < \infty$.

Conversely:

275. If the operator A of class K satisfies $\sup_{m,k} C_{m,k}(A) < \infty$, then A is an operator of scalar type.

276. If A is an operator of class K and $\neq 0$, then $C_{m,k}(A) \geq 1$.

Which operators attain the equal sign in the last inequality? We shall investigate this question only for operators A in Euclidean space. Under this assumption we find:

277. If A is a normal operator, then $C_{m,k}(A) = 1$ for all m and k.

278. If $C_{m,k}(A) = 1$ for any pair m,k, then A is an operator of scalar type.

279. Let λ and μ be two different eigenvalues of A such that the associated characteristic subspaces are not orthogonal. If $C_{m,k}(A) = 1$ for some pair m and k, then $\lambda^{d(m,k)} = \mu^{d(m,k)}$, where $d(m,k)$ denotes the greatest common divisor of m and k.

280. If for some pair m,k the number $C_{m,k}(A) = 1$ and if the function $\varphi(\lambda) = \lambda^{d(m,k)}$ is injective on the spectrum $\sigma(A)$,

then the operator A is normal.

281. If $C_{m,k}(A) = 1$ for some pair m,k, then the operator $A^{d(m,k)}$ is normal.

Thus A is normal, if m and k in (281) are relatively prime.

282. For any pair m,k ($m \geqq 2$, $1 \leqq k \leqq m-1$) there exists an operator A of class K in Euclidean space such that $C_{m,k}(A) = 1$, but none of the powers A^q, $q = 1,2,\ldots,d(m,k)-1$, is a normal operator.

absolute convergence of vector series D13
absolute basis D57
absolute norm D60
absolutely convex body D26
absolutely convex set D25
absorbing set (64)
anisometric measure D41
aperture of two subspaces D49
Ascoli-Mazur Theorem (83)
Auerbach basis D36

balanced set (52)
Banach-Mazur Distance D41
Banach space D2
basis of best approximation D48
Bernstein's theorem (139)
best approximation D45

Cauchy-Hadamard criterion, modified (26)
compression (operator) D56
c-norm D3
convergence in norm D12
c-space D6
cyclic group (188)

derived norm D61
derived norm relative to an ideal D62
deviation from being Euclidean D8
distance from a subspace D19
dominated norm D10
dominated seminorm D15
Dom p (58)
dual indices of ℓ^p-spaces D40

écart D28
ergodic (operator) D58
ergodic theorem for compression (202)
ergodic theorem (typical) (196)
equivalent norms D11
equivalent seminorms D18
Euclidean norm D7
Euclidean space D7
ε-universal space D44

extending a homomorphism (66)
extended seminorm D28

Fekete's Theorem (212)

Gel'fand's Formula (215)
globular aperture of two subspaces D51
Grassmann manifold D53
Gurariĭ-Kadec-Macaev Formula (113)

Hahn-Banach Theorem (88)
Hausdorff set (224)
Hermitian isometry D39
Hilbert norm D60
Hölder's Inequality (104)
hyperlayer D43

induced norm (36)
isometric normed spaces D38
isometric operator D55
isometry D38

Kreĭn-Mil'man-Rutman Theorem (176)

ℓ-norm D4
ℓ^p-norm D5
lower limit of sequence of subspaces D52a
ℓ-space D6
ℓ^p-space D6

metric D9
metric space D9
minimal ring norm D63
Minkowski Functional D27
Minkowski's Inequality D6
Minkowski Space D2
Minkowski's theorem (89)
multiplicative group (179)

natural metric in normed space (15)
nonlinear orthoprojector (134)
norm D1
norm convergence D12

normed space D2
normed vector system D17
norm-generated topology (17)
norm of bilinear functional D23
norm of functional D22
norm of Hermitian-bilinear functional D23
norm of Hermitian-quadratic functional D24
norm of homomorphism D20
norm of operator D21
norm of quadratic functional D24
norm topology (17)

order relation for norms on $\mathbb{M}(E)$ (248)
operator norm D59
operator of best approximation D47
operator of class K D64
orthogonal basis in normed space D36
orthogonality in normed space D31
orthogonality of subspace to subspace D32
orthogonality of subspace to vector D33
orthoprojector in normed space D30

parallelogram law (10)
projection constant of subspace D29

quadratically close bases (173)

rationally independent D58
ring norm (41), D59

Schwarz' Inequality D7
seminorm D14
semi-orthogonal basis D35
spherical aperture of two subspaces D50
Stiefel manifold D54
strict compression D56
strictly normed space D46
supporting functional D34
symmetric polyhedron D43
Sz.-Nagy's Theorem (152)

uniform norm D3
unit ball (53)
unit sphere D17
universal space D42
unity-preserving norm D59
upper limit of sequence of subspaces D52b

weak topology (17)
Wiener-Paley Theorem (175)

CHAPTER V

MULTILINEAR AND EXTERIOR ALGEBRA

1. Multilinear Mappings and Tensors.

In this section we return to a topic that was first introduced in Ch. I,D77. Its development was based on the idea of a bilinear functional defined on the space $E_1 \times E_2$. Here we shall start with the mapping (not necessarily into the common scalar field) of p such spaces. The universality of the construction (see (6)) reappears several times.

Definition 1. For given spaces $E_1, \ldots, E_p$ and E, a mapping F from the Cartesian product $E_1 \times \ldots \times E_p$ into E, i.e. a function $F(x_1, \ldots, x_p)$, $x_k \in E_k$, with values in E, is called *multilinear* if it is a homomorphism in each argument x_j $(j = 1,2,\ldots,p)$.

Definition 2. A multilinear mapping into the arithmetic space C^1 is called a *multilinear functional* or a *tensor*. The number p is the *valence* of the tensor. Thus a multilinear functional of p arguments can be called a p-valent tensor. A one-valent tensor is a linear functional, a two-valent tensor is a bilinear functional, and any multilinear mapping with p=2 is called bilinear.

1. The mapping $F(h,x) = hx$ $(h \in \mathrm{Hom}(E_1,E_2),\ x \in E_1)$ is bilinear (h and x vary independently over their domains).

2. The mapping $F(h_1,h_2) = h_1h_2$, where h_1 and h_2 are homomorphisms for which the multiplication makes sense, is bi-

linear.

If E is an algebra and if h_1, h_2 are endomorphisms, then the multiplication $h_1 h_2$ is a bilinear mapping $E \times E \to E$. Multilinear mappings are closely connected with tensor products, to be defined now.

Definition 3. Tensor product of the spaces $E_1, \ldots, E_p$, denoted by

$$E_1 \otimes \ \ldots \ \otimes E_p ,$$

is the name for the space of all multilinear functionals defined on the Cartesian product of the conjugate spaces, viz.

$$E_1' \times \ldots \times E_p' .$$

Definition 4. If $x_1, \ldots, x_p$ are any vectors from $E_1, \ldots, E_p$ respectively (which we may consider as fixed to begin with, cf. I,D77), then the multilinear functional

$$f_1(x_1) \cdots f_p(x_p) ,$$

where $f_1, \ldots, f_p$ vary independently over $E_1', \ldots, E_p'$ respectively, is called the *tensor product of the vectors* $x_1, \ldots, x_p$ and is denoted by

$$x_1 \otimes \ldots \otimes x_p .$$

Thus the tensor product of p fixed vectors $\in E_1 \times \ldots \times E_p$ is one multilinear functional defined on $E_1' \times \ldots \times E_p'$ (cf. I,(169)). It is an element of $E_1 \otimes \ldots \otimes E_p$. Note that $x_1 \otimes \ldots \otimes x_p = 0$, iff zero occurs among the x_k.

Let now the vectors $\{x_k\}_1^p$ vary independently over $\{E_k\}_1^p$. We obtain a set of corresponding multilinear functionals on $E_1' \times \ldots \times E_p'$.

3. The mapping of $E_1 \times \ldots \times E_p$ effected by the tensor product

$$\tau(x_1, \ldots, x_p) = x_1 \otimes \ldots \otimes x_p \quad \text{or} \quad \tau : (x_1, \ldots, x_p) \mapsto f_1(x_1) \cdots f_p(x_p)$$

is multilinear.

4. If F is a multilinear mapping into some space E_o and $h \in \mathrm{Hom}(E_o, E)$, then hF is a multilinear mapping into E.

Consequently,

5. For each homomorphism $h \in \mathrm{Hom}(E_1 \otimes \ldots \otimes E_p, E)$ the mapping $F = h\tau$, i.e.

$$F(x_1,\ldots,x_p) = h(x_1 \otimes \ldots \otimes x_p) \qquad (x_k \in E_k)$$

is multilinear.

It turns out that by this device (viz. $h(x_1 \otimes \ldots \otimes x_p)$) one can obtain all multilinear mappings. In this sense it is correct to say that the tensor multiplication is the *universal* multilinear mapping:

6. For each multilinear mapping F from $E_1 \times \ldots \times E_p$ into E there exists a unique homomorphism $F_\tau \in \mathrm{Hom}(E_1 \otimes \ldots \otimes E_p, E)$ such that $F = F_\tau\, \tau$.

For $p=1$, F_τ obviously is the same as F.

That property of universality characterizes the tensor product:

7. Let there be given spaces $E_1,\ldots,E_p$, a space E_o, and a multilinear mapping σ from the Cartesian product $E_1 \times \ldots \times E_p$ into E_o. Assume that

1. there exists a homomorphism $F_\sigma \in \mathrm{Hom}(E_o, E)$ such that $F = F\sigma$ for each multilinear mapping F from $E_1 \times \ldots \times E_p$ into some arbitrary space E;

2. the image of the mapping σ contains a system of vectors which is complete in E_o (see I,D17). Then E_o is isomorphic to the tensor product $E_1 \otimes \ldots \otimes E_p$, and under this isomorphism the mapping σ becomes τ of (3), i.e. the tensor product.[§]

We shall denote by $\Pi(E_1,\ldots,E_p; E)$ the linear space of multilinear mappings $E_1 \times \ldots \times E_p \to E$ with the natural definition of addition and multiplication by scalars.

8. $$\Pi(E_1,\ldots,E_p; E) \simeq \mathrm{Hom}(E_1 \otimes \ldots \otimes E_p, E)\ .$$

[§] A brief constructive account of uniqueness [n.i.] and existence of $E_1 \otimes \ldots \otimes E_p$ can be found in Greub, W. H., Multilinear Algebra, Springer, 1967, pp.9-11 and 27-30.

Hence

9. $\dim(E_1,\ldots,E_p;\ E) = \dim E_1\cdots\dim E_p\cdot\dim E$.

If we wish to make the linear space E into an algebra, the universality of the tensor product together with the usual requirements on multiplication permits us to consider any proposed definition of multiplication as a bilinear mapping $E\times E\to E$. Then (9) implies that

10. The maximal number of linearly independent multiplications in a space E equals n^3.

Lemma (8) also permits the natural conclusion

11. $E_1\otimes\ldots\otimes E_p \simeq (\Pi(E_1,\ldots,E_p;\ C^1))'$.

Definition 5. Suppose we introduce a collection of "basic" spaces $H_1,\ldots,H_p$, associated with the collection of the given spaces $E_1,\ldots,E_p$ such that either $E_k = H_k$ or $E_k = H_k'$ for some of the $k=1,\ldots,p$ (there are 2^p such choices possible). Then a multilinear functional (= tensor) $F(x_1,\ldots,x_p)$ on $E_1\times\ldots\times E_p$ is called *covariant* in the argument x_k if $E_k = H_k$; and *contravariant* in this argument if $E_k = H_k'$. A tensor covariant (contravariant) in all arguments is called covariant (contravariant). (Thus the notations x_k and E_k do here no longer imply that E_k is a "vector" space. E_k may be the conjugate of a vector space and x_k a functional.)

12. A linear functional defined on E is a one-valent covariant tensor ($H_1 = E_1$ (= E)).

We know that $(E')' = E$ [c.i.]. This permits us to look at each vector $\in E$ as a functional on E' (= E_1). Hence $x \in E$ is a contravariant tensor ($H_1' = E_1$ (= E'), so that $H_1 = E$).

13. A bilinear functional defined on E is a two-valent covariant tensor ($H_1 = H_2 = E_1 = E_2$ (= E)).

14. The bilinear functional $f(x)$ ($x \in E$, $f \in E'$) is a two-valent tensor, which is covariant in the argument x and contravariant in the argument f ($H_1 = H_2' = E$).

There exists a canonical method of identifying an arbitrary multilinear mapping into E with a tensor of suitable type. Let $F : E_1\times\ldots\times E_p \to E$ be the multilinear mapping. Consider the multilinear functional $f(F(x_1,\ldots,x_p))$ of the p+1 arguments $x_k \in E_k$, $f \in E'$ $(k = 1,\ldots,p)$, covariant in the arguments x_k and contravariant in the argument f. This defines a mapping

$$\kappa : \Pi(E_1,\ldots,E_p; E) \to \Pi(E_1,\ldots,E_p,E'; C^1) .$$

15. The mapping κ is an isomorphism.

Thus every multilinear mapping may be called a tensor.

16. A homomorphism from E_1 into E is a two-valent tensor $(H_1 = E_1, H_2' = E)$, covariant in the first argument and contravariant in the second, as implied by (15).

17. Multiplication in E is a trivalent tensor $(H_1 = H_2 = H_3' = E)$, covariant in the first two arguments and contravariant in the third (cf. remark before (10)).

These examples should illustrate the value of the tensor language, which is its ability to describe uniformly all the basic objects of linear analysis.

Let us now extend the coordinate-matrix representation of III,D1 to arbitrary tensors. Choose in each of the basic spaces H_k (dim $H_k = n_k$) a basis $\Delta_k = \{e_j^k\}_{j=1}^{n_k}$. If any of the spaces coincide, take also coinciding bases. In each of the conjugate spaces H_k' take the basis Δ_k' conjugate to Δ_k.

18. Let F be a p-valent tensor, covariant in the arguments x_k and contravariant in the arguments f^s. Let $\{\xi_k^j\}_{j=1}^{n_k}$ be the coordinates of the vector x_k relative to the basis Δ_k, and $\{\eta_\ell^s\}_{\ell=1}^{n_s}$ the coordinates of the functional f^s relative to the basis Δ_s'. Then

$$F(\ldots,x_k,\ldots;\ \ldots,f^s,\ldots) = \sum_{(\ldots j\ldots\ell\ldots)} \alpha^{\ldots\ell\ldots}_{\ldots j\ldots}\,\xi_k^j\ldots\eta_\ell^s\ldots, \qquad (*)$$

where $\alpha^{\ldots\ell\ldots}_{\ldots j\ldots}$ is a numerical function of the set of indices $(\ldots j\ldots\ell\ldots)$ occurring in a particular term of the sum, and the sum is taken over all sets of indices (all operations under the summation sign are multiplications).

The set of functions $\alpha^{\ldots\ell\ldots}_{\ldots j\ldots}$ is called the (multidimensional) "*matrix*" *of the tensor* F relative to the chosen bases.

In the particular case of multiplication in E such as in (17), the elements of that matrix are called the *structural constants* of the corresponding algebra.

The right member of (*) is called a *multilinear form*.

Theorem (18) determines a basis in the space of tensors of the type under consideration; it is induced by the bases chosen in the "basic" spaces. The tensors forming that basis are those in which all of the matrix elements equal zero except one which equals 1.

In conclusion let us consider the elementary operations on tensors.

Definition 6. Let $\varphi = \varphi(x_1,\ldots,x_p)$, $\psi = \psi(y_1,\ldots,y_q)$ be two tensors (multilinear functionals) with disjoint sets of variables. The *tensor multiplication of multilinear functionals* $\varphi\otimes\psi$ is the multilinear functional $\omega(x_1,\ldots,x_p, y_1,\ldots,y_q) = \varphi(x_1,\ldots,x_p)\cdot\psi(y_1,\ldots,y_q)$. This general definition is in agreement with the earlier definition which related to the particular case of linear functionals and vectors.

Under multiplication of tensors the valences are added, and the (co- or contravariant) type of a particular argument is taken from the factor to which that argument belongs.

Definition 7. The *tensor multiplication of multilinear mappings* can be defined in a natural way by means of the multiplication of the corresponding multilinear functionals (see (15)). The result is: if F_1 maps $E_{11}\times\ldots\times E_{1p}$ into E_1, and F_2 maps $E_{21}\times\ldots\times E_{2q}$ into E_2, then $F_1\otimes F_2$ maps $E_{11}\times\ldots\times E_{1p}\times E_{21}\times\ldots\times E_{2q}$ into $E_1\otimes E_2$.

19. In the tensor multiplication of two multilinear mappings F and G the corresponding homomorphisms F_τ and G_τ (cf.(6)) are subjected to tensor multiplication.

An important but more specialized operation is the contraction of a tensor with respect to conjugate arguments, described in the following problem and definition.

20. If $\varphi(\ldots,x,\ldots,f,\ldots)$ is a tensor in which x,f are conjugate arguments, i.e. $x\in E$ and $f\in E'$, then there exists a unique multilinear function T_φ of the remaining arguments with values in Hom(E,E) such that

$$\varphi(\ldots,x,\ldots,f,\ldots) = f(T_\varphi x) .$$

Definition 8. The tensor tr T_φ (the trace of the operator T_φ considered as a function of the remaining arguments) is called the *contraction of the tensor* φ with respect to the arguments x,f. In a contraction, the valence of a tensor is reduced by 2. (Here it is expedient to introduce the convention that a tensor of zero valence is simply a scalar.)

21. The contraction of a linear operator $T \in$ Hom(E,E) equals tr T (II,D44).

22. The contraction of a bilinear functional f(x) is dim E.

The relation between contractions and tensor multiplication is described by the following theorem:

23. Let φ and ψ be two tensors which admit tensor multiplication. The contraction of the tensor product $\varphi\otimes\psi$ with respect to a pair of arguments belonging to one factor equals the tensor product of the contraction of that factor by the other factor.

This is connected with the fact that the trace of a tensor multiplication can be expanded:

24. For any two operators $A_1 \in \mathbb{M}(E_1)$, $A_2 \in \mathbb{M}(E_2)$

$$\mathrm{tr}(A_1 \otimes A_2) = \mathrm{tr}\, A_1 \cdot \mathrm{tr}\, A_2 .$$

Other multiplications, which are not tensorial as such,

may be considered as combinations of tensor multiplication and contraction. For example:

25. If $h \in Hom(E,E_1)$ and $h_1 \in Hom(E_1,E_2)$, then the product h_1h coincides with the contraction of the tensor $h_1 \otimes h$ with respect to the corresponding arguments.

26. If $h \in Hom(E,E_1)$ and $x \in E$, then the "product" hx coincides with the contraction of the tensor $h \otimes x$ with respect to the corresponding arguments.

A general theorem on reduction follows:

27. Let F be a multilinear mapping of $E_1 \times \ldots \times E_p$ into E, considered as a tensor. Then $F(x_1,\ldots,x_p)$ may be looked at as a "product" of F and the vectors $x_1,\ldots,x_p$; it coincides with the p-fold contraction of the tensor $F \otimes x_1 \otimes \ldots \otimes x_p$ with respect to the corresponding arguments.

2. Symmetric and Antisymmetric Tensors. Theory of Determinants.

In this section we shall consider only tensors (= multilinear functionals) with arguments in the same space E; they are covariant in all arguments.

Definition 9. A tensor is called *symmetric* if it does not change under permutations of its arguments:

$$(x_{j_1},\ldots,x_{j_p}) = (x_1,\ldots,x_p) .$$

28. In the space of p-valent tensors the symmetric tensors form a subspace. Its dimension equals

$$\binom{n+p-1}{p} = \frac{n(n+1)\ldots(n+p-1)}{p!}$$

The dimension of the space of p-valent tensors equals n^p (cf.(9)).

29. The operator Q, defined on the space of p-valent tensors by the formula

$$(Q\varphi)(x_1,\ldots,x_p) = \frac{1}{p!}\sum_{(j_1,\ldots,j_p)} \varphi(x_{j_1},\ldots,x_{j_p}),$$

(where the summation is extended over all permutations of the indices $(1,\ldots,p)$) projects onto the subspace of symmetric tensors.

Definition 10. The operator Q described above is called the *symmetrization operator.*

Definition 11. A tensor is called *antisymmetric* if it is unchanged under an even permutation of the arguments and is multiplied by -1 under an odd permutation.

30. In the space of p-valent tensors the antisymmetric tensors form a subspace. Its dimension equals

$$\binom{n}{p} = \frac{n(n-1)\ldots(n-p+1)}{p!}$$

and vanishes for $p > n$.

31. The subspace of antisymmetric tensors of valence $p > 1$ is contained in the kernel of the symmetrization operator, but it does not coincide with that kernel for $p > 2$.

An especially important role is played by the subspace $\mathbb{D}(E)$ of antisymmetric tensors of valence n.

32. $\dim \mathbb{D}(E) = 1$.

Definition 12. According to a remark following (18) each basis Δ in E induces a basis in $\mathbb{D}(E)$ which by virtue of (32) consists of one elements d_Δ. The value of the multilinear functional $\mathbb{D}(E)$ on a system of arguments (= system of vectors) $\Gamma = \{x_1,\ldots,x_n\}$ is called the *determinant* of that system relative to the basis Δ and is denoted by $\det_\Delta\Gamma$ or $\det_\Delta\{x_1,\ldots,x_n\}$. If the choice of basis is clear from the context or makes no difference, then we write simply $\det\Gamma$ or $\det\{x_1,\ldots,x_n\}$.

33. $\det_\Delta \Delta = 1$.

34. For any pair of bases Δ and Δ_1

$$d_{\Delta_1} = (\det_{\Delta_1}\Delta)d_\Delta.$$

Hence

35. For any bases Δ, Δ_1, Δ_2

$$\det_{\Delta_2} \Delta = \det_{\Delta_1} \Delta \cdot \det_{\Delta_2} \Delta_1 .$$

This is the theorem on multiplication of determinants of vector systems. Its connection with the theorem on multiplication of determinants of operators will show up below, after we have described the determinant of an operator in the language of determinants of systems.

36. For any pair of bases Δ and Δ_1

$$\det_{\Delta_1} \Delta \cdot \det_{\Delta} \Delta_1 = 1 .$$

Observe in this connection that

37. For a linearly independent system Γ, $\det \Gamma \neq 0$.

Conversely:

38. If the system Γ is linearly dependent, then $\det \Gamma = 0$.

The last theorem is easily reduced to the following assertion:

39. If any two arguments of an antisymmetric multilinear functional coincide, then the value of the functional equals zero.

From (38) follows a corollary often useful in computation:

40. If to any of the vectors of the system $\Gamma = \{x_1,\ldots,x_n\}$ a linear combination of the remaining vectors is added, then the determinant of the system does not change.

Definition 13. We shall call the system $\Gamma = \{x_k\}_1^n$ *triangular relative to the basis* Δ if the matrix $(\alpha_{jk})_{j,k=1}^n$ of Γ is triangular in the basis Δ.

41. If the system Γ is triangular in the basis Δ, then $\det_\Delta \Gamma = \prod_{k=1}^{n} \alpha_{kk}$.

It is useful to compare this result with the definition of the determinant of an operator, II,D53.

Let $(j_1,\ldots,j_p)$ be a permutation of the indices $(1,\ldots,p)$

and define the *character* (or signum or sign) of that permutation by

$$\chi(j_1,\ldots,j_p) = \begin{matrix} 1, \text{ if the permutation is even,} \\ -1, \text{ if the permutation is odd.} \end{matrix}$$

42. If $\Delta = \{e_k\}_1^n$ is a basis and $(j_1,\ldots,j_n)$ is any permutation of the indices $(1,\ldots,n)$, then

$$\det\Delta\{e_{j_1},\ldots,e_{j_n}\} = \chi(j_1,\ldots,j_n).$$

43. Let $(\alpha_{jk})_{j,k=1}^n$ be the matrix of the system Γ in the basis Δ. Then

$$\det{}_\Delta\Gamma = \sum_{(j_1,\ldots,j_n)} \chi(j_1,\ldots,j_n)\alpha_{j_1 1}\ldots\alpha_{j_n n}.$$

In the classical theory of determinants the right-hand side of this equation is called the *determinant of the matrix* $(\alpha_{jk})_{j,k=1}^n$.

44. If the operator $A \in \mathbb{M}(E)$ is regular, then for any two bases Δ, Δ_1 we have the equality $\det_{A\Delta}A\Delta_1 = \det_\Delta\Delta_1$.

45. For any operator A the quantity $\det_\Delta A\Delta$ does not depend upon the basis Δ and coincides with the determinant of the operator A, i.e.: $\det_\Delta A\Delta = \det A$.

In other words, $\det_\Delta\Gamma = \det A$, where A is that uniquely defined operator, which transforms the basis Δ into the system Γ.

From (45) and (35) follows again the theorem II,(272) on the multiplication of determinants of operators, but this time with less effort.

Definition 14. Returning to tensors of arbitrary valence p, let us introduce the *antisymmetrization operator* $\bar{Q}$:

$$(\bar{Q}\varphi)(x_1,\ldots,x_p) = \frac{1}{p!}\sum_{(j_1,\ldots,j_p)} \chi(j_1,\ldots,j_p)\varphi(x_{j_1},\ldots,x_{j_p}).$$

46. The operator $\bar{Q}$ projects onto the subspace of antisymmetric tensors.

47. The subspace of the symmetric tensors of valence $p > 1$ is contained in the kernel of the antisymmetrization operator, but for $p > 2$ the subspace does not coincide with the kernel.

Comparison of theorems (46) and (47) with theorems (29) and (31) reveals a certain duality between symmetric and antisymmetric tensors.

3. Exterior Products and Exterior Forms.

Consider the p-*th* tensor power of the space E:

$$E\otimes\ldots\otimes E \text{ or } \overset{p}{\otimes}E$$

This space is conjugate to the space of tensors of valence p with arguments from E (see (11)).

Definition 15. The space, which is conjugate to the space of antisymmetric tensors of valence p with arguments from E, is called the p-*th exterior power* of the space E and is denoted by

$$E\wedge\ldots\wedge E \quad \text{or } \overset{p}{\wedge}E\ .$$

If $p > n$, then $\overset{p}{\wedge}E = 0$.

48. The operator $\overline{Q}$, which is the conjugate to the antisymmetrization operator, embeds a space isomorphic to $\overset{p}{\wedge}E$ into the space $\overset{p}{\otimes}E$:

$$\text{Im } \overline{Q}' \simeq \overset{p}{\wedge}E\ .$$

49. If $x_1,\ldots,x_p$ are vectors from E, then the element of $\overset{p}{\otimes}E$

$$\sum_{(j_1,\ldots,j_p)} \chi(j_1,\ldots,j_p)\ x_{j_1}\otimes\ldots\otimes x_{j_p}$$

belongs to $\overset{p}{\wedge}E$.§

§If $x_1\otimes\ldots\otimes x_p$ is written in the expanded form $f_1(x_1)\cdots f_p(x_p)$ according to D4, then only the x_k are permuted.

Definition 16. The linear combination of tensor products in (49) is denoted by $x_1 \wedge \ldots \wedge x_p$ and called the *exterior product* of the vectors $x_1, \ldots, x_p$.

50. The exterior product is a multilinear antisymmetric mapping of the Cartesian power $E \times \ldots \times E$ or $\overset{p}{\times} E$ into $\overset{p}{\wedge} E$; viz.

$$x_{j_1} \wedge \ldots \wedge x_{j_p} = \chi(j_1, \ldots, j_p)(x_1 \wedge \ldots \wedge x_p)$$

for all permutations $(j_1, \ldots, j_p)$ of the system of indices $(1, 2, \ldots, p)$.

51. $x_1 \wedge \ldots \wedge x_p = 0$ holds iff the vectors $x_1, \ldots, x_p$ are linearly independent.

Like tensor multiplication, exterior multiplication is universal. It is the *universal antisymmetric multilinear mapping*, and it is characterized by this property (cf. (4)-(7)):

52. If F is an antisymmetric multilinear mapping of the Cartesian power $\overset{p}{\wedge} E$ into some space E_o and $h \in \mathrm{Hom}(E_o, E_1)$, then hF is an antisymmetric multilinear mapping of $\overset{p}{\times} E$ into E_1.

53. For each homomorphism $h \in \mathrm{Hom}(\overset{p}{\wedge} E, E_1)$ the mapping $F = h\Lambda$ is an antisymmetric multilinear mapping of $\overset{p}{\times} E$ into E_1; here Λ denotes the mapping effected by the exterior product

$$\Lambda(x_1, \ldots, x_p) = x_1 \wedge \ldots \wedge x_p \qquad \text{(cf. (3))}.$$

54. For each antisymmetric multilinear mapping F from $\overset{p}{\times} E$ into E_1 there exists a unique homomorphism $F_\Lambda \in \mathrm{Hom}(\overset{p}{\wedge} E, E_1)$ such that $F = F_\Lambda \Lambda$.

55. Let there be given spaces E, E_o and an antisymmetric multilinear mapping M from the Cartesian power $\overset{p}{\times} E$ into E_o. Assume that

1. there exists a homomorphism $F_M \in \mathrm{Hom}(E_o, E_1)$ such that $F = F_M M$ for each antisymmetric multilinear mapping F from $\overset{p}{\times} E$ into some arbitrary space E_1;

2. the image of the mapping M contains a system which is complete in E_o. Then E_o is isomorphic to the exterior power $\overset{p}{\wedge}E$ and under this isomorphism the mapping M becomes Λ, i.e. the exterior product.

The theory of determinants, developed in the preceding section, fits naturally into the framework of the exterior product. The collection of antisymmetric multilinear mappings can in fact be made into an algebra called the (graded) *exterior algebra* of E. For this see Greub, op. cit., Ch. V.

56. If $\Delta = \{e_k\}_{k=1}^{n}$ is some basis in E, then for any system $\Gamma = \{x_j\}_{j=1}^{n}$

$$x_1\wedge\ldots\wedge x_n = \det_\Delta(x_1,\ldots,x_n)\cdot(e_1\wedge\ldots\wedge e_n).$$

This formula can be significantly generalized, when the concepts of minor and algebraic complement are introduced.

Definition 17. Let $J = \{j_1<\ldots<j_m\}$, $K = \{k_1<\ldots<k_m\}$ be any two subsystems of $\{1,\ldots,n\}$ with equal numbers of terms. For an arbitrary system of vectors $\Gamma = \{x_k\}_1^n$ denote by Γ_k the subsystem $\{x_{k_1},\ldots,x_{k_m}\}$, and by Δ_J the subsystem $\{e_{j_1},\ldots,e_{j_m}\}$ of the basis $\Delta = \{e_k\}_{k=1}^{n}$. Let $P_{J;\Delta}$ be the projector onto the linear hull of the subsystem $\Delta_J \subset \Delta$ parallel to the linear hull of the remaining vectors of the basis Δ. The determinant of the system $P_{J;\Delta}\Gamma_K$ relative to the basis Δ_J of the subspace sp Δ_J is called the *minor* of the system Γ relative to the basis Δ, corresponding to the pair of subsystems $\{J,K\}$; it will be denoted by $M_{J,K;\Delta}(\Gamma)$. And so

$$M_{J,K;\Delta}(\Gamma) \underset{\text{def}}{=} \det_{\Delta_J}\{P_{J;\Delta}\,\Gamma_K\}\ .$$

Now let us put $s(J) = \Sigma_{k=1}^{m} j_k$ and define $s(K)$ correspondingly.

Definition 18. The quantity

$$A_{J,K;\Delta}(\Gamma) \underset{\text{def}}{=} (-1)^{s(J)+s(K)} M_{J,K;\Delta}(\Gamma)$$

is called the *algebraic complement* of the system Γ relative to the basis Δ and corresponding to the pair of subsystems $\{J,K\}$.

Using these notational devices, we now write the general form of (56):

57. $$\bigwedge_{k\in K} x_k = \Sigma_J M_{J,K;\Delta}(\Gamma)\cdot(\bigwedge_{j\in J} e_j)$$

A consequence of the formula is the *Theorem of Laplace:*

58. $$\det_\Delta \Gamma = \Sigma_J M_{J,K;\Delta}(\Gamma) A_{J',K';\Delta}(\Gamma) ,$$

where

$$J' = \{1,\ldots,n\} - J \quad \text{and} \quad K' = \{1,\ldots,n\} - K.$$

Another important aspect of exterior algebra is E. Cartan's theory of exterior forms. We shall consider the simplest special case, namely exterior forms with constant coefficients.

Definition 19. An expression of the form

$$\omega(x_1,\ldots,x_n) = \Sigma_{J_\omega} \alpha^{j1,\ldots,jp} x_{j_1}\wedge\ldots\wedge x_{j_p} ,$$

where J_ω is an arbitrary set of systems of indices $\{j_1,\ldots,j_p\}$, $1 \leq j_k \leq n$, and the $\alpha^{j1,\ldots,jp}$ are numerical coefficients, is called an *exterior form of the p-th degree* of the n variables $x_1,\ldots,x_n$. Each exterior form of the p-*th* degree in n variables defines a mapping of the Cartesian product $\overset{n}{\times}E$ into the exterior product $\overset{p}{\wedge}E$. In the general case, this mapping is neither multilinear nor antisymmetric. Equality of exterior forms is understood as equality of the corresponding mappings. We shall say that the exterior form ω is in standard notation if

J_ω coincides with the set of all systems $\{j_1,\ldots,j_p\}$ for which $j_1<\ldots<j_p$.

59. Each exterior form equals some exterior form in standard notation.

60. If exterior forms in standard notation are equal, then their corresponding coefficients are equal.

61. The exterior forms of the p-*th* degree form a linear space (with the natural definitions of addition and scalar multiplication). The dimension of this space equals $\binom{n}{p}$ and is zero for $p > n$. We shall denote that space by $\Omega_p(E)$ and put

$$\Omega_o(E) = C^1, \quad \Omega_p(E) = 0 \quad \text{for } p < 0,\ p > n.$$

Definition 20 introduces the *product of* the *exterior forms*

$$\omega(x_1,\ldots,x_n) = \Sigma_{J_\omega} \alpha^{j_1,\ldots,j_p} x_{j_1}\wedge\ldots\wedge x_{j_p},$$

$$\varepsilon(x_1,\ldots,x_n) = \Sigma_{J_\varepsilon} \beta^{k_1,\ldots,k_q} x_{k_1}\wedge\ldots\wedge x_{k_q}:$$

$$(\omega\wedge\varepsilon)(x_1,\ldots,x_n) \underset{\text{def}}{=}$$

$$\Sigma_{J_\omega\times J_\varepsilon} \alpha^{j_1,\ldots,j_p} \beta^{k_1,\ldots,k_q} x_{j_1}\wedge\ldots\wedge x_{j_p}\wedge x_{k_1}\wedge\ldots\wedge x_{k_q}$$

Obviously,

62. The degrees of exterior forms add in multiplication.

63. $\varepsilon\wedge\omega = (-1)^{pq}\omega\wedge\varepsilon$

Exterior forms play an important role in *algebraic topology*, a synthesis of algebra and topology.

In conclusion of this section, we would like to explain some of the fundamental concepts used in that field of mathematics, but we must restrict ourselves to discuss only the algebraic formalism inasmuch as it is closely connected with the topic of this section. Even so, our special case of exterior forms with constant coefficients is essentially trivial in algebraic topology (see (66)), but it sufficiently illustrates the general formal scheme. Let us begin with a renaming that introduces the standard terminology.

Definition 21. We shall call an exterior form of the p-th degree a *p-dimensional chain*. The *boundary* (or *differential*) of the p-dimensional chain

$$\omega(x_1,\ldots,x_n) = \Sigma_{J_\omega} \alpha^{j_1,\ldots,j_p} x_{j_1}\wedge\ldots\wedge x_{j_p}$$

is the (p-1)-dimensional chain

$$(d\omega)(x_1,\ldots,x_n) = \Sigma_{J_\omega} \alpha^{j_1,\ldots,j_p} d(x_{j_1}\wedge\ldots\wedge x_{j_p}) ,$$

where

$$d(x_{j_1}\wedge\ldots\wedge x_{j_p}) \underset{\text{def}}{=} \Sigma_{s=0}^{p-1} (-1)^s (x_{j_1}\wedge\ldots\wedge x_{j_s}\wedge x_{j_{s+2}}\wedge\ldots\wedge x_{j_p}).$$

For p = 1 we regard $dx_{j_1} = 1$. For $p \leqq 0$ or $p > n$, we shall take $d\omega = 0$.

64. The mapping $d_p : \Omega_p(E) \to \Omega_{p-1}(E)$, where d_p acts on $\omega \in \Omega_p(E)$ according to $d_p\omega \underset{\text{def}}{=} d\omega$, is a homomorphism.

Definition 22. The mapping d_p is called the *boundary homomorphism*.

The fundamental property of the boundary homomorphism is expressed by the formula:

65. $$d_p d_{p+1} = 0.$$

This formula is often written symbolically in the form $d^2 = 0$.

Formula (65) is equivalent with the inclusion $\operatorname{Im} d_{p+1} \subset \operatorname{Ker} d_p$. We may therefore consider the factor space

$$H_p(E) = \operatorname{Ker} d_p / \operatorname{Im} d_{p+1} .$$

Definition 22a. $H_p(E)$ is called the p-dimensional *homology space*, and its elements are the p-dimensional *homology classes*. The chains which belong to $\operatorname{Ker} d_p$ are called p-dimensional *cycles*. According to the definition, the boundary of a cycle equals zero. Cycles which belong to the same homology class are colled *homologous*. Two homologous cycles are thus those whose difference is the boundary of some chain. Formula (65) says that all boundaries are cycles.

Obviously $H_p(E) = 0$, if $p < 0$ or $p > n$; but the situation considered here is trivial, because also for $p = 0,1,\dots,n$

66. $H_p(E) = 0$.

At the same time (66) is an (elementary) special case of one important *Theorem of de Rham*.

Together, (65) and (66) mean that the sequence of homomorphisms $\{d_p\}$ is exact (see I, after D58).

4. Tensorial and Exterior Powers of an Operator.

Definition 23. The *tensorial power* $\overset{p}{\otimes}A$ *of the operator* A acting in the space E is defined as the operator in the space $\overset{p}{x}E$ which acts by the formula

$$[(\overset{p}{\otimes}A)\varphi](f_1,\dots,f_p) \underset{\text{def}}{=} (A'f_1,\dots,A'f_p),$$

where φ is any multilinear functional on $\overset{p}{x}E'$.

This implies as a special case

67. $(\overset{p}{\otimes}A)(x_1\wedge\dots\wedge x_p) = Ax_1\wedge\dots\wedge Ax_p$.

The operator $\overset{p}{\otimes}A$ is uniquely defined by the last formula.

68. $\text{rk}(\overset{p}{\otimes}A) = (\text{rk } A)^p$.

In particular, this implies that

69. The tensorial power of a regular operator is regular.

Thus it has an inverse which satisfies

70. $(\overset{p}{\otimes}A)^{-1} = \overset{p}{\otimes}(A^{-1})$.

This result is related to the two following formulae:

71. $\overset{p}{\otimes}(AB) = (\overset{p}{\otimes}A)(\overset{p}{\otimes}B)$.

72. $\overset{p}{\otimes} I_E + I_{\overset{p}{\otimes} E}$, i.e. the identity operator in $\overset{p}{\otimes} E$.

And so the mapping $\overset{p}{\otimes}$ is a "homomorphism relative to multiplication of operators".

It is also obvious that

73. $\overset{p}{\otimes}(\alpha A) = \alpha^p \cdot \overset{p}{\otimes} A$.

How do trace and determinant of a tensorial power behave? From (24) follows

74. $\operatorname{tr}(\overset{p}{\otimes} A) = (\operatorname{tr} A)^p$.

Using the theorem on the general form of a multiplicative functional in $\mathbb{M}(E)$ (II,(287)) we can also assert that

75. $\det(\overset{p}{\otimes} A) = (\det A)^{pn^{p-1}}$.

By the way, this fact as well as (74) is easily derived from the spectral properties of the tensorial power to be established below.

76. If $e_1, \ldots, e_p$ are (not necessarily distinct) eigenvectors of the operator A corresponding to the eigenvalues $\lambda_1, \ldots, \lambda_p$, then $e_1 \otimes \ldots \otimes e_p$ is an eigenvector of the operator $\overset{p}{\otimes} A$ corresponding to the eigenvalue $\lambda_1 \cdots \lambda_p$.

77. The tensorial power of an operator of scalar type is again of scalar type.

78. If $\{\lambda_k\}_1^n$ are all eigenvalues of the operator A taken according to multiplicity, then all possible products of the form

$$\lambda_{k_1} \cdots \lambda_{k_p} \qquad (1 \leq k_s \leq n,\ s = 1, \ldots, p)$$

are the set of all eigenvalues of the operator $\overset{p}{\otimes} A$, again taken according to multiplicity.

One evident consequence is that the operator $\overset{p}{\otimes} A$ for $p > 1$ always has multiple eigenvalues.

Definition 24. The *exterior power* $\overset{p}{\wedge} A$ *of* the *operator* A acting in the space E is defined as the restriction of the operator $\overset{p}{\otimes}A$ to the subspace $\overset{p}{\wedge}E$ of $\overset{p}{\times}E$.

79. $(\overset{p}{\wedge}A)(x_1\wedge\ldots\wedge x_p) = Ax_1\wedge\ldots\wedge Ax_p$.

In particular:

80. $\overset{n}{\wedge}A$ is the operator that multiplies by det A.

By virtue of (79) we find:

81. The subspace $\overset{p}{\wedge}E$ of $\overset{p}{\otimes}E$ is invariant relative to $\overset{p}{\wedge}A$.

Thus $\overset{p}{\wedge}A$ is an operator in the space $\overset{p}{\wedge}E$.

The properties of exterior powers of operators differ only in some details from the properties of tensorial powers.

82. $\overset{p}{\wedge}(AB) = (\overset{p}{\wedge}A)(\overset{p}{\wedge}B)$.

83. $\overset{p}{\wedge}I_E = I_{\overset{p}{\wedge}E}$.

84. The exterior power of a regular operator is regular, and we have $(\overset{p}{\wedge}A)^{-1} = \overset{p}{\wedge}A^{-1}$.

85. $\overset{p}{\wedge}(\alpha A) = \alpha^p\overset{p}{\wedge}A$.

86. $\det(\overset{p}{\wedge}A) = (\det A)^{\binom{n-1}{p-1}}$, still similar to (75).

The trace of the operator $\overset{p}{\wedge}A$, however, cannot be expressed in terms of the trace of the operator A. The formula for the trace will be derived below.

87. Let $e_1,\ldots,e_p$ be linearly independent eigenvectors of the operator A, corresponding to the eigenvalues $\lambda_1,\ldots,\lambda_p$, not necessarily pairwise distinct. Then $e_1\wedge\ldots\wedge e_p$ is an eigenvector of the operator $\overset{p}{\wedge}A$ corresponding to the eigenvalue $\lambda_1\cdots\lambda_p$.

88. The exterior power of an operator of scalar type is again an operator of scalar type.

89. If $\{\lambda_k\}_1^n$ are all the eigenvalues of the operator A taken according to their multiplicity, then all possible products of the form $\lambda_{k_1}\cdots\lambda_{k_p}$ $(1\leq k_1<\ldots<k_p\leq n)$ form the set of all eigenvalues of the operator $\overset{p}{\wedge}A$, taken according to multiplicity.

90. Let the characteristic polynomial of the operator A equal

$$\mathcal{D}(\lambda;A) = \Sigma_{p=0}^{n} \alpha_p x^{n-p} .$$

Then $\mathrm{tr}(\overset{p}{\wedge}A) = (-1)^{n-p}\alpha_p \quad (p = 1,\ldots,n)$.

5. Volume of a System of Vectors. Existence of an Auerbach Basis.

Let E be a unitary (and thus a Euclidean) space. For such a space length (norm) and orthogonality of vectors are well-defined. Starting from these ideas, a theory of volume can be developed naturally. It turns out to be closely related to the theory of determinants, being in essence a metric interpretation of the latter.

Definition 25. We shall define the idea of the volume of a system of vectors $\Gamma = \{x_k\}_1^n$ by induction on m. The volume will be denoted by $\|\Gamma\|$ or $\|x_1,\ldots,x_m\|$. This notation is justified since we set the volume equal to the norm of the vector if m = 1. For m > 1, we put

$$\|x_1,\ldots,x_m\| \underset{\mathrm{def}}{=} \|x_1,\ldots,x_{m-1}\|\cdot\|Q(x_1,\ldots,x_{m-1})x_m\|,$$

where $Q(x_1,\ldots,x_{m-1})$ is the orthoprojector onto the orthogonal complement of the linear hull of the vectors $x_1,\ldots,x_{m-1}$. This definition is obviously suggested by the elementary theorems about the area of a parallelogram and the volume of a parallelepiped.

91. Always $\|\Gamma\| \geqq 0$; $\|\Gamma\| = 0$ iff the system Γ is linearly dependent.

Already here we notice a connection with the theory of determinants (cf.(51)).

We recall *Hadamard's Inequality*[§]

[§]The usual form of Hadamard's inequality is $\det(a_{ik})^2 \leqq \Sigma_k|a_{1k}|^2\cdots\Sigma_k|a_{mk}|^2$, where k runs from 1 to m in each sum, and (a_{ik}) is an m×m matrix.

92. $\|x_1,\ldots,x_m\| \leq \Pi_{k=1}^{m} \|x_k\|$.

93. Equality holds in (92) iff the system $\{x_k\}_1^m$ is either orthogonal or contains the zero vector.

In particular:

94. If Γ is an orthonormal system, then $\|\Gamma\| = 1$. Conversely, if Γ is a normalized system and $\|\Gamma\| = 1$, then Γ is an orthonormal system.

Since $\|\Gamma\| \leq 1$ for an arbitrary normalized system, the orthonormal systems have maximal volume among all normalized systems. This remark will be useful below ((106) - (107)).

95. If $\Delta = \{e_k\}_1^m$ is an orthonormal system, and $\Gamma = \{x_k\}_1^m$ is a triangular system relative to Δ, then

$$\|\Gamma\| = \Pi_{k=1}^{m} |\alpha_{kk}| = |\det_\Delta \Gamma|$$

(cf.(41)).

Note that for a given system Γ the system Δ, connected with Γ by the hypothesis of (95), can always be constructed by orthogonalization and normalization.

The general theorem connecting volume and determinant reads:

96. If $\Delta = \{e_k\}_1^m$ is an orthonormal system, and the system $\Gamma = \{x_k\}_1^m$ belongs to the linear hull sp Δ, then $\|\Gamma\| = |\det_\Delta \Gamma|$.

In a still more general form we have:

97. If the system $\Gamma = \{x_k\}_1^m$ belongs to the linear hull sp Δ of the linearly independent system $\Delta = \{e_k\}_1^m$, then $|\det_\Delta \Gamma| = \|\Gamma\|/\|\Delta\|$.

Hence:

98. For any linear operator A, $|\det A| = \|A\Delta\|/\|\Delta\|$.

This shows that unimodular operators and only these (II, D56) preserve the volume. In particular, all unitary operators preserve volume. Like the unitary operators, the unimodular operators form a group in multiplication.

We shall now consider the functional properties of volume.

99. The volume of a system of vectors does not depend upon the order:

$$\|x_{j_1},\ldots,x_{j_m}\| = \|x_1,\ldots,x_m\|.$$

This property cannot be established directly from the definition of volume.

100. The volume of a system is an absolutely homogeneous functional:

$$\|\alpha_1 x_1,\ldots,\alpha_m x_m\| = (\Pi_{k=1}^{m} |\alpha_k|)\cdot\|x_1,\ldots,x_m\|$$

101. The volume of a system does not change if to any vector of the system a linear combination of the remaining vectors is added.

The functional properties (100), (101) and the normalization (94) uniquely define the volume:

102. If the functional $v(\Gamma)$ defined on the set of systems $\Gamma = \{x_k\}_1^m$ satisfies

1. absolute homogeneity, i.e. $v(\alpha_1 x_1,\ldots,\alpha_m x_m) = (\Pi_{k=1}^{m} |\alpha_k|)\cdot v(x_1,\ldots,x_m)$;
2. invariance, if to any vector of the system a linear combination of the remaining vectors is added;
3. $v(\Gamma) = 1$ for orthonormal systems Γ;

then $v(\Gamma) = \|\Gamma\|$.

Theorem (97) does not speak of the two volumes themselves in the language of determinants, but of their ratios. An "absolute" description will now be pointed out.

Let Γ be some linearly independent system and Γ^* the biorthogonal system belonging to sp Γ (III,D29).

103. If Δ is an orthonormal basis in sp Γ, then

$$\det_\Delta \Gamma^* \cdot \overline{\det_\Delta \Gamma} = 1 .$$

104. We also have the formula $\|\Gamma\| = \sqrt{\det_{\Gamma^*}\Gamma}$.

This implies in particular that $\det_{\Gamma^*}\Gamma > 0$.

Definition 26. This last determinant is called the *Gram*

determinant of the system Γ and coincides with the determinant of its Gram matrix, III,D28.

As a consequence of (104)

.05. $$\|\Gamma^*\| \cdot \|\Gamma\| = 1.$$

The theory of volumes can be used to construct an existence proof for an Auerbach basis in an arbitrary normed space (IV,D36). Consider an arbitrary norm (not connected with the given unitary structure in the space E), and denote by N the set of bases which are normalized relative to that norm. By virtue of the usual considerations of compactness,

06. The functional $\|\Delta\|$ $(\Delta \in \mathbb{N})$ attains its maximum on $\mathbb{N}$.

.07. If $\Delta_o \in \mathbb{N}$ and $\|\Delta_o\| = \max_{\mathbb{N}}\|\Delta\|$, then Δ_o is an Auerbach basis.

But there can be normalized Auerbach bases which do not maximize the volume.

6. Norms on the Space of Tensor Products.

This section touches on the topic of Section 10 in Ch. IV. Just as in the space of operators, we can consider induced norms on spaces of the type $E_1 \otimes \ldots \otimes E_p$.

Definition 27. Each collection of norms on $E_1,\ldots,E_p$ induces a *norm* on $E_1 \otimes \ldots \otimes E_p$: If $\varphi(f_1,\ldots,f_p)$ is a multilinear functional on $E_1' \times \ldots \times E_p'$, then

$$\|\varphi\| \underset{\text{def}}{=} \sup \frac{|\varphi(f_1,\ldots,f_p)|}{\|f_1\|\ldots\|f_p\|}$$

(the norms on the spaces E_k' are induced by the chosen norms on the spaces E_k (IV,D22)).

08. $\|x_1 \otimes \ldots \otimes x_p\| = \|x_1\|\ldots\|x_p\|$ holds for any system of vectors $x_1,\ldots,x_p$ (cf.IV,(257)).

Definition 28. Every norm on $E_1 \otimes \ldots \otimes E_p$ which satisfies property (108) for given norms on $E_1,\ldots,E_p$ is called a *cross-norm*.

For the purpose of describing cross-norms let us simply put $p = 2$.

For each element $u \in E_1 \otimes E_2$ consider all of its representations of the form

$$u = \Sigma_{k=1}^{m} x_1^{(k)} \otimes x_2^{(k)} \qquad (x_1^{(k)} \in E_1, \; x_2^{(k)} \in E_2),$$

and denote the set of such representations by $F(u)$.§

109. Any cross-norm satisfies

$$\|u\| \leq \Sigma_{k=1}^{m} \|x_1^{(k)}\| \cdot \|x_2^{(k)}\|.$$

110. The functional $N(u) \underset{\text{def}}{=} \inf_{F(u)} \Sigma_{k=1}^{m} \|x_1^{(k)}\| \cdot \|x_2^{(k)}\| \quad (u \in E_1 \otimes E_2)$ is a cross-norm.

Thus N is the greatest cross-norm on $E_1 \otimes E_2$ under the usual notion of order in the set of norms, introduced in IV,(264). A smallest cross-norm does not exist. However, for a certain natural class of cross-norms a smallest cross-norm will appear, and, as one expects, it turns out to be an induced norm.

Definition 29. Let an arbitrary norm be given on $E_1 \otimes E_2$. It induces a norm on $(E_1 \otimes E_2)'$ and, with the aid of the canonical isomorphism, also on $E_1' \otimes E_2'$. The norm on $E_1' \otimes E_2'$ obtained in this way is called the *dual* of the original norm. A repetition of this process (with the use of the canonical isomorphisms $E_1'' \simeq E_1$, $E_2'' \simeq E_2$) leads back to the original norm.

Now $E_1' \otimes E_2'$ can be identified with the space $\mathbb{B}(E_1, E_2)$ of bilinear functionals. Because of this the dual norm can be described explicitly:

---

§There is indeed such an m for linearly independent systems $\{x_1^{(k)}\}_k$, $\{x_2^{(k)}\}_k$; see Greub, op. cit., p. 7.

111. The dual norm of the bilinear functional $B \in \mathbb{B}(E_1,E_2)$ equals the norm of the linear functional $F_B \in (E_1 \otimes E_2)'$ connected with B by the formula

$$B(x_1,x_2) = F_B(x_1 \otimes x_2) \qquad (x_1 \in E_1,\ x_2 \in E_2) .$$

From this interpretation we easily obtain that

112. The norm on $E_1' \otimes E_2'$ which is dual to the induced norm on $E_1 \otimes E_2$ is a cross-norm (relative to the induced norms on E_1' and E_2').

More generally:

113. If a given cross-norm $E_1 \otimes E_2$ majorizes the induced norm, then the norm on $E_1' \otimes E_2'$, which is dual to it, is also a cross-norm.

Conversely:

114. Let the norm ν on $E_1\ E_2$ be such that its dual norm on $E_1' \otimes E_2'$ is a cross-norm. Then ν majorizes the induced norm.

And so

115. The induced norm on $E_1 \otimes E_2$ is the least cross-norm whose dual is also a cross-norm (*Theorem of Schatten*).

Calculation of the norm, which is dual to the maximum cross-norm N, results in

116. The norm dual to N coincides with the induced norm on $E_1' \otimes E_2'$.

Thus cross-norms whose duals are also cross-norms are exactly those cross-norms which range between the induced norm and the norm N.

7. Formal Polynomials and Power Series in Several Variables.

Definition 30. The *formal polynomials* occurring in the beginning of this section are expressions of the form

$$\pi(z_1,\ldots,z_m) \underset{\text{def}}{=} \Sigma\alpha^{j_1,\ldots,j_s}\, z_{j_1} \ldots z_{j_s} .$$

The symbols z_k should not be considered as independent variables but as "letters" taken from a finite "alphabet" $z_1,\dots,z_m$. Disregarding the numerical coefficient $\alpha^{j_1,\dots,j_s}$, the literal "factor" $z_{j_1}\dots z_{j_s}$ is a "word of length s", i.e. a sequence of s letters taken from the alphabet in some definite order. Hence j_k is one of the numbers $1,2,\dots,m$, and repetitions ($j_p=j_q$) may occur. Each system of indices $(j_1,\dots,j_s)$ characterizes a "term" (= a word and its coefficient), and any permutation of a system of indices is considered as a different word, that is, the z_j do not commute. The formal sum extends over some finite set J_π of such systems. The word length s is called the degree of that particular term.
It is expedient to include the "free" number α among the terms of a formal polynomial and to assign to α the length $s = 0$ (the system of indices corresponding to α is empty). We also admit $J_\pi = \emptyset$, in which case we define π as 0.
Definition 31. The number $\max\{s \mid \alpha^{j_1,\dots,j_s} \neq 0\}$ is called the *degree of a formal polynomial*, if $\pi \neq 0$; it is denoted by deg π. If all terms of the formal polynomial have the same degree, then it is called *homogeneous*. Two formal polynomials are called equal if their corresponding coefficients are equal. Addition and multiplication by scalars are defined in the natural way.

117. The formal polynomials of m letters (or symbols) of degree no greater than p form a linear space of dimension $(m^{p+1} - 1)/(m - 1)$. It is the direct sum of the subspaces of homogeneous formal polynomials of degrees $0,1,\dots,p$.

Definition 32. The *product of* two *formal polynomials*

$$\pi(z_1,\dots,z_m) = \Sigma_{J_\pi} \alpha^{j_1,\dots,j_s} z_{j_1}\dots z_{j_s},$$

$$\rho(z_1,\dots,z_m) = \Sigma_{J_\rho} \beta^{k_1,\dots,k_\ell} z_{k_1}\dots z_{k_\ell}$$

is the formal polynomial

$$(\pi\rho)(z_1,\ldots,z_m) = \Sigma_{J_\pi \times J_\rho} \alpha^{j_1,\ldots,j_s} \beta^{k_1,\ldots,k_\ell} z_{j_1}\ldots z_{j_s} z_{k_1}\ldots z_{k_\ell} .$$

118. deg $(\pi\rho)$ = deg π + deg ρ.

Multiplication of formal polynomials is distributive; it commutes with scalar factors; it is associative, but <u>not commutative</u>. Division of formal polynomials, as a rule, does not exist, because a formal polynomial cannot have a multiplicative inverse unless it reduces to a scalar. This follows from

119. If $\pi\rho = 1$, then deg π = deg $\rho = 0$.

The situation can be improved by adjoining new objects to the formal polynomials.

Definition 33. The new objects required are *formal power series*. They are defined in the same way as formal polynomials, but J_π, the set of systems of indices, now can be infinite. By introducing zero coefficients if necessary we may assume that J_π consists of all systems of indices $(j_1,\ldots,j_s)$ $(1 \leqq j_k \leqq m;\ k = 1,\ldots,s_j;\ s = 0,1\ldots)$. Equality of formal power series is defined in the same way as for formal polynomials, and so are the operations on formal power series. Moreover, for formal power series it makes sense to form the sum of infinitely many terms, provided that for each homogeneous component only finitely many terms must be added.

Let us denote the free (or "constant") term of a formal power series a by $\varphi(a)$.

20. $\varphi(ab) = \varphi(a)\varphi(b)$.

Therefore a formal power series with constant term different from zero cannot be divided by a series without constant term. However, any formal power series can be divided by a series with nonzero constant term:

121. If $\varphi(a) \neq 0$, there exists a unique formal power series a^{-1} satisfying the relation $aa^{-1} = a^{-1}a = 1$.

Thus the power series with nonzero constant term form a group. This group contains all series with constant term 1 as a subgroup.

On the basis of formal power series the most general form of operator calculus can be developed. Each relation valid for formal power series carries over automatically to nonformal power series (i.e., if they are convergent in some sense). This holds for convergent power series in any associative algebra, in particular in the algebra of operators $\mathbb{M}(E)$.

In what follows let us agree to say briefly "power series", and "polynomial", omitting the word "formal". Let us also write the abbreviation a^k in place of $\underbrace{a\ldots a}_{k}$, and $a^o=1$ as usual.

The general operator calculus is obtained by adjoining the operation of <u>superposition</u> to the algebraic operations on power series. Superposition means here the substitution of arbitrary power series <u>without</u> constant terms in place of the arguments $z_1,\ldots,z_m$ in a power series. This operation is carried out according to the usual rules. If, in particular, we want to determine a scalar function of a power series, we let $f(\lambda)$ be a scalar function, holomorphic in the neighborhood of zero:

$$f(\lambda) = \Sigma_{k=0}^{\infty} \alpha_k \lambda^k.$$

Then for any power series a without constant term

$$f(a) \underset{\text{def}}{=} \Sigma_{k=o}^{\infty} \alpha_k a^k.$$

122. Let $\varphi(a) \neq 0$, $f(\lambda) = (\varphi(a) + \lambda)^{-1}$. Then $f(a - \varphi(a)) = a^{-1}$.

123. If $f_1(\lambda)$, $f_2(\lambda)$ are two scalar functions holomorphic in a neighborhood of zero and $f(\lambda) = f_1(\lambda)f_2(\lambda)$, then $f(a) =$

$f_1(a)f_2(a)$ ($\varphi(a)=0$ is assumed).

124. If $f_1(\lambda)$, $f_2(\lambda)$ are two scalar functions holomorphic in a neighborhood of zero; if, further, $f_2(0) = 0$, and $f(\lambda) \underset{\text{def}}{=} f_1(f_2(\lambda))$, then $f(a) = f_1(f_2(a))$ ($\varphi(a) = 0$ is assumed).

125. If $f'(0) \neq 0$, then the equation

$$f(a) = c \qquad (*)$$

is solvable with respect to the power series a ($\varphi(a) = 0$ again assumed), iff $\varphi(c) = f(0)$. If the equation (*) is solvable, then its solution is unique.

In particular:

126. The formula $c=e^a$ defines a bijection between the set of power series without constant term and the group of series with constant term 1. The inverse mapping is given by the formula $a = \log(1 + (c-1))$.[§]

The second member of this formula will be simply written as log c.

127. If $\varphi(a) = \varphi(b) = 0$ and $ab = ba$, then $e^a e^b = e^{a+b}$

It is a truly remarkable fact that we can discard the condition $ab = ba$ and still write an explicit formula for the product of exponential functions. We shall do that after giving some necessary information on the *Lie algebra of power series*. What we have to say has also independent interest.

Definition 34. The *commutator of the power series* a,b is defined in the standard form of II,D5:

$$[a,b] = ab - ba .$$

128. $\varphi([a,b]) = 0$.

Definition 35. We shall call a power series (polynomial) without constant term a *Lie power series* (polynomial) if

---

[§] e^λ and $\log(1+\lambda)$ are here shorthand symbols for the corresponding power series.

it can be represented in the form

$$\sum_{(k_1,\ldots,k_s)} \gamma^{k_1,\ldots,k_s}[z_{k_1},[\ldots,[z_{k_{s-1}},z_{k_s}]\ldots]]$$

(for s = 1 the commutations indicated by the brackets are absent).

To each power series $a = \Sigma\alpha^{j_1,\ldots,j_s} z_{j_1}\ldots z_{j_s}$ we can assign the Lie series

$$Qa = \Sigma\alpha^{j_1,\ldots,j_s}[z_{j_1},[\ldots,[z_{j_{s-1}},z_{j_s}]\ldots]] .$$

Obviously,

$$Q(a + b) = Qa + Qb, \quad Q(\alpha a) = \alpha Qa.$$

The behavior of the mapping Q in the multiplication of series is more complicated:

129. If a is a Lie series and b is an arbitrary power series, then $Q(ab) = [a,Qb]$.

Hence

130. If a and b are Lie series, then

$$Q([a,b]) = [Qa,b] + [a,Qb] .$$

We need a criterion for the "Lie character" of a power series. To prepare for the proof of such a criterion, we introduce

Π_s, the space of homogeneous polynomials of degree $s > 0$,

Q_s, the restriction of Q to the space Π_s,

$P_s \underset{\text{def}}{=} \frac{1}{s} Q_s$, a convenient modification of the mapping Q_s.

Using (130) we can easily prove that

131. The operator P_s projects the space Π_s onto the subspace L_s of homogeneous Lie polynomials of degree s.

Now we modify the mapping Q by multiplying every homogeneous group of terms of degree $s > 0$ by $1/s$ and obtain the mapping P of the space of power series without constant term on the space L of Lie series with constant term $\neq 0$:

$$Pa = \Sigma \frac{1}{s} \alpha^{j_1,\dots,j_s}[z_{j_1},[\dots,[z_{j_{s-1}},z_{j_s}]\dots]] .$$

The criterion for the Lie character of the power series without constant term is now:

.32. The power series a is a Lie power series iff Pa = a.

To justify the remark after (127) we require still another important criterion for the Lie character of a. In order to formulate it we must introduce a tensor multiplication in the linear space of (formal) power series. However, this space is not finite-dimensional, hence the methods used so far in this chapter are of little use. On the other hand, we only will need tensor multiplications of the type $a_1 \otimes \dots \otimes a_p$, where the a_k are (formal) power series, i.e., we need the elements of the tensor power $\overset{p}{\otimes} A$ (A denotes here the space of power series).

We wish to preserve the validity of the usual rules

1. $(a_1+a_2)\otimes b = a_1\otimes b + a_2\otimes b$, $a\otimes(b_1+b_2) = a\otimes b_1 + a\otimes b_2$;
2. $(\alpha a)\otimes b = a\otimes(\alpha b) = \alpha(a\otimes b)$.

This will enable us to reduce a $\otimes$-product of two power series to a sum of expressions such as

$$\alpha(z_{j_1}\dots z_{j_s}) \otimes (z_{k_1}\dots z_{k_t}), \qquad (*)$$

but $\otimes$-products, where one or both factors are scalars, will also occur.

Recalling that A is not only a linear space, but also an associative algebra by D32, we introduce (cf. Greub, loc. cit., p. 43) the rule

3. $(a_1\otimes a_2)(b_1\otimes b_2) = (a_1 b_1) \otimes (a_2 b_2)$,

which holds for tensor products of algebras A and B and will permit the reduction of $\otimes$-products of the form (*) to simpler structures.

If p = 2, then for the complete determination of all

elements of $\overset{2}{\otimes}A$ only the following four types of products

$$z_j \otimes z_k,\ 1\otimes z_j,\ z_j\otimes 1,\ 1\otimes 1,\ 1 \leq j,k \leq m$$

require new definitions.

Definition 36. We put $1\otimes 1 \underset{\text{def}}{=} 1$, and consider the three remaining product types as $(m+1)^2 - 1$ *new letters*. They are the symbols on which the product $a\otimes b$ depends, provided that a and b depend on the same symbols $\{z_k\}_1^m$. Note, however, that some of the new symbols commute: $(z_1\otimes 1)(1\otimes z_2) = (1\otimes z_2)(z_1\otimes 1)$, although $z_1 z_2 \neq z_2 z_1$.

Now let us compare each power series

$$a = \Sigma\alpha^{j_1,\ldots,j_s} z_{j_1}\cdots z_{j_s}$$

with the series

$$ha \underset{\text{def}}{=} \Sigma\alpha^{j_1,\ldots,j_s}(hz_{j_1})\cdots(hz_{j_s}),$$

where we have put

$$hz_k = z_k\otimes 1 + 1\otimes z_k \qquad (k = 1,\ldots,m).$$

Then the second criterion is:

133. A power series is a Lie series iff $ha = a\otimes 1 + 1\otimes a$.

Hence

134. The exponential mapping $b = e^a$ is a bijection between the set of Lie series and the set of power series with constant term 1, which satisfy the relation $hb = b\otimes b$.

By direct calculation it can first be shown that

135. $$\log(e^{z_1}e^{z_2}) = \Sigma_{s=1}^{\infty}\frac{(-1)^{s-1}}{s}\Sigma_{p_j+q_j>0}\frac{z_1^{p_1}z_2^{q_1}\cdots z_1^{p_s}z_2^{q_s}}{p_1!q_1!\cdots p_s!q_s!}$$

It is then easy to verify that

136. The power series $\log(e^{z_1}e^{z_2})$ is a Lie series.

Hence we can apply the operator P of (132) and obtain by substituting $z_1\to a$, $z_2\to b$ the *Campbell-Hausdorff-Dynkin Formula*:

137. $\log(e^a e^b) = \Sigma_{s=1}^{\infty} \frac{(-1)^{s-1}}{s}$ x

$$\Sigma_{p_j+q_j>0} \frac{1}{\Sigma_{j=1}^{\infty}(p_j+q_j)} \frac{[a^{p_1},[b^{q_1},\ldots,[a^{p_s},b^{q_s}]\ldots]]}{p_1!q_1!\ldots p_s!q_s!} .$$

Here we have used the following notation:

$$[u^p,v] \underset{\text{def}}{=} \underbrace{[u,[u,\ldots,[u,v]\ldots]]}_{u \text{ occurs } p \text{ times}},$$

$$[u^p,v^q] = 0\,(q>1),\ [u^p,v^o] = \begin{cases} u, & p=1 \\ 0, & p>1 \end{cases} .$$

Let us write the initial segment of the series (137):

$$\log(e^a e^b) = a + b + \tfrac{1}{2}[a,b] + r ;$$

here r is a power series which does not contain terms of lower degree than the third and vanishes if ab = ba.

138. In the algebra of operators $\mathbb{M}(E)$ there exists a neighborhood W of zero such that the Campbell-Hausdorff-Dynkin series with the substitution $a \to A$, $b \to B$ $(A,B \in W)$ converges and its sum equals the operator $\log(e^A e^B)$.

Proofs of criteria (132) and (133) can be found in Nr. 3 of the Special Literature to Ch. V, Thms. 8 and 9 on pp. 169 and 170, but the proof of Thm. 9 uses some advanced results of Lie theory.

An essentially elementary proof of (137) is the content of Dynkin's paper "On the representation of the series $\log(e^x e^y)$ with non-commuting x and y by commutators", Matematičeskiĭ Sbornik 25(1949), pp. 155-162.

algebraic complement D18
"alphabet" D30
antisymmetrization operator D14
antisymmetric tensor D11
Auerbach basis, existence (107)

boundary homomorphism D22

Campbell-Hausdorff-Dynkin Formula (137)
Cartan's theory of exterior forms D19ff.
character of permutation (42)
commutator of two power series D34
contraction of bilinear functional (22)
contraction of linear operator (21)
contraction of tensor D8
covariant tensor D5
criteria for Lie character of power series (132),(133)
cross-norm D28

degree of formal polynomial D31
determinant D12
determinant of matrix (43)
dual norm D29

exterior form of p-*th* degree D19
exterior forms in algebraic topology D21ff.
exterior power of operator D24
exterior product of vectors D16

formal polynomial D30

Gram determinant of system D26
group of formal power series (121)

Hadamard's Inequality (92)
homogeneous formal polynomial D31
homologous D22a

Laplace's Theorem (58)
Lie power series D35

"matrix" of a tensor (18)
minor D17
multilinear function D1
multilinear functional D2
multilinear mapping into E as tensor (15)
multiplication considered as tensorial operator (25)-(27)
multiplication in E as trivalent tensor (17)
multiplication of determinants of vector systems (35)

norm induced on $E_1\otimes\ldots\otimes E_p$ D27

p-dimensional cycle D22a
p-dimensional homology space D22a
product of exterior forms D20
product of two formal polynomials D32
p-*th* exterior power of space D15
p-valent tensor D2

Schatten's Theorem (115)
structural constants (18)
superposition of formal power series (121)
symmetric tensor D9
symmetrization operator D10

tensor D2
tensorial power of operator D23
tensor multiplication of multilinear functionals D6,D7
tensor product of spaces D3
tensor product of vectors D4
tensors in matrix representation (18)
Theorem of de Rham (66)
trace of $\overset{p}{\wedge}A$ (90)
triangular system relative to a basis D13

universality of exterior multiplication (52)
universality of tensor multiplication (6),(7)

valence of tensor D2

volume and determinant connected (96),(98)
volume of a vector system D25

"word of length s" D30

CHAPTER VI
REAL LINEAR SPACE

In the earlier chapters, our object of investigation was the complex linear space; the space elements (vectors) had to admit multiplication by complex numbers (scalars). In a more general setting, the elements of some _field_ may serve as scalars. In this case we speak of a space _over_ that field. In analysis, the most important cases are the fields of complex and of real scalars.
We have now reached the point where we must stop at the second case and discuss its special properties. Clearly, the great majority of definitions and theorems in previous chapters carry over without change to real spaces (and even to spaces over arbitrary fields). This holds especially for Chapters I and III-V. The construction of the spectral theory in Ch. II, however, is closely connected with the existence of roots of polynomial equations and, more generally, with the machinery of the theory of analytic functions. On the other hand, there is nothing special whatsoever about being complex e.g. in Ch. V, and that chapter carries over literally to real spaces.

1. Complexification.

The close connection between real and complex spaces is of course a consequence of the relation between the real and the complex fields. This relation permits a systematic reduction of investigations in a real space to questions already answered for a complex space. (This, of course, works also in the opposite direction.) Such a reduction is not always necessary nor is it always convenient, but in many cases it produces a quick answer.-- In this chapter R will denote the basic real space, and n its dimension as a real space. The formal definition of dim R is thus exactly the same as for a complex space (cf. I,D22), only the fields differ. If needed we shall use and distinguish the terms "real dimension" and "complex dimension".

1. The field of complex numbers C is a two-dimensional real linear space.

Definition 1. The *complex* n-dimensional *space* E can be formally related to the *real spaces* C and R by the real operation ⊗. First embed the real R in the complex E, e.g. by writing R and E as $sp_{r,c}\{x_i\}_{i=1}^{n}$ and using real and complex scalars in sp_r and sp_c, respectively. Now recall that theorem V,(8) associates every bilinear mapping $\pi: C\times R \to E$ with exactly one homomorphism $h: C\otimes R \to E$. Choose $\pi:(\alpha,x) \mapsto x$, $\alpha\in C$, $x\in R$, $\alpha x\in E$. There is then the homomorphism $h: \alpha\otimes x \mapsto x$, which is clearly bijective. Note that $\alpha x\in E$ is well-defined, since R was embedded in E.

2. The real tensor product $R^{(c)} \underset{def}{=} C\otimes R$ now becomes a complex linear space if we introduce the natural multiplication by a scalar: $\beta(\alpha x) \underset{def}{=} (\beta\alpha)x$.

Note, however, that $R^{(c)}$ can also be considered as a space over the real scalar field R.

Definition 2. The complex space $R^{(c)}$ is called the *complex hull* of R, and the transition from R to $R^{(c)}$ is known as the complexification of the space R.

In what follows we shall simplify the notation by writing αx ($\alpha\in C$, $x\in R$) instead of $\alpha\otimes x$. (This means "forgetting" about the isomorphism h in D1.) The initial space R can be identified with the subset $\{y|y=1x;\ x\in R\}$ of $R^{(c)}$. R is then a subspace of $R^{(c)}$, where $R^{(c)}$ is now considered as a real space.

3. If the system $\{x_1,\ldots,x_n\} \in R$ is linearly independent, then so is the system $\{x_1,\ldots,x_n, ix_1,\ldots,ix_n\}$ in $R^{(c)}$, considered here as a real space.

4. The complex dimension of $R^{(c)}$ equals n, the real dimension of R. But the real dimension of $R^{(c)}$ is evidently 2n.

5. Every vector $z \in R^{(c)}$ has a unique representation of the form

$$z = x + iy \qquad (x,y \in R).$$

Definition 3. The vectors x and y in (5) are said to be the *real* and *imaginary parts* of z. Thus (5) describes the real space $R^{(c)}$ as the direct sum of R and iR.

If $z = z+iy$ $(x,y \in R)$, then $\bar{z} \underset{\text{def}}{=} x - iy$ is called the *complex conjugate* of z.

6. If $\alpha = \rho + i\sigma$ (ρ,σ real scalars), and $z = x + iy$ $(x,y \in R)$, then

$$\alpha z = (\rho x - \sigma y) + i(\sigma x + \rho y) .$$

7. The mapping $jz = \bar{z}$ is a Hermitian isomorphism $R^{(c)} \to R^{(c)}$ (cf. I,D81), and it is an involution.

If L is a subspace of R, then $L^{(c)}$, the *complex hull* of L, is defined as the set of vectors αx $(\alpha \in C,\ x \in L)$.

8. The complex hull $L^{(c)}$ is a subspace of $R^{(c)}$,

and so

9. The complex dimension of $L^{(c)}$ equals the real dimension of L.

10. Let R_1 and R_2 be two real spaces, and let $h \in \mathrm{Hom}(R_1,R_2)$. There is a unique extension of the homomoprhism h to $h^{(c)}$: $R_1^{(c)} \to R_2^{(c)}$.

Definition 4. The homomorphism $h^{(c)}$ of (10) is known as the *complex continuation* of h, and the transition from h to $h^{(c)}$ is the *complexification of the homomorphism* h.

11. $$\mathrm{Ker}\, h^{(c)} = (\mathrm{Ker}\, h)^{(c)}.\ \mathrm{Im}\, h^{(c)} = (\mathrm{Im}\, h)^{(c)}.$$

Hence

12. $$\mathrm{def}\, h^{(c)} = \mathrm{def}\, h, \quad \mathrm{rk}\, h^{(c)} = \mathrm{rk}\, h.$$

Parallel to complexification of a homomorphism there is a

conjugate procedure, viz. the *Hermitian complexification*:

13. Let R_1 and R_2 be two real spaces and $h \in \mathrm{Hom}(R_1, R_2)$. There is a unique extension of the homomorphism h to a Hermitian homomorphism mapping $R_1^{(c)}$ into $R_2^{(c)}$.

These operations on homomorphisms apply to linear functionals as special cases.

In the case of bilinear functionals it is convenient to use the term complexification directly for the extension to a Hermitian-bilinear functional (I,D84):

14. Let R_1 and R_2 be real spaces and B a bilinear functional on $R_1 \times R_2$. B permits a unique extension to a Hermitian-bilinear functional $B^{(c)}$ on $R_1^{(c)} \times R_2^{(c)}$.

15. Rank and defect of a bilinear functional do not change in complexification.

16. For a symmetric bilinear functional B on R,

$$B^{(c)}(z,z) = B(x,x) + B(y,y), \qquad (z = x+iy).$$

17. The complexification of a symmetric bilinear functional produces a Hermitian-symmetric bilinear functional (III,D12).

18. Let Q be a quadratic functional on R. Q can be uniquely extended to a "real" (in the sense of III,D13) Hermitian-quadratic functional $Q^{(c)}$ on $R^{(c)}$.

19. For every symmetric bilinear functional B on R there is a vector system $\{e_k\}_1^r \subset R$ with $r = \mathrm{rk}\, B > 0$ such that

$$B(x,y) = \Sigma_{k=1}^{r}\ \varepsilon_k B(x,e_k)B(e_k,y) \qquad (\varepsilon_k = \pm 1).$$

20. For every quadratic functional Q on R there is a linearly independent system of linear functionals $\{f_k\}_1^r$ with $r = \mathrm{rk}\, Q > 0$ such that

$$Q(x) = \Sigma_{k=1}^{r}\ \varepsilon_k [f_k(x)]^2 \qquad (\varepsilon_k = \pm 1).$$

Propositions (19) and (20) can be proved by complexification (they reduce then to III,(41) and (43)) but they can of course be shown "without leaving R" by following the pattern of proof for the corresponding proposition in complex space.

The law of inertia for "real" Hermitian-quadratic functionals carries over without change to such functionals defined on R.

21. The indices of inertia of a quadratic functional on R do not change in complexification (cf. III,D14).

In particular, complexification of a positive functional produces a positive functional.

22. If R is a unitary space, then the complexification of the scalar product makes $R^{(c)}$ into a complex unitary space.

Let us now look at the complexification of the norm.

Definition 5. We introduce as *norm on* $R^{(c)}$ a *complex extension* of the norm on R, which coincides on the subspace $R \subset R^{(c)}$ with the initial norm on R. (Cf. here the remark preceding (3).)

23. Let R now be Euclidean and let $\|.\|$ denote the norm on R. A Euclidean complex extension of $\|.\|$ can be defined as

$$\|x + iy\|_1 \underset{\text{def}}{=} \sqrt{\|x\|^2 + \|y\|^2} .$$

This turns out to be the only extension under the condition that serves as hypothesis in

24. If R is a normed space and the norm on $R^{(c)}$ satisfies

$$\|x + iy\|_1 = \nu(\|x\|, \|y\|) \qquad (x,y \in R),$$

then $(\alpha,\beta) = (\alpha^2+\beta^2)^{\frac{1}{2}}$ and R is Euclidean (IV,(12)).

Therefore the complexification of a non-Euclidean norm cannot be brought into concordance with the representation $z = x+iy$. But does such a complexification in general exist? An affirmative answer to this question is obtained from the theory of the cross-norm in V, Sect. 6.

In what follows we shall consider the modulus of a complex number as a norm on $\mathbb{C}$.

25. For a normed space R the set of complex extensions of the norm coincides with the class of cross-norms on $\mathbb{C} \otimes R$

(V,D28).

Therefore:

26. The functional

$$\|z\| \underset{\text{def}}{=} \inf \Sigma|\alpha_k| \; \|x_k\| ,$$

where the infimum is taken over all representations $z = \Sigma\alpha_k x_k$ ($\alpha_k \in C$, $x_k \in R$), is the greatest complex extension of the norm on R.

27. The induced norm on $C\otimes R$ has the form (V,D27)

$$z = \sup_{f\in R'} \frac{|f^{(c)}(z)|}{\|f\|} \quad (z = x+iy).$$

Now Schatten's theorem (V,(115)) implies:

28. The induced norm is the least complex extension of the norm on R, which is in concordance with the complexification of linear functionals: for it must satisfy $\|f^{(c)}\| = \|f\|$ for $f \in R'$ by D4.

2. Decomplexification.

This term shall denote a procedure that is in a certain sense inverse to complexification.

32. Every complex n-dimensional space E is a 2n-dimensional real space.

Definition 6. If we imagine E in this role, we shall denote the real space by $E^{(r)}$; the transition from E to $E^{(r)}$ will be called the *decomplexification of the space* E.

33. Every m-dimensional subspace $L \subset E$ is a 2m-dimensional subspace of $E^{(r)}$.

In this role of L we shall again denote the real subspace by $L^{(r)}$.

34. Multiplying E by i is an automorphism of $E^{(r)}$.

Denote this automorphism by J.

35. $$J^2 = -I .$$

36. A subspace of $E^{(r)}$ coincides with some $L^{(r)}$ iff it is invariant in regard to J.

Definition 7. *Subspaces* of $E^{(r)}$ with that invariance property are called *complexifiable.*

37. If M is a subspace of the space $E^{(r)}$, then the smallest subspace of E which contains M is given by the sum $\tilde{M} = M+JM$ (notation in I,D28).

38. $\frac{1}{2} \dim M \leq \dim \tilde{M} \leq \dim M$. (dim M = the real dimension, dim $\tilde{M}$ = the complex dimension).

39. $\dim \tilde{M} = \frac{1}{2} \dim M$ is attained iff M is complexifiable.

40. $\dim \tilde{M} = \dim M$ is attained iff $JM \cap M = 0$.

Definition 8. Subspaces of $E^{(r)}$ with the property (40) are called *complexifying.* Every subspace of a complexifying subspace is clearly complexifying.

41. If M is a complexifying subspace, then $\dim M \leq n$.

42. If in the preceding proposition $\dim M < n$, then there is a complexifying subspace $M_1 \supset M$ such that $\dim M_1 = \dim M+1$.

43. For each $m \leq n$ there is a complexifying subspace of $E^{(r)}$ with dimension m.

44. A complexifying subspace M is maximal (I,D27) iff dim M = n.

45. If R is a maximal complexifying subspace, then every $z \in E$ has a unique representation $z = x+iy$ with $x,y \in R$.

This establishes a natural isomorphism between E and the complex hull $R^{(c)}$ of R. However, no unique determination of the subspace R has been given.

The inverse of (45) is:

46. If every $z \in E$ can be uniquely represented as $x + iy$ with $x,y \in R \subset E^{(r)}$, then R is a maximal complexifying subspace.

We can describe the complexifiable subspace of $E^{(r)}$ in the dual terminology of complexifying subspaces.

47. Every complexifiable subspace has even dimension.

48. If the subspace M is complexifiable, then it includes a complexifying subspace such that $M = K \oplus JK$.

Conversely,

49. The sum $K + JK$ is direct for any complexifying subspace K and is a complexifiable subspace.

50. If M is a complexifiable subspace and $\dim M > 0$, then there is a complexifiable subspace $M_1 \subset M$ such that $\dim M_1 = \dim M-2$.

51. For every $m \leqq n$ there is a complexifiable subspace of dimension $2m$.

52. M is a nontrivial minimal complexifiable subspace of $E^{(r)}$ iff $\dim M = 2$.

53. Every nontrivial complexifiable subspace (this includes $E^{(r)}$) can be split into a direct sum of two-dimensional complexifiable subspaces.

Let us now turn to the decomplexification of homomorphisms and functionals.

54. Let E_1 and E_2 be complex spaces and $h \in \mathrm{Hom}(E_1,E_2)$. Then also $h \in \mathrm{Hom}(E_1^{(r)},E_2^{(r)})$.

Definition 9. In this role we shall write $h^{(r)}$ instead of h and call the transition from h to $h^{(r)}$ the *decomplexification of the homomorphism* h. Now denote by J_k the operators $E_k^{(r)} \to iE_k^{(r)}$, $k = 1,2$.

55. $h^{(r)}J_1 = J_2h^{(r)}$.

Definition 10. A *homomorphism* $g \in \mathrm{Hom}(E_1^{(r)},E_2^{(r)})$ that satisfies $gJ_1 = J_2g$ is called *complexifiable.*

56. If $g \in \mathrm{Hom}(E_1^{(r)},E_2^{(r)})$ is a complexifiable homomorphism, then there is a unique $h \in \mathrm{Hom}(E_1,E_2)$ such that $g = h^{(r)}$.

57. If f is a linear functional on E, then $f^{(r)}(x) \underset{\mathrm{def}}{=} \mathrm{Re}\, f(x)$ is a linear functional on $E^{(r)}$.

Definition 11. The transition from f to $f^{(r)}$ is called the *decomplexification of the functional* f. It is not reducible to the decomplexification of a homomorphism.

58. The decomplexification of functionals establishes an injection from E' into $E^{(r)})'$.

59. If B is a Hermitian-bilinear functional on $E_1 \times E_2$, then

$B^{(r)}(x,y) = \mathrm{Re}\, B(x,y)$ is a bilinear functional on $E_1^{(r)} \times E_2^{(r)}$

60. Here decomplexification establishes an injection from the space of Hermitian-bilinear functionals on $E_1 \times E_2$ into the space of bilinear functionals on $E_1^{(r)} \times E_2^{(r)}$.

61. If E is a unitary space, then the space $E^{(r)}$, under the decomplexified scalar product§, is an inner product space.

62. If E is a normed space, then the norm on E is a norm on $E^{(r)}$.

In Euclidean space the decomplexification of the norm and that of the scalar product are in concordance.

Clearly,

63. Decomplexification of a homomorphism preserves the norm, i.e., $\|h^{(r)}\| = \|h\|$.

64. Decomplexification of a linear functional preserves the norm.

3. Operators in Real Space.

Operator theory in real space owes its special character to the fact that the field of real scalars is not algebraically closed: an equation with real coefficients does not necessarily have real roots. Hence a linear operator in real space need not have eigenvectors. But after complexification the eigenvectors will appear, and the task that remains is the "back projection" of the spectral theory into R.

Definition 12. In this section, A denotes a linear operator in R. The eigenvalues of its complex extension $A^{(c)}$, which operates in $R^{(c)}$, will be called the *eigenvalues of* A for simplicity. Also the term "spectral

---

§A real space R is an inner product space provided there is a bilinear functional $(\cdot,\cdot): R \times R \to \mathcal{R}$ ($\mathcal{R}$ is the real field), which satisfies the conditions of III,D17. Note that 3. of that definition becomes $(x,y) = (y,x)$

radius" and other terms of spectral theory will refer to $A^{(c)}$, if they are used in connection with A.

Let j denote the conjugation mapping in $R^{(c)}$: $jz \underset{def}{=} \bar{z}$.

65. The operator $A^{(c)}$ commutes with j : $jA^{(c)} = A^{(c)}j$.

Put $W^s(\lambda) \underset{def}{=} \operatorname{Ker}(A^{(c)} - \lambda I)^s$, $s = 1,2,\ldots$; $\lambda \in \sigma(A^{(c)})$.

66. The subspaces $W^s(\lambda)$ and $W^s(\bar{\lambda})$ of $E^{(c)}$ are isomorphic.

67. If λ is an eigenvalue of A (recall D12), then so is $\bar{\lambda}$, and $\bar{\lambda}$ has the same multiplicity as λ.

68. The characteristic polynomial of A is real.

69. The minimal polynomial of A is real.

70. If λ is a real eigenvalue of A, then each of the subspaces $W^{(s)}(\lambda)$ has a real basis, i.e. its vectors belong to R.

71. If all eigenvalues of A are real, then A has a Jordan basis (II,D33) in R.

The converse of (71) is trivial.

72. The operator A has a basis of eigenvectors, iff the roots of its minimal polynomial are real and simple.

73. Every operator A has an invariant subspace whose dimension is ≤ 2.

Definition 13. We shall say that an *invariant subspace* of A is *minimal*, if it is $\neq 0$ and does not contain other invariant subspaces. By (73) the dimension of a minimal invariant subspace is not larger than 2. If it is = 1, then the invariant subspace is characteristic.

74. A two-dimensional invariant subspace L of A is minimal, iff the matrix of the operator $A|L$ has the form

$$\begin{bmatrix} \alpha & -\beta \\ \beta & \alpha \end{bmatrix}, \quad \beta \neq 0$$

in some basis of L. For all such bases the scalar pairs α,β are the same.

75. The operator A has a chain of invariant subspaces

$$0 = J_o \subset J_1 \subset \ldots \subset J_m = R$$

such that $0 < \dim J_{k+1} - \dim J_k \leq 2$, $k = 0,1,\ldots,m-1$.

This is the real analog of II,(58) on the "triangular representation". In the present case one does not, in general, obtain a triangular matrix; the matrix is block-triangular, but the orders of the diagonal blocks do not exceed 2. If the order of a block in the diagonal is 2, then it can be reduced to the form indicated in (74). The relation of (75) and II,(58) is similar to the relation of the [decomposition of a real polynomial into linear and quadratic factors] to the [decomposition of a complex polynomial into linear factors].

76. If the minimal polynomial of the operator A has simple roots, then A splits R into a direct sum of invariant subspaces of dimension $\leqq 2$.

The converse assertion is also true.

77. If dim R is odd, then A possesses an eigenvector.

78. In an even-dimensional R there are operators without eigenvectors. (To comply here with the artificial convention D12, do not consider "eigenvector" as a "spectral term".)

We shall now assume that the space R is unitary. Then $R^{(c)}$ becomes also unitary by the canonical complexification described in (22).

Definition 14. The operator A is called *symmetric*, if

$$(Ax,y) = (x,Ay) \qquad (x,y \in R) .$$

79. If A is symmetric, then $A^{(c)}$ is self-adjoint (III,D37).

80. The spectrum of a symmetric operator is real.

81. A symmetric operator possesses an orthonormal basis of eigenvectors.

This result and the remaining theory of symmetric operators can be obtained without complexification by operations on R that are quite analogous to those in the complex case; but in doing so, the existence of eigenvalues must first be established by a "real" method. Without leaving R, one can indeed prove that

82. The number $\lambda = \min_{(x,x)=1} (Ax,x)$ is an eigenvalue of the symmetric operator A in R.

Definition 15. The *operator* A is called *orthogonal*, if

$$(Ax,Ay) = (x,y) \qquad (x,y \in R).$$

83. If A is orthogonal, then $A^{(c)}$ is unitary, and conversely.

84. The spectrum of an orthogonal operator is unitary.

85. Any orthogonal operator decomposes R into an orthogonal sum of invariant subspaces of dimension ≤ 2.

86. The restriction of an orthogonal operator to a one-dimensional invariant subspace L multiplies all $x \in L$ either by 1 or by -1.

87. In the restriction to a two-dimensional minimal invariant subspace the matrix of an orthogonal operator becomes in any orthonormal basis

$$\begin{bmatrix} \cos\varphi & -\sin\varphi \\ \sin\varphi & \cos\varphi \end{bmatrix},$$

where φ is between 0 and π and independent of the choice of basis.

88. The determinant of an orthogonal operator equals ±1.

An orthogonal operator is thus unimodular.

Definition 16. An orthogonal *operator* with determinant 1 is called *properly orthogonal* or a *rotation*. A normalized vector $x \neq 0$ satisfying $Ax = x$ is called the *axis of rotation*.

89. In odd-dimensional space every rotation possesses an axis.

90. In even-dimensional space there are rotations without an axis.

Note here that the

91. Orthogonal operators form a group in multiplication.

92. The rotations form a subgroup of the group of all orthogonal operators.

93. This rotation group is connected (in the topology of (91)).

94. The connected components in the group of orthogonal operators are the rotation group and its complement, i.e. the set of improperly orthogonal operators.

This result is in concordance with

95. The set of orthonormal bases in R splits into two connected components (cf. here IV,(167)).

The bases of one of the two components may be termed "right-handed" arbitrarily; then those of the other component are the "left-handed" bases.

Definition 17. The discrimination made in favor of one of the two components is what we call the *orientation* of the *space* R. Thus real space permits two "opposite" orientations.

Theorem (95) can be easily extended to any bases.

Definition 18. There is a close relation between orthogonal operators and *antisymmetric operators* defined by

$$(Ax,y) = -(x,Ay) .$$

96. If A is antisymmetric, then $iA^{(c)}$ is self-adjoint, and conversely.

97. The spectrum of an antisymmetric operator is imaginary.

98. In an odd-dimensional space, every antisymmetric operator is non-regular.

99. In an even-dimensional space there are regular antisymmetric operators.

100. For any antisymmetric operator A, $\det A \geqq 0$.

101. Any antisymmetric operator A splits R into an orthogonal sum:

$$R = \text{Ker } A \boxplus \text{Im } A .$$

Im A itself decomposes into an orthogonal sum of two-dimensional minimal invariant subspaces.

102. In the restriction to a two-dimensional minimal invariant subspace the matrix of an antisymmetric operator becomes in any orthonormal basis

$$\begin{bmatrix} 0 & -\beta \\ \beta & 0 \end{bmatrix},$$

where $\beta \neq 0$ does not depend on the choice of the basis.

A connection between orthogonal and antisymmetric operators can be established by a slightly modified form of the Cayley transformation II,D46:

103. If A is an antisymmetric operator, then $U = (I+A)(I-A)^{-1}$ is a rotation, and all rotations can be so represented.

Symmetric, antisymmetric, and orthogonal operators belong to the class of <u>real normal operators</u>, which could have been introduced in complete analogy with III, Sect. 7. From the spectral point of view <u>real</u> normal operators can be characterized as follows: they are orthogonal sums of irreducible operators acting in subspaces of dimension ≤ 2.

4. Differential Maps. Smooth Norms.

Definition 19. Let R_1 and R_2 be real spaces and let $G \subset R_1$ be an open set. A *map* $f : G \to R_2$ is called *differentiable* at $x_o \in G$, if there is a homomorphism $h \in \text{Hom}(R_1, R_2)$ such that

$$\lim_{x \to x_o} \frac{\|Fx - Fx_o - h(x-x_o)\|}{\|x-x_o\|} = 0 ,$$

or, expressed differently,

$$Fx = Fx_o + h(x-x_o) + o(\|x-x_o\|) \qquad (x \to x_o)$$

104. For given F, differentiable at the given x_o, the homomorphism h is uniquely determined.

Definition 20. The homomorphism h of (104) is called the *(Fréchet) derivative* of the mapping F at the vector (or point) x_o and denoted by $F'(x_o)$. Keeping the notation used in analysis, we shall write $dx \underset{def}{=} x-x_o$, $dF \underset{def}{=} F'(x_o)dx$, and use the term *differential*. Note, however, that dF means exactly the same as $h(x-x_o)$, i.e. h determined by F and x_o, taken at the running point $x-x_o$.

105. If $h_1 \in Hom(R_1,R_2)$ and $h_1 \neq F'(x_o)$, then

$$\overline{\lim_{x \to x_o}} \frac{\|Fx - Fx_o - h_1(x-x_o)\|}{\|x-x_o\|} > 0 .$$

In this sense the derivative is the best linear approximation to the map F at the point x_o.

106. A linear map F from R_1 into R_2 (i.e., $F \in Hom(R_1,R_2)$ is differentiable at every $x \in R_1$ and

$$F'(x) = F \quad \text{for all } x \in R_1.$$

107. Let F be a map into R_2 and Φ a map from R_2. If F is differentiable at the point x, and Φ is differentiable at the point Fx, then the product ΦF is differentiable at x and

$$(\Phi F)'(x) = \Phi'(Fx)F'(x)$$

(the theorem on the derivative of a composite function).

Definition 21. The map $F: G \to R_2$ is said to be *differentiable on G* if it is differentiable at every $x \in G$.

Definition 22. A map is called *constant* if the set $\{Fx | x \in G\}$ has only one point.

108. A constant map is differentiable and its derivative is zero (viz. the zero homomorphism).

Conversely,

109. A differentiable map with zero derivative is constant. Thus a differentiable map is uniquely determined by its derivative (homomorphism) and its value at one point - exactly as in scalar analysis. Also the recovery of the map from its derivative (= integration) is carried out as in the scalar case; it actually reduces to that case. To keep the formulation simple, let us only consider maps from R_1 into R_2.

110. To reobtain the differentiable map F, integrate over the scalar t:

$$Fx = Fx_o + {}_0\int^1 F'(x_o + t(x-x_o))(x-x_o)\ dt.$$

This formula yields an interesting estimate of the distance between the points Fx_1 and Fx_2:

111. Let F be a differentiable map. Then

$$\|Fx_1 - Fx_2\| \leqq M\|x_1 - x_2\| ,$$

where $M = \sup_{x\in[x_1,x_2]} \|F'(x)\|$ and $[x_1,x_2]$ is the segment given by $[x_1,x_2] = \{x|x = x_1 + t(x_2-x_1),\ 0 \leqq t \leqq 1\}$.

Proposition (111) plays here the role of the mean-value theorem of multidimensional calculus. To describe one of its applications, we fix

Definition 23. The map F is said to be *compressive* (cf. IV,D56), if

$$\|Fx_1 - Fx_2\| \leqq \|x_1 - x_2\| \quad \text{for all } x_1,x_2 \in R_1,$$

112. A differentiable map F is compressive iff $\|F'(x)\| \leqq 1$ $(x \in R_1)$, that is, if the derivative F' (a homomorphism) is compressive at every x.

Definition 24. A map F is *strictly compressive* if

$$\|Fx_1 - Fx_2\| < \|x_1 - x_2\|;$$

it is *uniformly compressive*, if there is a number q $(0 \leqq q < 1)$ such that

$$\|Fx_1 - Fx_2\| \leqq q\|x_1 - x_2\|.$$

We have then the analog to (112):

13. A differentiable map is uniformly compressive iff $\sup_{\in R_1} \|F'(x)\| < 1$.

Uniformly compressive maps that map from R into R are particularly interesting, because the *fixed-point principle* holds for them:

114. If $F: R \to R$ is uniformly compressive, then the equation $x = Fx$ has always a unique solution.

Let us now consider derivatives of higher order. If $F: R_1 \to R_2$ is differentiable, then it generates a map from R_1 into $\mathrm{Hom}(R_1, R_2)$, since $h = F'(x)$. Here x stands for x_o of (104) (as already in (106) and (107)), and $F'(x)$ assigns to every $x \in R_1$ an element of $\mathrm{Hom}(R_1, R_2)$. The collection of homomorphisms so created belongs indeed to the range of the map $R_1 \to \mathrm{Hom}(R_1, R_2)$.

Definition 25. If the map $R_1 \to \mathrm{Hom}(R_1, R_2)$ is differentiable at x, then its derivative is called the *second* (Fréchet) *derivative* of the map F at x and denoted by $F''(x)$. Clearly,

$$F''(x) \in \mathrm{Hom}(R_1, \mathrm{Hom}(R_1, R_2)) .$$

Repeating this construction, we can introduce a derivative $F^{(m)}(x)$ of any order m. $F^{(m)}(x)$ is a homomorphism from R_1 into the space H_{m-1} of homomorphisms, of which $F^{(m-1)}(x)$ is an element. The statement "F is m times differentiable at some x" needs now no further explanation.

Definition 26. The *differential of order m* at the point x_o is defined in terms of the derivative of the same order:

$$d^m F = (\dots((F^{(m)}(x_o)dx\dots)dx \qquad (dx = x - x_o).^{\S}$$

115. If the map F has the form

$$Fx = (\dots(\underbrace{\Phi x)x\dots)x}_{m} ,$$

where $\Phi(-x) = -\Phi(x)$ and $\Phi \in H_m$, then F is m times differentiable, $F^{(m)} = m!\Phi$, and all its derivatives of order $< m$

§It is not necessary to use the same x in all the differentials dx. $d^m F$ at x_o appears then as an m-linear operator from R_1 to R_2, symmetric in $x_1, \dots, x_n$.

vanish at $x = 0$.

116. If the map F is m times differentiable at x_o and each $F^{(j)}x_o$ is the zero homomorphism of the appropriate space for $j = 1,\dots,m$, then $F(x) - F(x_o) = o(\|x-x_o\|^m)$, as $x \to x_o$.

117. For any map F. which is m times differentiable at x_o, Taylor's theorem with Peano's form of the remainder holds:

$$Fx = Fx_o + dF + \tfrac{1}{2}d^2F + \dots + \frac{1}{m!}d^mF + o(\|x-x_o\|^m).^{\S}$$

This theory applies in particular to (nonlinear) functionals; in that case the derivative at x_o becomes a linear functional.

Definition 27. We shall call the derivative of a functional φ the *gradient* of φ and denote it by $\nabla\varphi$.§§ For φ differentiable at x_o we have therefore

$$\varphi(x) = \varphi(x_o) + (\nabla\varphi|_{x_o})(x - x_o) + o(\|x - x_o\|) .$$

The second derivative of a (not necessarily) linear functional assigns to every point a homomorphism from R into R', sometimes called the *Hessian* (of the functional).

118. Let Δ be a basis in R and Δ' be the conjugate basis in R' (I,D62). The matrix of the Hessian relative to the pair Δ,Δ' (I,D53) is symmetric (a *Theorem of H. A. Schwarz*).

If R is Euclidean, then the Hessian may be considered as an operator in E. According to Schwarz' theorem that operator is self-adjoint.

Consider now the differentiation of a norm on the space R (no norm is differentiable at $x = 0$).

Definition 28. A norm differentiable at every $x \neq 0$ is called a *smooth norm* for brevity.

---

§The various $d^{(j)}F$ operate in different spaces. See the discussion in Nr. 8 of the List of General References.

§§Confusion with the commutation symbol II,D5 is hardly possible.

119. All ℓ^p-norms with $1 < p < \infty$ are smooth, but

120. Neither the c- nor the ℓ-norm is smooth, if $n > 1$.

121. The gradient of any smooth norm has everywhere a norm equal to 1.

For instance,

122. The gradient of the Euclidean norm equals $x/\|x\|$ $(x \neq 0)$ (if linear functionals are canonically identified with vectors).

Even norms that are not smooth are still differentiable in a more general sense. To see this, let us introduce

Definition 29. The *derivative at x in the direction of the vector* e (the directional derivative of scalar calculus) is the vector

$$F'_e(x) = \lim_{t \to 0+} \frac{F(x + te) - F(x)}{t}$$

If it exists, we speak of differentiability at x in direction e (*Gâteaux derivative*).

123. If the map F is Fréchet-differentiable at x, then it is differentiable in every direction, and $F'_e(x) = F'(x)e$.

124. Every norm on R is differentiable in every direction at all $x \neq 0$.

The derivative of a norm in direction e at a given point $x \neq 0$, considered as a functional of the argument e, will be designated as gradient as before.

125. The gradient $\gamma(e)$ of a norm at any point $x \neq 0$ satisfies

1. $\gamma(\alpha e) = \alpha\gamma(e) \quad (\alpha > 0)$;
2. $\gamma(e_1 + e_2) \leq \gamma(e_1) + \gamma(e_2)$;
3. $\sup_{\|e\|=1} |\gamma(e)| = 1$.

126. A norm is Fréchet-differentiable at $x \neq 0$, iff its gradient $\gamma(e)$ at that point is antisymmetric, i.e. $\gamma(-e) = -\gamma(e)$ for every e.

The differentiability of a norm has a simple geometrical sense:

127. A norm is differentiable at $x \neq 0$, iff all supporting functionals at x (IV,D34) are proportional.

In other words, a norm is differentiable at x, iff the linear functional f_x, which satisfies both conditions $f_x(x) = 1$ and $\|f_x\|.\|x\| = 1$, is unique ($\|.\|$ is of course the norm under consideration and the corresponding induced norm). For a smooth norm, the map $y = f_x$ is therefore well-defined for all $x \in R-\{0\}$; it is known as the *inversion* mapping, because

128. The inversion has the usual form $f_x = x/\|x\|^2$ for a Euclidean norm.

129. Given a smooth norm on R and its gradient γ_x at x, the associated inversion is $\gamma_x/\|x\|$.

Hence it appears that one-time differentiability of the inversion is equivalent to two-times differentiability of the norm.

130. The inversion is a continuous map from $R-\{0\}$ onto $R'-\{0\}$. It maps the unit sphere of R onto the unit sphere of R'.

131. If the norm on R' is also smooth, then the iterated inversion returns every point to its initial position.

Thus the inversion is an example for the duality between normed spaces.

Concluding this section, let us calculate the gradients of two important functionals defined on the Euclidean space R.

Let S be a symmetric operator in R (D14), and let $y \in R$ be fixed.

132. The gradient of the inhomogeneous quadratic functional

$$\varphi(x) = \tfrac{1}{2}(Sx,x) - (y,x)$$

is given by $\nabla\varphi(x) = Sx - y$.

133. The gradient of the *Rayleigh functional*

$$\varphi(x) = (Sx,x)/(x,x)$$

is given by

$$\nabla\varphi(x) = (Sx - \varphi(x)x).2\|x\|^{-2} ,$$

i.e., $\nabla\varphi(x) = PSx$, where P is the orthoprojector upon the subspace orthogonal to x.

5. Differentiation of Functions of an Operator.

Let $\mathbf{S}(E)$ be the real space of all self-adjoint operators acting in a complex or real unitary space E (cf. III, (131)), and let $\varphi(x)$ be a <u>real</u> scalar function defined on some open set G of the real axis. Put

$$\mathbf{G} = \{S \mid S \in \mathbf{S}(E), \quad \sigma(S) \subset G\}.$$

The function $\varphi(\lambda)$ induces a mapping φ: $\mathbf{G} \to \mathbf{S}(E)$ by assigning to every $S \in \mathbf{G}$ the self-adjoint operator $T = \varphi(S)$ (see remark preceding III,(148), and (149)).

In this section we shall study the specialized application of our differential calculus to that induced mapping. Let both functions $\varphi(S)$ and $\psi(S)$, maps from $\mathbf{S}$ into $\mathbf{G}$, be differentiable at the "point" S. Then, clearly,

1. $d[\varphi(S) + \psi(S)] = d\varphi(S) + d\psi(S)$,
2. $d[\alpha\varphi(S)] = \alpha d\varphi(S)$.

Further,

3. $d[\varphi(S)\psi(S)] = \varphi(S)d\psi(S) + d\varphi(S)\psi(S)$

and finally

134. $$d[\varphi(S)]^{-1} = -[\varphi(S)]^{-1}d\varphi(S)[\varphi(S)]^{-1} .$$

These rules of differentiation allow us to work out the simplest examples for differentiable functions of an operator.

135. $\varphi(S) = S^m$ (m a natural number) is differentiable and

$$dS^m = S^{m-1}dS + S^{m-2}dS\cdot S + \cdots + dS\cdot S^{m-1}$$

136. If $\varphi(\lambda)$ is a polynomial, then $\varphi(S)$ is differentiable.

137. If $\varphi(\lambda)$ is a rational function, then $\varphi(S)$ is differentiable.

Starting from (136) we can characterize the class of

functions $\varphi(\lambda)$ that generate differentiable functions $\varphi(S)$. Note first of all that (135) implies:

138. For any polynomial $\varphi(\lambda)$ we have the formula[§]

$$d\varphi(S) = \Sigma_{j,k=1}^{n} [(\varphi(\lambda_j)-\varphi(\lambda_k))/(\lambda_j-\lambda_k)]P_j \cdot dS \cdot P_k ;$$

here P_j is the orthoprojector upon the one-dimensional characteristic subspace of S, generated by the eigenvectors $e_j \in \{e_k\}_1^n$, the orthonormal characteristic basis of S.

Further:

139. If $\varphi(\lambda)$ has a continuous derivative in some closed neighborhood of the spectrum $\sigma(S)$, then

$$\|\Sigma_{j,k=1}^{n}[(\varphi(\lambda_j)-\varphi(\lambda_k))/(\lambda_j-\lambda_k)]P_j \cdot dS \cdot P_k\| \leq \max_{\alpha\leq\lambda\leq\beta} |\tilde{\varphi}'(\lambda)| \cdot \|dS\|;$$

here $\|.\|$ on the left side is the operator norm (IV,D59), $\|.\|$ on the right side is the Hilbert norm (IV,D60), and $\tilde{\varphi}(\lambda)$ is some extension of $\varphi(\lambda)$ to an interval $(\alpha,\beta) \supset \sigma(S)$ under preservation of its continuous differentiability.

Finally,

140. If $\varphi(\lambda)$ has a continuous derivative in a neighborhood of $\sigma(S)$, then the function $\varphi(S)$ is differentiable at S, and (138) holds for $d\varphi(S)$.

141. Let S(t), an operator-function of the real parameter t with range in $\mathbf{S}(E)$, have a continuous derivative in a neighborhood of t_o, and let $\varphi(\lambda)$ have a continuous derivative in a neighborhood of the spectrum of the operator $S(t_o)$. Then $d\varphi(S(t))/dt|_{t=t_o}$ exists and is given by

$$\Sigma_{j,k=1}^{n}[(\varphi(\lambda_j)-\varphi(\lambda_k))/(\lambda_j-\lambda_k)]P_j S'(t_o) P_k .$$

It is not difficult to extend (140) and (141) to derivatives and differentials of higher order. In the following

[§]Here and later ratios of differences must be replaced by the corresponding derivatives if the "nodes" (here the eigenvalues) coincide.

Definition 30, we introduce by recursive definition the shorthand notations

$$\varphi^{[o]}(\lambda) = \varphi(\lambda)$$

$$\varphi^{[1]}(\lambda,\mu) = (\varphi^{[o]}(\lambda) - \varphi^{[o]}(\mu))/(\lambda-\mu)$$

$$\varphi^{[2]}(\lambda,\mu,\nu) = (\varphi^{[1]}(\lambda,\mu)-\varphi^{[1]}(\lambda,\nu))/(\mu-\nu)$$

$$\ldots = \ldots$$

in order to facilitate the formulation of the higher-order derivatives and differentials.

142. If the operator-function $S(t)$ has a continuous second derivative in a neighborhood of t_o, and if $\varphi(\lambda)$ has a continuous second derivative in a neighborhood of the spectrum of $S(t_o)$, then $d^2\varphi(S(t))/dt^2\,|_{t=t_o}$ exists and is given by

$$2\Sigma\varphi^{[2]}(\lambda_j,\lambda_k,\lambda_r)P_jS'(t_o)P_kS'(t_o)P_r + \Sigma\varphi^{[1]}(\lambda_j,\lambda_k)P_jS''(t_o)P_k \;,$$

where the arguments $\lambda_j,\lambda_k,\lambda_r$ (resp. λ_j,λ_k) run independently over $\sigma(S)$ in the first (resp. second) sum. (An analogous summation is meant in some of the following formulas.)

Similarly:

143. If $\varphi(\lambda)$ has a continuous second derivative in a neighborhood of $\sigma(S)$, then $\varphi(S)$ is twice differentiable at the point S and

$$d^2\varphi(S) = 2\Sigma\varphi^{[2]}(\lambda_j,\lambda_k,\lambda_r)P_j\cdot dS\cdot P_k\cdot dS\cdot P_r$$

144. For $d^m\varphi(S)$ we have under analogous conditions

$$d^m\varphi(S) = m!\Sigma\varphi^{[m]}(\lambda_{j_1},\ldots,\lambda_{j_{m+1}})P_{j_1}\cdot dS\cdot P_{j_2}\cdot dS\cdots P_{j_{m+1}} \;.$$

Thus the Taylor formula

$$(S + \Delta S) = \varphi(S) + d\varphi(S) + \ldots + 1/m!\; d^m\varphi(S) + o(\|\Delta S\|^m)$$

holds, provided $\varphi(\lambda)$ has a continuous m-*th* derivative in a neighborhood of $\sigma(S)$.

145. The remainder term in Taylor's formula vanishes iff $\varphi(\lambda)$ coincides with some polynomial of degree $\leq m$ in a neigh-

orhood of $\sigma(S)$.

.46. If ΔS commutes with S, then

$$(S + \Delta S) = \varphi(S) + \varphi'(S)\Delta S + \ldots (\varphi^{(m)}(S)/m!)\Delta S^m + o(\|\Delta S\|^m).$$

If we want the Taylor formula in a similarly detailed shape also in the general case, we set

$$\varphi^{[k]}(S) \underset{\text{def}}{=} \Sigma\varphi^{[k]}(\lambda_{j_1},\ldots,\lambda_{j_{k+1}})P_{j_1}\otimes\ldots\otimes P_{j_{k+1}}$$

and introduce

Definition 31. The *Schur multiplication*, here denoted by $\circ$, of an element of $\overset{k+1}{\otimes}\mathbf{S}(E)$ and an element of $\overset{k}{\otimes}\mathbf{S}(E)$, i.e.

$$(T_1\otimes\ldots\otimes T_{k+1})\circ(S_1\otimes\ldots\otimes S_k),$$

is the successive product of the operators $T_1, S_1, T_2, S_2, \ldots, S_k, T_{k+1}$ in this order.

The general form of (146) may now be written as

47. $$(S + \Delta S) =$$
$$\varphi(S) + \varphi^{[1]}(S)\circ\Delta S + \frac{1}{2!}\varphi^{[2]}(S)\circ\Delta S^{(2)} + \ldots$$
$$+ \frac{1}{m!}\varphi^{[m]}(S)\circ\Delta S^{(m)} + o(\|\Delta S\|^m)$$

here $\Delta S^{(k)}$ is the k-*th* derivative of the "incremental" perator ΔS.

In this context the *Schur* or *Hadamard product* of two matrices (here indicated by the same sign $\circ$) should be distinguished from D31. If we write the matrix of the first-order differential $d\varphi(s)$ in the characteristic basis Δ of the operator S, we obtain

$$(d\varphi(S))_\Delta = [(\varphi(\lambda_j)-\varphi(\lambda_k))/(\lambda_j-\lambda_k)]_{j,k=1}^{n}\ (dS)_\Delta .$$

Let us now consider the application of the Taylor formula to the perturbation theory of self-adjoint operators. (Other applications will occur in VIII, Sect. 9.)

Let S and $T \in \mathbf{S}(E)$, and $S(\varepsilon) \underset{\text{def}}{=} S + \varepsilon T$; choose $\varphi(\lambda) = 1$

in a neighborhood of the point $\lambda_1, \varphi(\lambda) = 0$ in a neighborhood of $\sigma(S)-\{\lambda_1\}$. On the remaining neighborhood of $\sigma(S)$ choose $\varphi(\lambda)$ as some smooth function. For sufficiently small $|\varepsilon|$, the operator $P(\varepsilon) \underset{\text{def}}{=} \varphi(S(\varepsilon))$ will be an orthoprojector upon the linear hull of those eigenvectors of $S(\varepsilon)$ which belong to eigenvalues of $S(\varepsilon)$, included in the chosen neighborhood of the point λ_1.

The key to the perturbation theory is the expansion of $P(\varepsilon)$ in a Taylor series.

148. If λ_1 is a simple eigenvalue of S, then

$$P(\varepsilon) = P_1 + \varepsilon\Sigma_{k=2}^{n}(\lambda_1-\lambda_k)^{-1}[P_1TP_k + P_kTP_1] + \ldots$$

Here the notation P_k is the same as in (138), and the expansion may be carried to terms of arbitrarily high order.

Under the same conditions, we have:

149. The eigenvalue $\lambda_1(\varepsilon)$ is for sufficiently small $|\varepsilon|$ again a simple eigenvalue of $S(\varepsilon)$, and the corresponding eigenvector is given by

$$e_1(\varepsilon) = e_1 + \varepsilon\Sigma_{k=2}^{n}(\lambda_1-\lambda_k)^{-1}(Te_1,e_k) + \ldots$$

150. $$\lambda_1(\varepsilon) = \lambda_1 + \varepsilon(Te_1,e_1) + \ldots$$

151. For any function $\varphi(\lambda)$ which has a continuous m-*th* derivative,

$$\text{tr}\,\varphi(S+\varepsilon T) = \text{tr}\,\varphi(S) + \varepsilon\Sigma_{k=1}^{n}\varphi'(\lambda_k)(Te_k,e_k) + \\ + \varepsilon^2\Sigma_{j,k=1}^{n}[(\varphi'(\lambda_j)-\varphi'(\lambda_k))/(\lambda_j-\lambda_k)]|(Te_j,e_k)|^2 + \ldots \\ + o(\varepsilon^m)\ .$$

6. Steinitz' Theorem for Vector Series.

Steinitz' Theorem has already been formulated in I, after (327). It is no restriction of generality if we assume E as a real Euclidean space, and we may even presuppose that the sum of the conditionally convergent series under consideration is zero. Then Steinitz' theorem may be formulated as follows:

52. If $\Sigma_{k=1}^{\infty} x_k = 0$, but $\Sigma_{k=1}^{\infty}\|x_k\| = \infty$, then the set F of sums f convergent series, obtained from the given series by re-rdering of terms, is the orthogonal complement of its domain f unconditional convergence N.

In accordance with I,D100 and (322),(323), and since E was assumed as a <u>real</u> Euclidean space (implying $E' \approx E$ by III,D19), N may be described as the largest of the sub-spaces $L \subset E$, for which $\Sigma_{k=1}^{\infty}\|Px_k\|$ converges, where P is the orthoprojector upon L.

Note that (152) becomes the classical Riemann reordering theorem for dim E = 1 (i.e., either N = 0 or N = E; hence $N^{\perp}$ is either E ($\Rightarrow \Sigma_{k=1}^{\infty} x_{j_k}$ may be any number), or $N^{\perp} = 0$ ($\Rightarrow \Sigma_{k=1}^{\infty} x_{j_k} = 0$). For n = 2, (152) becomes the extension of Riemann's theorem to series of complex numbers (<u>three</u> possibilities!).

We now sketch one of the methods of proof of Steinitz' theorem; it is based on the following lemma:

53. Any vector system $\Gamma = \{x_k\}_1^m$ which satisfies

$$\Sigma_{k=1}^{m} x_k = 0, \quad \|x_k\| \leq 1 \quad \text{for all } k,$$

n be so reordered that for any r, $1 \leq r \leq m$,

$$\|\Sigma_{k=1}^{r} x_{j_k}\| \leq \ell.$$

re ℓ is independent of the chosen Γ.

Lemma (153) means geometrically: if the sides of a closed polygonal line are ≤ 1, then they can be reordered in such a way that the diameter of the (new) polygonal line so obtained is $\leq 2\ell$ (proof by induction with respect to dim $\mathrm{sp}\{x_k\}_1^m$).

54. If the vector s is the limit of some subsequence of the rtial sums of the series in (152), then $s \in F$.

155. If s_1 and $s_2 \in F$, then $(1 - \alpha)s_1 + \alpha s_2 \in F$ $(-\infty < \alpha < \infty)$.
F is therefore a subspace of E.

156. $\dim F \neq 0$.

157. $N = 0 \Rightarrow F = E$.

Now it is no longer difficult to obtain the general theorem (152) from (157).

antisymmetric real operator D18
antisymmetric real operator and rotation (103)
axis of rotation D16

block-triangular matrix (75)

complex dimension of $R^{(c)}$ (4)
complex dimension, preceding (1)
complex extension of norm D5
complex hull D2
complex space as tensor product D1
complexifiable homomorphism D10
complexifiable subspace D7
complexification of bilinear functionals (14)
complexification of homomorphism D4
complexification of non-Euclidean norm (24) ff.
complexification of real space D2
complexification of scalar product (22)
complexification of symmetric bilinear functional (17)
complexifying subspace D8
compressive map D23
constant map D22

decomplexification D6
decomplexification of functional D11
decomplexification of homomorphism D9
derivative in direction of vector D29
derivative of composite map (107)
differentiable map D19
differentiable norm (126)
differential of map D20
differential of order m D26
differentiation of function of operator, preceding (134)
differentiation of operator-function (141)
differentiation of operator polynomial (138)

fixed-point principle (114)
Fréchet derivative D20

Gâteaux derivative D29
gradient of map D27
gradient of norm (125)

Hermitian complexification (13)
Hessian of functional D27

inversion mapping (127)

"left-handed" basis (95)

mean-value theorem for map (111)
minimal invariant subspace D13

operator in real space, preceding D12
orientation of real space D17
orthogonal operator D15

perturbation of self-adjoint operator (148) ff.
properly orthogonal D16

Rayleigh functional (133)
real dimension of $R^{(c)}$ (4)
real dimension, preceding (1)
real normal operators (103)
Riemann's theorem, following (152)
"right-handed" basis (95)
rotation D16

Schur multiplication D31
Sçhur product of matrices, following (147)
second Fréchet derivative D25
smooth norm D28
Steinitz' Theorem (152)
strictly compressive map D24

symmetric operator in real space D14

Taylor's Theorem for differentiable map (117)

uniformly compressive map D24

unimodularity of orthogonal operator (88)

CHAPTER VII

CONVEX SETS IN A REAL SPACE

1. Wedges and Cones.

Throughout this chapter unless stated otherwise E will denote a real linear space. We shall have to consider (finite) linear combinations of vectors $\in$ E with real coefficients together with nonnegative and, in particular, positive linear combinations.

Definition 1. A *nonnegative (positive) linear combination* is a linear combination with nonnegative (positive) coefficients.

Definition 2. The *nonnegative linear hull* of a finite or infinite set of vectors $F \subset E$ is the set of all (finite) nonnegative linear combinations of vectors of F. The *positive linear hull* of a finite set of vectors $F \subset E$ is the set of all positive linear combinations of these vectors and zero.

Definition 3. The vectors $\{x_k\}_1^m$ are *positively linearly independent* if no nontrivial nonnegative linear combination of them is zero.

Definition 4. The set $K \subset E$ is a *wedge* if

1. $x_1, x_2 \in K$ implies $x_1+x_2 \in K$,
2. $x \in K$, $\alpha \geqq 0$ implies that $\alpha x \in K$.

A wedge K is a *cone*[§] if $x \in K$ ($x \neq 0$) implies $-x \notin K$[§§]. Obviously, any subspace (in particular {0} and E) is a wedge, but it is not a cone unless it is zero.

Definition 5. The *dimension* (or upper dimension) dim K of the wedge K is the dimension of the smallest subspace containing the wedge. A wedge of maximum dimension dim K = n is called *solid*.

A simple example of a wedge which is not a subspace is $K = \{\alpha x \mid \alpha \geqq 0\}$ where $x \in E$ ($x \neq 0$) is fixed. This wedge has dimension one and is called a halfline (or ray). A halfline is a cone. The dual to this example is the set $K = \{x \mid f(x) \geqq 0\}$ where the fixed functional $f \in E'$ ($f \neq 0$). This wedge has dimension n and is called a closed halfspace.

The set F of all linear functionals f supporting the unit sphere of the normed space E at the point x_o (IV,D34) is also a wedge (in E') provided $f(x_o) > 0$ for all $f \in F$.

Definition 6. The *vector sum* (or simply *sum*) of a finite set of subsets $\{F_k\}_1^m$ of the space E is the set of all sums of the form $\Sigma_{k=1}^m x_k$ ($x_k \in F_k$).

If the collection of subsets $F_\nu \in E$ is infinite, then the definition of the sum is purposeful only under the condition that all subsets F_ν (except, possibly, finitely many) contain the zero vector. In this case the sum is defined as the set of all finite sums of the form

Σx_{ν_k} ($x_{\nu_k} \in F_{\nu_k}$, ν_k pairwise distinct); the sum will

then contain every F_ν as a subset.

1. The sum $K = \Sigma K_\nu$ of a set of wedges $\{K_\nu\}$ is a wedge. This wedge is the smallest wedge containing all K_ν.

[§] The terms "convex wedge" and "convex cone" would be more appropriate, but they are rarely used.

[§§] Wedge and cone are defined in exactly the same way in a complex space.

Dually,

. The intersection $K = \cap K_\nu$ of a set of wedges $\{K_\nu\}$ is a edge. This wedge is the largest wedge contained in all K_ν.

. The sum of a finite or infinite set of halflines is a edge.

Dually,

. The intersection of a finite or infinite set of closed alfspaces is a wedge.

. The positive linear hull of any finite set of vectors $\subset E$ is a wedge.

Dually,

. Let F' be a finite set of functionals from E'. Then the et

$$\{x|f(x) > 0, f \in F'\} \cup \{0\}$$

s a wedge (in E).

If the set F' in (6) consists of a single functional $f \neq 0$, then the corresponding wedge is called an <u>open halfspace</u>§.

. Let $\{x_k\}_1^m$ be a linearly independent system of vectors rom E ($m \geq 2$). The set

$$K = \left\{x|x = \Sigma_{k=1}^{m} \xi_k x_k,\ \xi_m^2 - \Sigma_{k=1}^{m-1} \xi_k^2 \geq 0,\ \xi_m \geq 0\right\}$$

s a wedge. The set

$$\left\{x|x = \Sigma_{k=1}^{m} \xi_k x_k,\ \xi_m^2 - \Sigma_{k=1}^{m-1} \xi_k^2 > 0,\ \xi_m \geq 0\right\}$$

ogether with zero is also a wedge. For $m = n$ both wedges re solid cones (*Minkowski cones*).

. If the vectors which determine the halflines in (3) are inearly independent, then the corresponding wedge is a cone. f the vectors of the set F in (5) are linearly independent,

---

§ This is not an open set, but it becomes so after removal of he zero vector.

then the corresponding wedge is a cone.

9. The intersection of any set of cones is a cone.

10. Any wedge K contains a largest subspace L.

Definition 7. The largest linear subspace L contained in the wedge K is called the *linear part* of K, and the number dim L may be called the *lower dimension* of K; it is not denoted by dim K.

11. If K is a wedge and L is a subspace, then the set $\{[x]_L | x \in K\}$ is a wedge in the factor space E/L (notation in I,D36).

12. If L is the linear part of the wedge K, then the set $\{[x]_L | x \in K\}$ is a cone in E/L.

From (12) it follows that

13. Each wedge is representable as the sum of its linear part and some cone.

14. If a wedge is represented as the sum of a subspace and a cone, then the subspace is its linear part.

Thus a wedge is a cone iff its linear part is zero.

Definition 8. A wedge which is representable as the sum of a finite number of halflines is called a *finite wedge* or a *polyhedral wedge*. The zero subspace is by definition a finite wedge.

An example of a finite wedge is the *positive hyperoctant*, which is defined relative to some basis $\{e_k\}_1^n$ by the system of inequalities $\xi_k \geqq 0$ $(k = 1,2,\ldots,n)$. The positive hyperoctant is clearly a cone.

Definition 9. Any set of vectors whose nonnegative linear hull coincides with a finite wedge K is called a *frame* of this wedge. A frame with a minimal number of vectors is called a *basis* of the finite wedge. The number of vectors in a basis of the wedge K is called the *order* (or *affine rank*) of the wedge K and is denoted by ord K. Obviously ord K $\geqq$ dim K.

15. Any subspace L is a finite wedge of order dim L + 1.

16. Any closed halfspace is a finite wedge of order n + 1.

Note that the wedge defined in (5) is finite only for rk F $\leq$ 1, and the wedge defined in (6) is finite only when its dimension is $\leq$ 1. In (7) the first cone is finite only for m = 2, and the second is not finite.

Definition 10. The vector $x_o \in K$ ($x_o \neq 0$) is called an *extremal vector* of the wedge K if it cannot be represented in the form $x_o = x_1 + x_2$, where x_1 and x_2 are linearly independent vectors of K. Together with x_o any vector αx_o ($\alpha > 0$) is extremal.

17. A finite cone K is the nonnegative linear hull of the set of its extremal vectors. They form a positively linearly independent set, and the order of K equals the number of its extremal vectors.

18. If L is the linear part of a finite wedge K, then

$$\text{ord } K = \text{ord } L + \text{ord } K/L .$$

19. The sum of a finite number of finite wedges is a finite wedge.

20. The intersection of a finite number of finite wedges is a finite wedge.

In particular,

21. The intersection of a finite number of closed half-spaces is a finite wedge.

The converse of this theorem will be considered in Section 3.

Let L be the smallest subspace containing the wedge K. In the following, terms such as interior or boundary point of the wedge will be used in the sense of the topology of L. Note that a finite wedge is always closed.

22. A wedge K is closed iff its intersection with any two-dimensional subspace is a finite wedge.

The set of interior points of a wedge K is denoted by Int K.

23. If the wedge K is closed, then the equality K = Int K holds iff K is a subspace.

24. The intersection of any family of closed wedges is a closed wedge.

25. The sum of a finite number of closed wedges is a closed wedge.

26. The closure $\overline{K}$ of any wedge is a wedge. For example, in (7) the closure of the second wedge is the first.

But it is easy to see that the closure of a cone is not always a cone.

27. If $h \in \mathrm{Hom}(E, E_1)$ and K is a wedge in E, then the set hK is a wedge in E_1.

28. If the wedge K is finite, then hK is also a finite wedge.

29. $\dim hK \leq \dim K$.

30. If the wedge K is finite, then $\mathrm{ord}\, hK \leq \mathrm{ord}\, K$.

31. Let K be a cone. Then $\mathrm{Ker}\, h = 0$ is sufficient for hK to be a cone.

32. Let K be a wedge and L be its linear part. Then $L \subset \mathrm{Ker}\, h$ is necessary for hK to be a cone.

33. If K is a wedge in E, then the set

$$K' = \{f \mid f(x) \geq 0 \quad \text{for all } x \in K\}$$

is a wedge in E'.

Definition 11. The wedge K' in (33) is called the *conjugate* (wedge) of K.

34. The conjugate wedge is always closed: $(K') = K'$.

35. $(\overline{K})' = K'$.

36. If the wedge K is a subspace L, then $K' = L^{\perp}$.

37. If the wedge K is finite, then the wedge K' is also finite.

38. For any wedge K

$$K \subset K'' \quad \text{[c.i.]} \quad \text{(cf. I,(194))}.$$

39. $(K_1 + K_2)' = K_1' \cap K_2'$.
40. $(K_1 \cap K_2)' \supset K_1' + K_2'$.

(39)–(40): (cf. D6 for notation)

If in (38) and (40) the wedges K, K_1, K_2 are subspaces, then the inclusions can be replaced by equalities (cf. I,

(182),(184)). But this is in general not possible. Counterexamples are easy to construct by using non-closed wedges. We shall see in Section 3 that equality holds in (38) and (40) for closed wedges. Then the duality of (39) and (40) and the self-duality of (38) become manifest.

41. The mapping (') is monotone decreasing: if $K_1 \subset K_2$, then $K_1' \supset K_2'$ (cf. I,(183)).

42. If K' is solid, then K is a cone. Conversely, if K is a cone and K" = K [c.i.], then K' is solid.

We shall see later (cf.(88)) that K" = K iff K is closed. Thus, if K is a closed cone, then K' is solid.

In general,

43. If m is the upper dimension of the wedge K and m' is the lower dimension of the conjugate wedge K', then m+m' = n (cf. I(181), complementation formula).

44. If the cone K is given in some basis Δ by the inequality

$$\xi_n \geqq (\Sigma_{k=1}^{n-1} |\xi_k|^p)^{1/p} \qquad (1 \leqq p \leqq \infty) ,$$

then the cone K' in the conjugate basis Δ' is defined by the inequality

$$\eta_n = (\Sigma_{k=1}^{n-1} |\eta_k|^q)^{1/q}$$

where 1/p + 1/q = 1 (cf. IV,(104)).

In closing this section let us have a quick look at the case when E is a <u>Euclidean space</u>; we may then assume $K' \subset E$.

45. K + (-K') = K.

46. If $K \cap K' = \{0\}$, then K is a subspace.

Definition 12. A wedge K in a Euclidean space is *acute* if $(x,y) \geqq 0$ for any $x,y \in K$, and *obtuse* if for any $y \notin K$ there exists an $x \in K$ such that $(x,y) < 0$. A wedge is *straight* if it is both acute and obtuse.

47. An acute wedge is a cone.

48. An obtuse wedge is solid.

49. A wedge K is acute iff $K \subset K'$.

50. A wedge K is obtuse iff $K \supset K'$.

Definition 13. A wedge is *self-conjugate* if $K = K'$. Obviously a wedge is straight iff it is self-conjugate. A simple example of a self-conjugate wedge is the positive hyperoctant relative to an orthonormal basis. It is called the positive orthant.

51. The wedge K is obtuse only if the wedge K' is acute. If K' is acute and closed, then K is obtuse.

Let us define the cone K_p $(1 \leqq p \leqq \infty)$ in an orthonormal basis by the inequality $\xi_n \geqq (\Sigma_{k=1}^{n-1} |\xi_k|^p)^{1/p}$. Then

52. $K_p' = K_q$ $(1/p + 1/q = 1)$.

53. The cone K_p is acute for $1 \leqq p \leqq 2$ and obtuse for $2 \leqq p \leqq \infty$.

Thus the cone K_2 is straight. It is called the *circular Minkowski cone.*

2. Convex Sets.

Wedges and absolutely convex sets in a real space[§] are particular cases of convex sets defined as follows.

Definition 14. A set $W \subset E$ is *convex* if $x_1, x_2 \in W$ implies

$$(1 - \theta)x_1 + \theta x_2 \in W \quad (0 < \theta < 1).^{\S\S}$$

Definition 15. An *affine manifold* G in E is a sum

$$G = x_o + L ,$$

where x_o is a vector and L is a subspace. The dimension

[§]Absolutely convex sets are defined in real and complex spaces in the same way. (Admitting $\theta=0,1$ in D14 would not change anything.)

[§§]This definition remains the same for convex sets in a complex space.

of the affine manifold G is the number $\dim G \underset{\text{def}}{=} \dim L$. When $\dim G = n-1$ the affine manifold is called a <u>hyperplane</u>.

Each hyperplane can be given by an equation $f(x) = \alpha$ ($f \in E'$, $f \neq 0$) and, conversely, each such equation defines some hyperplane. The hyperplane $f(x) = \alpha$ divides the space E into two open halfspaces[§] $f(x) > \alpha$ and $f(x) < \alpha$. An affine manifold is obviously convex.

Definition 16. The *linear dimension* (or briefly dimension) $\dim F$ of some set $F \subset E$ (and in particular of a convex set) is the dimension of the smallest affine manifold containing F.

Definition 17. A convex set of dimension n is a *convex body*.

54. A convex set W is absolutely convex iff it is symmetric about the origin, i.e., together with x it contains $-x$ (cf. IV,D25).

55. If the set W is convex, then the set αW is also convex.

56. A convex set W is a wedge iff $\alpha W \subset W$ ($\alpha \geqq 0$).

57. If W is a convex set in E and $h \in \mathrm{Hom}(E,E_1)$, then the set hW is convex in E_1.

58. If W is a convex set, then the set $\bigcup_{\alpha \geqq 0} \alpha W$ is a wedge, and it is a cone iff $0 \notin W$.

59. If W is a convex set, then for every $x_0 \in E$ the set $x_0 + W$ is also convex.

From (58) and (59) follows

60. If $\tilde{E} = E \times E_1$ with $\dim E_1 = 1$ and if W is a convex set in E, then the set $K = \bigcup_{\alpha \geqq 0} \alpha(y_0 + W)$, where $y_0 \in E$, is a wedge in $\tilde{E}$ and a cone if $y_0 \neq 0$.

The <u>extension</u> of the convex set $W \subset E$ to the wedge K described in (60) permits the reduction of some theorems

---

[§]The meaning of the term "open halfspace" here is different from that mentioned in the preceding section.

on convex sets to corresponding theorems on wedges. The contraction of a wedge K to the initial convex set W can be used for the converse reduction.

61. The intersection $W = \cap W_\nu$ of an arbitrary collection $\{W_\nu\}$ of convex sets is a convex set. Thus W is the largest convex set contained in all members W_ν of the collection.

62. The sum $W = \Sigma W_\nu$ of an arbitrary collection W_ν of convex sets is a convex set.

Note that unlike (1) the sum W in general does not contain the sets W_ν.

Definition 18. A *convex combination* of the vectors $\{x_k\}_1^m$ is a sum of the form

$$\Sigma_{k=1}^m \alpha_k x_k \quad (\alpha_k \geqq 0, \ \Sigma_{k=1}^m \alpha_k = 1).$$

Definition 19. The *convex hull* Co F of a set $F \subset E$ is the set of all convex combinations of vectors of F.

63. The convex hull Co F is the smallest convex set containing F.

In particular,

64. The convex hull Co $(\cup F_\nu)$ of the union of arbitrary sets $F_\nu \subset E$ is the smallest convex set containing all F_ν.

65. $\text{Co}\,(F_1 + F_2) = \text{Co}\,F_1 + \text{Co}\,F_2$,

$\text{Co}\,(\alpha F) = \alpha \text{Co}\,F$.

66. $\text{Co}\,(F_1 \cup F_2) = \text{Co}\,(\text{Co}\,F_1 \cup \text{Co}\,F_2)$,

$\text{Co}\,(F_1 \cap F_2) \subset \text{Co}\,F_1 \cap \text{Co}\,F_2$.

Let G be the smallest affine manifold containing the convex set $W \subset E$. We shall understand all topological terms referring to W in the sense of the topology naturally induced in G by the topology of the Space E. In this sense we shall speak of the set of interior points of W, denoted by Int W, relative to G. If G = E, we shall omit "relative to E" for brevity. The vector $x \in$ Int W, if every segment $[x, x+\varepsilon y] \in W$ for some ε and arbitrary $y \in E$.

Definition 20. The *closed convex hull* of the set $F \subset E$

is the closure of the set Co F.

67. The intersection $\cap W_\nu$ of any collection of closed convex sets is a closed convex set.

68. The sum ΣW_k of a finite number of closed convex sets is a closed convex set.

69. The convex hull Co($\cup F_k$) of the union of a finite collection $\{F_k\}$ of compact sets is compact.

70. If dim F = m, then each point $x \in$ Co F is a convex combination of at most m+1 points of F, depending in general upon x (*Carathéodory's Theorem*).

Under some additional restrictions theorem (70) allows more precision:

71. If the set F is connected or has at most m connected components[§], then each point $x \in$ Co F is a convex combination of m points of F.

72. If the set F contains at least n + 2 points, then it can be decomposed into non-empty parts

$$F_1, F_2 \quad (F = F_1 \cup F_2,\ F_1 \cap F_2 = \emptyset)$$

such that

$$\text{Co } F_1 \cap \text{Co } F_2 \neq \emptyset .$$

73. If the intersection of any n + 1 of $k \geqq n + 2$ compact convex bodies of the space E is non-empty, then the intersection of all k bodies is also non-empty (*Helly's Theorem*).

One proof utilizes induction on k. Helly's theorem remains true for $k = \infty$.

The following proposition, which results from (73), is dual to (73).

74. If a compact convex body $W \subset E$ is covered by some finite collection of open halfspaces, then m + 1 halfspaces cover-

---

[§] A suitable criterion for connectedness is: S is connected, iff any two members of any open covering $\{O_\alpha\}$ of S are first and last members of a finite chain $O_1, O_2, \ldots, O_p$ such that $O_k \cap O_{k+1} \neq \emptyset$.

ing W can be chosen from that collection.

Propositions 70-74 belong to combinatorial geometry, which we shall not touch further on; but we shall again encounter a particular case of theorem (73) in algebraic form in VIII,(52).

Definition 21. A *convex polyhedron* is the convex hull W of a finite set $F \subset E$, and the set F is its *frame*.

A convex polyhedron is analogous to a finite wedge (but no wedge $k \neq 0$ is a convex polyhedron).

75. A convex polyhedron is a compact set.

76. If W is a convex polyhedron then αW is also a convex polyhedron.

77. If W is a convex polyhedron in E and $h \in \mathrm{Hom}(E,E_1)$, then the set hW is a convex polyhedron in E_1.

78. Sum, intersection, and convex hull of a finite set of convex polyhedra are convex polyhedra.

79. A convex set W is a convex polyhedron iff its extension to a wedge is a cone.

80. If the intersection W of a finite number of closed half-spaces is a bounded set, then it is a convex polyhedron.

The converse of this theorem will be given below in (99).

In conclusion let us consider the problem of characterizing Euclidean spaces in terms of projection constants (cf. IV,D29 and (71)).

Let E be a normed space and let the norm in E be twice differentiable, its second derivative being an isomorphism $E \to E'$.

81. If the projection constants (IV,D29) of all subspaces of codimension 1 equal unity, then the space E is Euclidean.

For the proof it is sufficient to consider the case when E is a three-dimensional space. One can introduce some scalar product in E and establish the following geometrical lemmas, which hold under the conditions of (81).

82. If one of the plane sections of a convex body W is a disc, then all parallel sections are discs.

83. For each boundary point x_o of W there exists a linear operator A_o satisfying the condition $Ax_o = x_o$ and transforming W into a convex body, for which x_o is an umbilical point of the boundary (i.e., at the point x_o the curvatures of all normal sections of the boundary are equal).

Note that theorem (81) remains true without any assumptions about the norm[§].

3. Separation Theorems.

Definition 22. Let K be a wedge in E and $L \subset E$ a subspace such that $\dim(K \cap L) \geqq 1$. A linear functional f defined on L is *nonnegative* relative to K if $f(x) \geqq 0$ for all $x \in K \cap L$ and $f(x) > 0$ for some $x \in K \cap L$.

84. If a subspace L contains an interior point of the wedge K, then any nonnegative linear functional defined on L permits an extension to a nonnegative linear functional $\tilde{f}(x)$ defined on all of E (*Theorem of M.G. Krein*).

The proof[§§] can be carried out by induction on dim L, if we make use of the following remark.

85. If $x \notin L$, then there exist vectors x' and x" in L such that $x - x' \in K$, $x'' - x \in K$; and this implies

$$\sup_{x'} f(x') \leqq \inf_{x''} f(x'') .$$

It is also possible to use the generalization of the Hahn-

[§]Cf. S. Kakutani, Some characterizations of Euclidean space, Japan J. Math. 16 (1940), 93-97.

[§§]This is the proof of Thm. 1.1 in Kreĭn, M.G., and Rutman, M.A., Linear operators leaving invariant a cone in a Banach space, Am. Mathem. Society Translation Nr. 26, 1950, if the transfinite induction for ∞-dimensional E is omitted. H. Hahn used this method for his proof of the Hahn-Banach Thm. (IV,(88)) (= Satz III of paper Nr.34 in Sp.Lit.IV). In the present case, E need not be a normed space. The weak topology of I,(274) suffices to show the existence of x' and x" in (85).

Banach theorem to seminormed spaces (cf. IV,(96)).

From Kreĭn's theorem follows in particular

86. For any wedge $K \neq E$ with interior point there exists at least one functional $f \in E'$ nonnegative relative to K[§].

Therefore, if $K \neq E$, then $K' \neq 0$.

From Kreĭn's theorem also follows a separation theorem for hyperplanes:

87. If K is a closed wedge in E and $x \notin K$, then there exists a functional $f \in K'$ such that $f(x) < 0$.

Now we can supplement theorems (38), (21), and (40).

88. $K = K''$, iff the wedge K is closed (cf. (38)).

89. For any wedge K

$$K'' = \overline{K} .$$

Therefore:

90. A closed wedge $K \neq E$ is the intersection of the set of all closed halfspaces which contain K. For any wedge K this intersection is $\overline{K}$.

In particular, the inversion of theorem (21) holds:

91. Any finite wedge $\neq E$ is the intersection of a finite number of closed halfspaces.

92. If the wedges K_1 and K_2 are closed, then

$$(K_1 \cap K_2)' = K_1' + K_2' \quad \text{(cf. (40)).}$$

Note that the separation theorem (87) contains the nontrivial part of the following propositions about finite wedges.

93. For any finite wedge K

$$K'' = K \quad \textit{(Farkas' Theorem)}.$$

[§] Thms. (84) and (86) also hold if K contains only interior points relative to $L_m \supset K$, where $m = \dim K < n$. The nonnegative starting functional f may first be extended to $\tilde{f} \in L_m'$. The further extension to $\tilde{\tilde{f}} \in E'$ is trivial: along $n-m$ linearly independent vectors $\{x_k\}_{m+1}^{n}$ assign freely chosen scalars ξ_k and put $\tilde{\tilde{f}}(x_k) = \xi_k$.

94. For any finite wedge $Q \subset E'$ there exists a wedge $K \subset E$ such that $K' = Q$ (*Minkowski's Theorem*).

95. For any finite wedge $Q \subset E'$ there exists a wedge $K \subset E$ such that $K' = Q$ and $Q' = K$ (*Theorem of H. Weyl*).

96. The conjugation mapping defined by D11 generates a bijective and monotone mapping of the set of all finite wedges of the space E onto the set of all finite wedges of the space E'.

As already observed in some cases, all these assertions are true for the set of all closed (and not just finite) cones.

In addition, the following analog of theorem (90) holds.

97. A closed convex set W is the intersection of all closed halfspaces containing W. For any convex set W this intersection is $\overline{W}$.

Further:

98. The closed convex hull of any set $F \subset E$ coincides with the intersection of all closed halfspaces containing F.

The following proposition is the converse of (80):

99. Any convex polyhedron is the intersection of a finite number of closed halfspaces (cf. (91)).

The theorem about the separation by a hyperplane permits the development

100. If the wedges K_1 and K_2 satisfy

$$\text{Int } K_1 \cap \text{Int } K_2 = \emptyset ,$$

then there exists a functional $f \in E'$ $(f \neq 0)$ such that $f(x) \geq 0$ for $x \in K_1$ and $f(x) \leq 0$ for $x \in K_2$.

In particular:

101. If the subspace L does not contain an interior point of the wedge K, then there exists a nonnegative functional $f \in E'$ such that $f(x) = 0$ for $x \in L$.

This assertion can be generalized to an arbitrary convex set. Formulated in geometric language, it says:

102. If the affine manifold G does not contain an interior point of the convex set W, then there exists a hyperplane $H \supset G$ such that W lies on one side of H.

This is a more general formulation of the theorem of Ascoli-Mazur (cf. IV,(82),(83)). If G is reduced to the single point $x_o \in W$, then H becomes a supporting hyperplane to the set W at the point x_o.

We finish this elaboration of Kreĭn's theorem with the following proposition about the separation of convex sets.

103. If the convex sets W_1 and W_2 satisfy

$$\operatorname{Int} W_1 \cap \operatorname{Int} W_2 = \emptyset ,$$

then there exists a hyperplane separating W_1 from W_2, i.e., there exists a functional $f \in E'$ ($f \neq 0$) for which

$$\sup_{x_1 \in W_1} f(x_1) \leq \inf_{x_2 \in W_2} f(x_2) .$$

Let us now consider some applications of the separation theorem to the problem of moments.

Let $\{x_\nu\}$ be some (finite or infinite) set of vectors of the normed space E and $\{\alpha_\nu\}$ a set of real numbers.

104. There exists a functional $f \in E'$ with norm $\|f\| \leq \rho$ satisfying the inequalities

$$f(x_\nu) \geq \alpha_\nu$$

for all ν iff for any finite choice of indices $\nu_1, \nu_2, \ldots, \nu_m$ and any positive $\lambda_1, \lambda_2, \ldots, \lambda_m$

$$\rho \| \Sigma_{k=1}^{m} \lambda_k x_{\nu_k} \| \geq \Sigma_{k=1}^{m} \lambda_k \alpha_k .$$

105. Let K be a wedge. If for some $\nu = \nu_o$

$$x_{\nu_o} \in \operatorname{Int} K, \ \alpha_{\nu_o} > 0 ,$$

then there exists a linear functional $f \in E'$ nonnegative relative to K and satisfying the condition

$$f(x_\nu) = \alpha_\nu$$

for all ν, iff for any finite choice of indices $\nu_1, \nu_2, \ldots, \nu_m$ and of reals $\lambda_1, \lambda_2, \ldots, \lambda_m$ satisfying $\Sigma_{k=1}^{m} \lambda_k \alpha_k = 0$

$$\Sigma_{k=1}^{m} \lambda_k x_{\nu_k} \notin \text{Int } K .$$

106. Let $\{S_\nu\}$ be a set of self-adjoint operators in a unitary space and $\{\alpha_\nu\}$ a set of real numbers. If for some $\nu = \nu_o$ the operator S_{ν_o} is positive and $\alpha_{\nu_o} > 0$, then there exists a nonnegative self-adjoint operator C satisfying the condition

$$\text{tr } (CS_\nu) = \alpha_\nu$$

for all ν iff for any finite choice of indices $\nu_1, \nu_2, \ldots, \nu_m$ and of reals $\lambda_1, \lambda_2, \ldots, \lambda_m$, for which $\Sigma_{k=1}^{m} \lambda_k \alpha_{\nu_k} = 0$,

$$\Sigma_{k=1}^{m} \lambda_k S_{\nu_k}$$

is not a positive operator.

We consider in passing the so-called *L*-problem of moments.

107. Let $\{x_k\}_{k=1}^{m}$ $(m < n)$ be a linearly independent system of vectors of a normed space E and $\alpha_1, \alpha_2, \ldots, \alpha_n$ be real numbers. Put

$$\mu = \inf_{\{\lambda_k\}} \|\Sigma_{k=1}^{m} \lambda_k x_k\|, \text{ where } \Sigma_{k=1}^{m} \alpha_k \lambda_k = 1.$$

There exists a functional $f \in E'$ satisfying the conditions

$$f(x_k) = \alpha_k \quad (k = 1,2,\ldots,m), \ \|f\| < L,$$

iff $L \geq 1/\mu$.

108. If $L > 1/\mu$, then the solutions of the L-problem of moments form a convex set which is the intersection of the unit ball with an $(n - m)$-dimensional affine manifold.

109. If the space E' is strictly normed (cf. IV,D46),

then the L-problem of moments has a unique solution iff $L = 1/\mu$.

4. Extreme Points.

Definition 23. Let W be a convex set. The point $x \in W$ is called an *extreme* point of W if the representation

$$x = (1 - \theta)x_1 + \theta x_2, \quad 0 < \theta < 1, \; x_1, x_2 \in W$$

is possible only for $x_1 = x_2$ (cf. D10 for (convex) wedges). If all boundary points of W are extremal, then W is a *strictly convex* set.

110. The point $x \in W$ is extremal iff

$$x = \tfrac{1}{2}(x_1 + x_2) \qquad x_1, x_2 \in W$$

implies $x_1 = x_2 = x$.

111. Each extreme point of a convex polyhedron belongs to any frame (D21) of it.

112. A convex polyhedron coincides with the convex hull of its extreme points.

Thus the set of extreme points of a convex polyhedron is a minimal frame (basis) of it.

Definition 24. The number of extreme points of the convex polyhedron P is called its *order* (or *affine rank*) and is denoted by ord P.

113. Every compact set W has an extreme point. For example:

114. If $y \notin W$, then among the points $x \in W$ furthest from y there is an extreme point.

Note here that the set of extreme points of a compact convex set need not be closed.

115. Any compact convex set coincides with the closed convex hull of its extreme points (*Kreĭn-Mil'man Theorem*).

This theorem and theorems (70) and (71) immediately imply

116. If W is a compact convex set and dim W = m, then each point $x \in W$ is a convex combination of m + 1 of its extreme points. If the set of extreme points is connected (or has no more than m connected components), then each point $x \in W$ is a convex combination of m extreme points.

Below we shall consider some applications of the theorems about convex sets. For preparation we introduce two special classes of operators and investigate their spectral properties.

Definition 25. The operator $T \in \mathbb{M}(E)$ is a *permutation operator* relative to the fixed basis Δ if it effects an interchange of the vectors of the basis Δ. The set of permutators relative to the basis Δ will be denoted by $\mathbb{P}$.

Definition 26. The operator $A \in \mathbb{M}(E)$ is *stochastic* relative to the basis Δ if the elements of its matrix α_{jk} $(j,k = 1,\ldots,n)$ in this basis are nonnegative and satisfy the relations $\Sigma_{k=1}^{n} \alpha_{jk} = 1$ $(j = 1,2,\ldots,n)$.

If, in addition, $\Sigma_{j=1}^{n} \alpha_{jk} = 1$ $(k = 1,2,\ldots,n)$, then the operator A is *doubly stochastic* (or *bistochastic*) relative to the basis Δ. The matrix $(\alpha_{jk})_{j,k=1}^{n}$ of a stochastic (doubly stochastic) operator in the basis Δ is also called stochastic (doubly stochastic or bistochastic). We shall denote the set of doubly stochastic operators relative to a basis Δ by $\mathbb{W}$; obviously, $\mathbb{P} \subset \mathbb{W}$.

Let $T \in \mathbb{P}$, $\Delta = \{e_k\}_1^n$, $Te_k = e_{j_k}$ $(k = 1,2,\ldots,n)$, and denote the decomposition of the permutation

$$\sigma = \begin{pmatrix} 1, 2, \ldots, n \\ j_1, j_2, \ldots, j_n \end{pmatrix}$$

into disjoint cycles by

$$\sigma = (r_1, r_2, \ldots, r_{p_1}) \ldots (s_1, s_2, \ldots, s_{p_m}) .$$

117. Relative to an orthonormal Δ, the operator T^* is a

permutation operator, and the substitution σ^* effected by T^* is the inverse of σ.

118. $T^p = I$ for p equal to the least common multiple of the cycles of the substitution σ.

119. Each eigenvalue of a permutation operator is a natural root of unity.

120. The multiplicity of the eigenvalue $\lambda = 1$ of the permutation operator T equals the number of cycles of the substitution σ.

By II,(44) a permutation operator is an operator of scalar type. The following proposition contains (119) and (120).

121. The characteristic polynomial of the permutation operator T is

$$\mathcal{D}(\lambda;T) = (\lambda^{p_1} - 1)(\lambda^{p_2} - 1) \ldots (\lambda^{p_m} - 1) .$$

Conversely,

122. If the characteristic polynomial of the operator T of scalar type has the form (121), then T is a permutator relative to some basis in $E^{(c)}$ (see notation in VI,(2)).

We shall now point out the spectral properties of stochastic operators.

123. The spectral radius of a stochastic operator is one.

124. Those eigenvalues of a stochastic operator that lie on the circle $|\lambda| = 1$ are roots of unity.

Studying the structure of the set $\mathbb{W}$ of doubly stochastic operators, we find:

125. The set $\mathbb{W}$ is compact and convex, and $\dim \mathbb{W} = (n-1)^2$.

126. The set of extreme points of $\mathbb{W}$ is $\mathbb{P}$. Thus $\mathbb{W}$ is a polyhedron and ord $\mathbb{W} = n!$. Therefore

127. The set of doubly stochastic operators is the convex hull of the set of permutation operators (*G. Birkhoff's Theorem*).

128. Each doubly stochastic operator is the convex combination of $n^2 - 2n + 2$ permutation operators.

Definition 27. An operator T is said to be a *quasi-permutation* relative to the basis $\Delta = \{e_k\}_1^n$ if

$$Te_k = \pm e_{j_k} \quad (k = 1,2,\ldots,n) ,$$

where $(j_1,j_2,\ldots,j_n)$ is a permutation of $(1,2,\ldots,n)$.

Note the following analog of Birkhoff's theorem:

129. If the operator T is a compression (IV,D56) in the c-norm and in the ℓ-norm relative to a basis Δ, then it belongs to the convex hull of the quasipermutation operators relative to Δ, and conversely.

For any sequence of real numbers $\{\alpha_j\}_1^n$ let $\{\overset{\geq}{\alpha}_j\}_1^n$ denote the numbers α_j numbered in nonincreasing order.

Let $x,y \in R^n$, $x = \{\xi_j\}_1^n$, $y = \{\eta_j\}_1^n$, and write $x \lll y$, if

$$\Sigma_{j=1}^k \overset{\geq}{\xi}_j \leqq \Sigma_{j=1}^k \overset{\geq}{\eta}_j \quad (k = 1,2,\ldots,n) ;$$

if, in addition, $\Sigma_{j=1}^n \xi_j = \Sigma_{j=1}^n \eta_j$, write $x \prec y$. Obviously, $\lll$ and $\prec$ are order relations.

For a fixed vector $y = \{\eta_j\}_1^n$ of R^n, denote by W the set of all vectors $x \in R^n$ which satisfy the condition $x \prec y$.

130. The set W is convex and compact.

131. The point $x \in W$ is an extreme point of W iff

$$x = \left\{\eta_{j_k}\right\}_{k=1}^n$$

where $(j_1,j_2,\ldots,j_n)$ is some permutation of $(1,2,\ldots,n)$.

Consequently the set W is a polyhedron and ord W $\leqq$ n! Thus

132. For any vector $y \in R^n$ the set W of vectors $x \in R^n$ which satisfy the condition $x \prec y$ coincides with the convex hull of the set of vectors obtained from the vector y by permuting its coordinates (*Rado's Theorem*).

There is the following connection between the theorems of Birkhoff and Rado, which can also be used to prove

Rado's theorem.

133. $x \prec y$ iff $x = Ay$ where A is a doubly stochastic operator relative to the canonical basis of the space R^n (which is formally the same as the basis in I,D18 for C^n).

The results (130)-(132) permit the following extension.

134. The set W of vectors $x = \{\xi_j\}_1^n$, which for a fixed $y \in R^n$ satisfy the condition $|x| \prec\prec |y|$, is compact and convex§.

135. The point $x \in W$ is an extreme point of the subset W iff $x = \left\{\varepsilon_k |\eta_{j_k}|\right\}_{k=1}^n$, where $(j_1, j_2, \ldots, j_n)$ is some permutation of $(1,2,\ldots,n)$, and $\varepsilon_k = \pm 1$.

Consequently, the set W is a polyhedron and ord $W \leq 2^n\, n!$

Thus

136. For any $y \in R^n$ the set W of vectors $x \in R^n$ which satisfy the relation $|x| \prec\prec |y|$ coincides with the convex hull of the set of vectors obtained from the vector $|y|$ by permuting its coordinates and multiplying them by ± 1 (*A.S. Markus' Theorem*).

5. Inequalities between Eigenvalues and Singular Numbers.

Here we shall consider inequalities between eigenvalues and s-numbers arising in connection with the order relations $\prec$ and $\prec\prec$.

Let $K \subset R^n$ be a wedge defined by

$$K = \{x = \{\xi_k\}_1^n \mid \xi_1 \geq \xi_2 \geq \ldots \geq \xi_n\} .$$

If $x, y \in K$, $x = \{\xi_k\}_1^n$, $y = \{\eta_k\}_1^n$, then the relation $x \prec\prec y$ is equivalent to the system of inequalities

$$\Sigma_{k=1}^m \xi_k \leq \Sigma_{k=1}^m \eta_k \qquad (m = 1,2,\ldots,n) .$$

§Here $|\cdot|$ means: replace the coordinates of a vector by their absolute values.

We shall consider the system of eigenvalues $\lambda(S) = \{\lambda_k(S)\}_1^n$ of the self-adjoint operator S as a vector in the space R^n. The vector s(A) with components $\{s_k(A)\}_1^n$,

$$s(A) = \lambda((A^*A)^{\frac{1}{2}}) ,$$

is introduced analogously (III,D53). Thus, for any operators $S \in \mathbb{S}(E)$, $A \in \mathbb{M}(E)$ in the complex Euclidean space E,

$$\lambda(S) \in K, \ s(A) \in K .$$

For any operator A, enumerate the eigenvalues taking account of their multiplicity, so that their moduli are nonincreasing, and denote by $\omega_k(A)$ the modulus of the k-*th* eigenvalue ($k = 1,2,\ldots,n$). In particular, $\omega_1(A) = \rho(A)$, the spectral radius. The vector $\omega(A) = \{\omega_k(A)\}_1^n \in R^n$, and, obviously $\omega(A) \in K$.
We shall use the following <u>convention</u> for a nondecreasing function $\varphi(\xi)$:

$$\varphi(x) \underset{\text{def}}{=} \{\varphi(\xi_k)\}_1^n \in K, \quad \text{if } x \in K \quad (x = \{\xi_k\}_1^n) .$$

The following criterion for monotonicity of a functional defined on the wedge K is the source of a series of inequalities which connect the eigenvalues with the s-numbers.

137. If the functional $\Phi(x)$ defined on K is differentiable and

$$\nabla\Phi(x) = \left\{\frac{\partial\Phi}{\partial\xi_k}\right\}_1^n \in K \qquad (x \in K) ,$$

then the relation $x \ll y$ $(x,y \in K)$ implies the inequality

$$\Phi(x) \leq \Phi(y) .$$

The proof can be effected by the change of variables

$$\zeta_k = \Sigma_{j=1}^k \xi_k \qquad (k = 1,2,\ldots,n) .$$

Applications of theorem (137) are based on the following

remarkable *Inequalities of H. Weyl.*

138. $$\prod_{k=1}^{m} \omega_k(A) \leq \prod_{k=1}^{m} s_k(A) \qquad (m=1,2,\ldots,n).$$

For the proof the exterior powers of the operator can be utilized (cf. V,(89)).

139. In (138) equality holds for $m \geqq \mathrm{rk}\, A$. Equality holds in all the inequalities (138) iff the operator A is normal (cf. III,D50 and (252)).

Weyl's inequalities are the best possible ones in the sense that

140. For any complex numbers $\lambda_1,\ldots,\lambda_n$ $(|\lambda_1| \geqq \ldots \geqq |\lambda_n|)$ and nonnegative numbers $s_1 \geqq \ldots \geqq s_n$ satisfying the relations

$$\left|\prod_{k=1}^{m} \lambda_k\right| \leq \prod_{k=1}^{m} s_k \quad (m = 1,2,\ldots,n-1), \quad \left|\prod_{k=1}^{n} \lambda_k\right| = \prod_{k=1}^{n} s_k ,$$

there exists an operator A with eigenvalues $\{\lambda_k\}_1^n$ and singular numbers $\{s_k\}_1^n$ (*Horn's Theorem*).

Weyl's inequalities can also be obtained from the following theorem which is an immediate result of the intertwining of the eigenvalues (cf. III,(232)).

141. For any orthonormal system of vectors $\{e_k\}_1^m$

$$\det((Ae_j,Ae_k))_{j,k=1}^{m} \leq \prod_{k=1}^{m} s_k(A).$$

More generally,

142. For any system of vectors $\{e_k\}_1^m$ $(m \leq n)$,

$$\det((Ae_j,Ae_k))_{j,k=1}^{m} \leq \prod_{k=1}^{m} s_k(A) \cdot \det((e_j,e_k))_{j,k=1}^{m}.$$

For $m = n$ equality holds.

We now consider some applications of theorem (137).

143. Let the functional $F(y)$, which is defined on the intersection of the wedge K with the positive hyperoctant, generate the functional

$$\Phi(x) = F(e^x) \qquad (x \in K;\ \text{recall the convention})$$

and assume that it satisfies the conditions of theorem (137).

Then

$$F(\omega(A)) \leqq F(s(A))$$

(*Ostrowski's Theorem*).

From Ostrowski's theorem follows:

144. If the function

$$\varphi(\xi) = f(e^{\xi}) \qquad (-\infty < \xi < \infty)$$

is convex, then

$$\Sigma_{k=1}^{m} f(\omega_k(A)) \leqq \Sigma_{k=1}^{m} f(s_k(A)) \qquad (m = 1,2,\ldots,n)$$

(*H. Weyl's Theorem*).

In particular

145. $(\omega(A))^p \prec\!\prec (s(A))^p \qquad (p > 0)$.

In expanded form relation (145) means that

$$\Sigma_{k=1}^{m} \omega_k^p(A) \leqq \Sigma_{k=1}^{m} s_k^p(A) \qquad (m = 1,2,\ldots,n) \ . \qquad (*)$$

For $p = 1$ and $p = 2$ the last of these inequalities has already been met (cf. III,(257), (259)). Let us also note that for $p = 2$ all the indicated inequalities can be established immediately by utilizing a Schur basis (III,D52).

146. Let $F_r(\tau_1,\tau_2,\ldots,\tau_m)$ be the r-*th* elementary symmetric function:

$$F_r(\tau_1,\ldots,\tau_m) = \Sigma\tau_{j_1}\tau_{j_2}\ldots\tau_{j_r}, \text{ summed over } 1 \leqq j_1<\ldots<j_k \leqq m.$$

Then

$$F_r(\omega_1(A),\omega_2(A),\ldots,\omega_m(A)) \leqq$$
$$F_r(s_1(A),s_2(A),\ldots,s_m(A)) \qquad (1 \leqq r \leqq m \leqq n).$$

The inequalities (*) for $p = 1$ permit a generalization of theorem III,(265):

147. For any $m \leqq n$

$$\sup\Sigma_{k=1}^{m}|(UAe_k,e_k)| = \Sigma_{k=1}^{m} s_k(A) \ ,$$

where the supremum is taken over all $U \in \mathbb{U}(E)$ and all orthonormal systems $\{e_k\}_1^m$.

In particular, the inequalities

$$\Sigma_{k=1}^{m} |(Ae_k, e_k)| \leqq \Sigma_{k=1}^{m} s_k(A) \qquad (m = 1,2,\ldots,n)$$

hold. From these inequalities and Weyl's theorem (144) follows:

148. For any orthonormal system

$$\{|(Ae_k, e_k)|^p\}_1^n \prec\prec \{s_k^p(A)\}_1^n \qquad (p > 0).$$

Theorem (147) also implies the following *Inequality of Ky Fan* (cf. III,(264)).

149. $s(A + B) \prec\prec s(A) + s(B)$.

In particular,

150. For $S,T \in \mathbb{S}(E)$

$$\omega(S + T) \prec\prec \omega(S) + \omega(T) .$$

For nonnegative S,T (cf. III,D38) this relation can be obtained independently of (137) from the following theorem.

151. For $S,T \in \mathbb{S}(E)$

$$\lambda(S + T) - \lambda(T) \prec \lambda(S)$$

(*Wielandt's Theorem*).

Wielandt's theorem can be easily derived from III, (239)-(241).

152. Let $A \in M(E)$, and $\tilde{A} \in M(\tilde{E})$ where $\tilde{E} = E \times E$ (I,D70); define $\tilde{A}$ by the block matrix

$$\begin{bmatrix} 0 & A \\ A^* & 0 \end{bmatrix}, \text{ i.e., } \tilde{A}\begin{bmatrix} x \\ y \end{bmatrix} = \begin{bmatrix} A\,y \\ A^*x \end{bmatrix} \qquad (x,y \in E) .$$

Then the eigenvalues of the operator $\tilde{A}$ equal $\pm s_k(A)$ ($k = 1,2,\ldots,n$).

From this remark and Wielandt's theorem follows:

153. For $A,B \in \mathbb{M}(E)$

$$\{|s_k(A + B) - s_k(B)|\}_1^n \lll s(A)$$

(*Mirsky's Theorem*).

From Mirsky's theorem follows once again the inequality of Ky Fan.

The following theorems are the multiplicative analogs of assertions (149), (150), (153), (151).

154. For regular operators $A,B \in \mathbb{M}(E)$

$$\log s(AB) \prec \log s(A) + \log s(B) \qquad \textit{(Horn's Inequality)}.$$

155. For positive operators $S,T \in \mathbb{S}(E)$,

$$\log \lambda(ST) \prec \log \lambda(S) + \log \lambda(T).$$

156. For regular operators $A,B \in \mathbb{M}(E)$

$$\{\log s_k(AB) - \log s_k(B)\}_1^n \prec \log s(A)$$

(*Amir-Moez' Theorem*).

157. For positive operators $S,T \in \mathbb{S}(E)$

$$\{\log \lambda_k(ST) - \log \lambda_k(T)\}_1^n \prec \log \lambda(S).$$

6. Convex Sets in Problems Concerning the Localization of Spectra of Self-adjoint Operators

Consider the real vector $\lambda = \{\lambda_k\}_1^n$ as a function of the self-adjoint operator S. Obviously, the image of the space $\mathbb{S}(E)$ under the mapping $\lambda: \mathbb{S}(E) \to R^n$ coincides with the wedge K defined in the preceding section. Localizing the spectrum of an operator S relative to some set of operators $\mathbb{H} \subset \mathbb{S}(E)$ means describing the image $\lambda(\mathbb{H})$.

Definition 28. Let us introduce the *outer diameter* of a finite sequence of real numbers as the difference between the largest and the smallest of them, and the *inner diameter* as the least of the absolute values of the pairwise differences of numbers of this sequence. --

We begin with the localization of the spectrum of the sum of operators with fixed spectra.

Let $a = \{\alpha_k\}_1^n$, $b = \{\beta_k\}_1^n$ be fixed vectors from K and

$$\Sigma = \{C \mid C = S + T;\ S,T \in \mathbf{S}(E),\ \lambda(S) = a,\ \lambda(T) = b\}$$

158. The set $\lambda(\Sigma)$ is connected.

Wielandt's and Rado's theorems imply the following *Theorem of Lidskiĭ* on the localization of the spectrum of the <u>sum</u> of two self-adjoint operators:

159. Let U_1 and U_2 be the convex hulls of the sets of vectors of the form $\{\alpha_k + \beta_{j_k}\}_{k=1}^n$ and $\{\beta_k + \alpha_{j_k}\}_{k=1}^n$ respectively, where $(j_1, j_2, \ldots, j_n)$ runs through the set of permutations of the indices $(1,2,\ldots,n)$. Then

$$\lambda(\Sigma) \subset U_1 \cap U_2 .$$

Among other things it follows at once:

160. If the outer diameter of the spectrum of one of the self-adjoint operators S,T is less than the inner diameter of the spectrum of the other operator, then the operator S + T has no multiple eigenvalue.

In Lidskiĭ's theorem inclusion cannot in general be replaced by equality (even when $U_1 \cap U_2 \subset K$). However

161. If the outer diameter of one of the sequences a,b is less than the inner diameter of the other, then

$$\lambda(\Sigma) = U_1 \cap U_2 .$$

Under the conditions of theorem (161) the set $\lambda(\Sigma)$ is thus a convex polyhedron.

We pass now to the localization of the spectrum of operator products.

Let $a = \{\alpha_k\}_1^n$, $b = \{\beta_k\}_1^n$ be again fixed vectors but now with <u>positive</u> coordinates belonging to the cone K and $\Pi = \{C \mid C = ST;\ S,T \in \mathbf{S}(E),\ \lambda(S) = a,\ \lambda(T) = b\}$. Then

162. The set $\lambda(\Pi)$ is connected.

From theorems (157) and (132) follows the *Theorem of Lidskiĭ* on the localization of the spectrum of the product of two operators:

63. Let V_1 and V_2 be the convex hulls of the sets of ectors of the form $\{\log \alpha_k + \log \beta_{j_k}\}_{k=1}^{n}$ and $\log \alpha_{j_k} + \log \beta_k\}_{k=1}^{n}$ respectively, where $(j_1, j_2, \ldots, j_n)$ ıns through the set of permutations of the indices (1,2, ..,n). Then

$$\log \lambda(\Pi) \subset V_1 \cap V_2 .$$

54. Let S and T be positive operators. If the exterior iameter of the sequence of coordinates of one of the vecors $\log \lambda(S)$, $\log \lambda(T)$ is less than the inner diameter of ıe sequence of coordinates of the other, then the operator T does not have a multiple eigenvalue.

55. If the outer diameter of the coordinates of one of the ectors log a, log b is less than the inner diameter of the oordinates of the other, then

$$\log \lambda(\Pi) = V_1 \cap V_2 .$$

Under the conditions of theorem (165) the set $\log \lambda(\Pi)$ is thus a convex polyhedron.

Let us note two more *Theorems of Markus* about the localization of s-numbers of the sum and product of operators, analogous to the theorems of Lidskiĭ.

66. Let $a = \{\alpha_k\}_1^n$, $b = \{\beta_k\}_1^n$ be fixed vectors of K with onnegative coordinates and

$$\Sigma = \{C \mid C = A + B;\ A,B \in \mathbb{M}(E),\ s(A) = a,\ s(B) = b\}.$$

ıen

$$S(\Sigma) \subset U_1 \cap U_2 ,$$

ıere U_1 and U_2 are the convex hulls of the vector sets $\alpha_k + \varepsilon_k \beta_{j_k}\}_{k=1}^{n}$ and $\{\beta_k + \varepsilon_k \alpha_{j_k}\}_{k=1}^{n}$ respectively; here $j_1, j_2, \ldots, j_n)$ runs through the set of permutations of the ıdices $(1,2,\ldots,n)$ and $\varepsilon_k = \pm 1$.

167. Let $a = \{\alpha_k\}_1^n$, $b = \{\beta_k\}_1^n$ be fixed vectors of K with positive coordinates and

$$\Pi = \{C \mid C = AB;\ A,B \in \mathbb{M}(E),\ s(A) = a,\ s(B) = b\}.$$

Then

$$\log s(\Pi) \subset V_1 \cap V_2 ,$$

where V_1 and V_2 are the convex hulls of the vector sets $\{\log \alpha_k + \log \beta_{j_k}\}_{k=1}^n$ and $\{\log \beta_k + \log \alpha_{j_k}\}_{k=1}^n$, respectively; here $(j_1,j_2,\ldots,j_n)$ runs through the set of permutations of the indices $(1,2,\ldots,n)$.

There is also an analog of Lidskiĭ's theorem (163) for unitary operators, but we shall not touch on this.

We shall consider now a converse problem of localization. The converse problem of localization in general requires a description of the set of operators whose spectra satisfy a given restriction.

Let Λ be an arbitrary set in the space R^n. We shall denote by $\mathbb{W}(\Lambda)$ the set of all self-adjoint operators whose eigenvalues in any order of enumeration represent a point of Λ. Together with $\mathbb{W}(\Lambda)$ let us introduce the set $\mathbb{V}_o(\Lambda)$ of self-adjoint operators whose eigenvalues, if numbered in a suitable order, represent a point of Λ.

168. If the set Λ is closed and convex, then the set $\mathbb{W}(\Lambda)$ is also closed and convex.

This theorem follows from (169) and (170).

169. If Λ is the intersection of some set $\{\Lambda_\nu\}$ of closed halfspaces in R^n,

$$\Lambda = \cap \Lambda_\nu ,$$

then $\mathbb{W}(\Lambda) = \cap \mathbb{W}(\Lambda_\nu)$.

170. If Λ is a closed halfspace of the space R^n defined by the inequality

$$\alpha_1\lambda_1 + \alpha_2\lambda_2 + \ldots + \alpha_n\lambda_n \geqq \alpha,$$

hen $\mathbb{W}(\Lambda)$ coincides with the set of self-adjoint operators satisfying the inequality

$$\operatorname{tr}(AS) \geqq \alpha \qquad (A \in \mathbb{S}(E),\ \lambda(A) = \{\alpha_k\}_1^n) .$$

The nontrivial part of this theorem can be established with the aid of Birkhoff's theorem.

Theorem (168) permits the following development:

71. If the set $\Lambda \subset R^n$ is invariant relative to all permu-ations of the coordinates of the point $\{\lambda_k\}_1^n$ and is the losed convex hull of some set $M \subset R^n$, then $\mathbb{W}(\Lambda) = \operatorname{Co} \mathbb{V}_o(M)$.

This together with Carathéodory's theorem implies

72. If M is a conical set of the space R^n (i.e., $\gamma M \subset M$ or $\gamma \geqq 0$), then $\mathbb{W}(\Lambda)$ is the set of operators which can be epresented as the sum of n operators from $\mathbb{V}_o(M)$.

This contains in particular an obvious corollary of the spectral theorem for self-adjoint operators:

73. We can represent every nonnegative operator as the sum f n nonnegative operators of rank $\leqq 1$ (I,D47).

7. Unitary-Invariant Norms and Symmetrically Convex Bodies.

In this section, E will denote a complex Euclidean space. (The theory to be presented applies without change to the case of a real space E, but the role of unitary operators would then be played by the orthogonal operators.)

Definition 29. A norm in the space $\mathbb{M}(E)$ is called *unitary-invariant* if for any $U \in \mathbb{U}(E)$

$$\|UA\| = \|AU\| = \|A\| \qquad (A \in \mathbb{M}(E)).$$

74. The operator norm $\|A\| = s_1(A)$ is unitary-invariant notation in III,D53).

75. The Hilbert norm

$$\|A\| = (\operatorname{tr} A^*A)^{\frac{1}{2}} = [\Sigma_{k=1}^n s_k^2(A)]^{\frac{1}{2}}$$

is unitary-invariant.

176. The functional $p(A) = \Sigma_{k=1}^{n} s_k(A)$ $(A \in \mathbb{M}(E))$ is a unitary-invariant norm (cf. III,(264)).

Below $\|.\|$ denotes an arbitrary unitary-invariant norm. When simultaneously considering a norm $\|.\|$ and an operator norm, i.e., a norm induced by the norm on E, we shall denote the latter by $|\cdot|$.

The unitary-invariant norm of an operator depends only upon its operator modulus (III,D47):

177. $\|A\| = \|(AA^*)^{\frac{1}{2}}\| = \|(A^*A)^{\frac{1}{2}}\|$.

Further:

178. If for some $\gamma \geqq 0$

$$s_k(A) \leqq \gamma s_k(B) \qquad (k = 1,2,\ldots,n),$$

then $\|A\| \leqq \gamma\|B\|$.

Hence it follows that a unitary-invariant norm of an operator depends only upon its s-numbers.

179. If $s_k(A) = s_k(B) \qquad (k = 1,2,\ldots,n)$,

then $\|A\| = \|B\|$.

From (178) and III,(266) it follows that

180. $\|BAC\| \leqq |B|\cdot|C|\cdot\|A\|$.

By the way, any norm in $\mathbb{M}(E)$ which satisfies this inequality is unitary-invariant.

181. If rk $A = 1$, then $\|A\| = \alpha|A|$, where α is some positive number which does not depend upon A.

Definition 30. A *cross-norm* on $\mathbb{M}(E)$ is a norm which satisfies the condition

$$N(A) = |A|, \quad \text{if rk } A = 1.$$

This definition coincides with the general definition of cross-norm in V,D18 on account of the canonical isomorphism $\mathbb{M}(E) \simeq E'\otimes E$ (cf. V,(14,(15)).

182. Any unitary-invariant cross-norm satisfies the inequality (cf. III,(250))

$$s_1(A) \leq \|A\| \leq \Sigma_{k=1}^{n} s_k(A) \qquad (A \in \mathbb{M}(E))$$

Any norm satisfying inequality (182) is clearly a cross-norm.

Thus the set of unitary-invariant cross-norms possesses a least and a greatest element

$$\|A\|_{min} = s_1(A), \quad \|A\|_{max} = \Sigma_{k=1}^{n} s_k(A) .$$

Note that (180) and (182) imply

183. All unitary-invariant cross-norms are ring norms (IV,D52).

As already said, any unitary-invariant norm is a function of the s-numbers:

$$\|A\| = p(s(A)) \tag{*}$$

where p is some functional defined on the cone

$$K^+ = \{x = \{\xi_k\}_1^n | \xi_1 \geq \xi_2 \geq \ldots \geq \xi_n \geq 0\}.$$

There arises the problem of describing the class of functionals p which generate unitary-invariant norms by formula (*). The key to the solution is given by theorem (189), which we prepare by some necessary definitions and simple propositions.

Denote by X the set of vectors obtained from the vector $x = \{\xi_k\}_1^n$ through permutation and multiplication by ±1 of its coordinates and denote by $\hat{x}$ that vector of X which belongs to the cone K^+.

Definition 31. A seminorm p(x) in R^n is called *symmetric* if it is invariant relative to permutations of the coordinates of the vector x and multiplication of them by ±1, i.e., $p(x) = p(\hat{x})$.

Definition 32. A closed convex set $W \subset R^n$ is said to be *symmetrically convex* if $x \in W$ implies $X \subset W$.

184. A symmetrically convex set distinct from {0} is a convex body (cf. D17).

Instead of "symmetrically convex set" we shall thus say below "symmetrically convex body".

185. A seminorm p is symmetric iff the unit p-ball is a symmetrically convex body.

186. The norms in the real spaces ℓ^p $(1 \leqq p \leqq \infty)$ are symmetric seminorms.

Therefore:

187. The unit ball of the real space ℓ^p $(1 \leqq p \leqq \infty)$ is a symmetrically convex body.

Any symmetric seminorm is monotone in the following sense:

188. If p is a symmetric seminorm and $\hat{x} \leqq \hat{y}$ (that is, if each coordinate of the vector $\hat{x}$ does not exceed the corresponding coordinate of the vector $\hat{y}$), then $p(x) \leqq p(y)$.

With the aid of Rado's theorem (132) this proposition can be considerably strengthened:

189. If p is a symmetric seminorm and $x \lll y$, then $p(x) \leqq p(y)$ (*Theorem of Ky Fan*).

If a functional $p \neq 0$ is defined only on the cone K^+, then its symmetric extension to all of R^n, denoted by $\tilde{p}$, is uniquely determined by the condition $\tilde{p}(x) = p(\hat{x})$ $(x \in R^n)$.

190. The functional $\tilde{p}$ is a symmetric seminorm iff the functional p satisfies for $x,y \in K^+$ the conditions

1) $p(x) > 0 \quad (x \neq 0);$
2) $p(\alpha x) = \alpha p(x) \quad (\alpha \geqq 0);$
3) $p(x+y) \leqq p(x) + p(y);$
4) $p(x) \leqq p(y)$ for $x \prec y$.

191. In order that a functional p defined on the cone K^+ can be generated according to formula (*) by a unitary-invariant norm it is necessary and sufficient that its symmetric extension $\tilde{p}$ be a seminorm (another theorem of Schatten).

The triangle inequality, which constitutes the nontrivial part of Schatten's theorem, follows from (189) and

III,(264).

Definition 33. A symmetrically convex body is said to be *normal* if the corresponding seminorm $p(x)$ satisfies the condition

$$p(e_1) = 1 \quad (e_1 = \{1,0,\dots,0\}).$$

By virtue of Schatten's theorem the unitary-invariant cross-norms in $\mathbb{M}(E)$ are in a one-to-one correspondence with the normal symmetrically convex bodies in R^n. By this correspondence the following assertion, which could also be established independently, becomes the counterpart of theorem (182):

192. Every normal symmetrically convex body $W \subset R^n$ satisfies the relation

$$W_1 \subset W \subset W_\infty ,$$

where

$$W_1 = \{x = \{\xi_k\}_1^n \mid \Sigma_{k=1}^n |\xi_k| \leq 1\} ,$$

$$W_\infty = \{x = \{\xi_k\}_1^n \mid \max_k |\xi_k| \leq 1\} .$$

Thus the set of normal symmetrically convex bodies possesses a least and a greatest element: $W_{min} = W_1$, $W_{max} = W_\infty$.

In the particular case of the symmetric seminorm $p(x) = (\Sigma_{k=1}^n |\xi_k|^p)^{1/p}$, Schatten's theorem implies immediately

193. $[\Sigma_{k=1}^n s_k^p(A+B)]^{1/p} \leq [\Sigma_{k=1}^n s_k^p(A)]^{1/p} + [\Sigma_{k=1}^n s_k^p(B)]^{1/p}$ $(1 \leq p \leq \infty)$.

For $p = 1,2,\infty$ this inequality has already been met earlier.

We shall denote the space $\mathbb{M}(E)$ endowed with the unitary-invariant norm

$$\|A\| = [\Sigma_{k=1}^n s_k^p(A)]^{1/p} \qquad (1 \leq p \leq \infty) ,$$

by $S_p(E)$. It is the operator analog of the space ℓ^p.

One would expect that the spaces $S_p(E)$ and $S_q(E)$ with $1/p + 1/q = 1$ are mutually conjugate.

In verifying this assertion let us consider the more general problem of finding the norm in the space conjugate to the space $\mathbb{M}(E)$ with a given unitary-invariant norm.

For any symmetric seminorm $\tilde{p}$ in R^n put

$$\tilde{p}^*(y) = \sup_{x \in K^+} [(\tilde{p}(x))^{-1} \Sigma_{k=1}^{n} \eta_k \xi_k],$$

where $y \in K^+$, $y = \{\eta_k\}_1^n$, $x = \{\xi_k\}_1^n$.

Using proposition (190) we can establish that

194. The functional $\tilde{p}^*(y)$ is a symmetric seminorm in R^n.

195. Let the normed spaces $\tilde{R}$ and $\tilde{R}^*$ be the space R^n endowed with norms $\tilde{p}$ and $\tilde{p}^*$, respectively. Then the conjugate space to $\tilde{R}$ is isometric with $\tilde{R}^*$.

196. If $\tilde{p}$ is a symmetric seminorm in R^n, then the space conjugate to the normed space $\tilde{\mathbb{M}}(E)$ with norm $\tilde{p}(s(A))$ is isometric with the normed space $\tilde{\mathbb{M}}^*(E)$ with norm $\tilde{p}^*(s(A))$ (cf. II,(203)).

In particular,

197. The spaces $S_p(E)$ and $S_q(E)$ are mutually conjugate, if $1/p + 1/q = 1$ $(1 \leq p \leq \infty)$.

In conclusion we shall formulate two <u>interpolation theorems</u>. Let A be a linear operator in R^n and $\|A\|_p$ its operator norm in ℓ^p relative to the canonical basis.

Introduce the scalar product in R^n in the usual way.

198. If $\|A\|_1 \leq 1$ and $\|A\|_\infty \leq 1$, then for any unitary-invariant norm for which $\|I\| = 1$ we have $\|A\| \leq 1$ (*Theorem of Mityagin*).

In particular,

199. If $\|A\|_1 \leq 1$ and $\|A\|_\infty \leq 1$, then $\|A\|_p \leq 1$ for all $p \geq 1$ (a theorem of *M. Riesz*).

For the proofs one can make use of theorem (129).

Instead of unitary invariance it is sufficient in (198)

to assume only that the norms of all quasi-permutations (relative to the canonical basis) equal 1. Note that theorems (198) and (199) carry over to complex spaces.

absolutely convex set (54)
acute wedge in Euclidean space D12
affine manifold D15
affine rank of convex polyhedron D24
affine rank of wedge D9
Amir-Moez' Theorem (156)
Ascoli-Mazur Theorem, generalized (102)

basis of wedge D9
G. Birkhoff's Theorem (127)

Carathéodory's Theorem (71)
circular Minkowski cone (53)
closed halfspace, preceding D6
closed convex hull D20
cone D4
conjugate wedge D11
convex body D17
convex combination D18
convex hull (Co F) D19
convex polyhedron D21
convex set D14
cross-norm on $\mathbb{M}(E)$ D30

dimension of affine manifold D15
dimension of wedge D5
doubly stochastic operator D26

extension of convex set, following (60)
extremal vector of wedge D10
extreme point of convex set D23

finite wedge D8
frame of convex polyhedron D21
frame of wedge D9

halfline, preceding D6
Helly's Theorem (73)
Horn's Inequality (154)
Horn's Theorem (140)
hyperplane D15

inner diameter D28
interior points of W (Int W), preceding D20

Kreĭn-Mil'man Theorem (115)
M. G. Kreĭn's Theorem (84)
Ky-Fan's Inequality (149)
Ky-Fan's Theorem (189)

Lidskiĭ's Theorem (product) (163)
Lidskiĭ's Theorem (sum) (159)
linear dimension of set D16
linear part of wedge D7
lower dimension of wedge D7
L-problem of moments (107)ff.

Markus' Theorem (136)
Markus' Theorems (sum and product) (166),(167)
Minkowski cone (7)
Minkowski's Theorem (94)
Mirsky's Theorem (153)
Mityagin's Theorem (198)

nonnegative linear combination D1
nonnegative linear functional D22
nonnegative linear hull D2
normal symmetrically convex body D33

obtuse wedge in Euclidean space D12
open halfspace, following (6)
order of convex polyhedron D24
order of wedge D9
order symbol $\prec$ defined, preceding (130)
Ostrowski's Theorem (143)
outer diameter D28

permutation operator D25
polyhedral wedge D8
positive hyperoctant, following D8
positive linear combination D1

positive linear hull D2
positively linearly independent D3
problem of moments (104)ff.

quasipermutation D27

Rado's Theorem (132)
M. Riesz' Theorem (199)

self-conjugate wedge D13
separation theorem for hyperplanes (87)
solid wedge D5
spectral properties of stochastic operators (123) ff.
stochastic operator D26
straight wedge in Euclidean space D12
strictly convex set D23
symmetrically convex body (184)
symmetrically convex set D32
symmetric seminorm D31

unitary-invariant norm D29

vector sum of subsets D6

wedge D4
Weyl's Inequalities (138)
Weyl's Theorem (95),(144)
Wielandt's Theorem (151)

CHAPTER VIII

PARTIALLY ORDERED SPACES

1. The Order Relation in a Linear Space.

Definition 1. A real linear space E is called *quasi-ordered* if we choose in E a wedge K (called the *positive wedge*) and define a quasiorder by the rule: $x \geq y$ (or $y \leq x$), if $x-y \in K$.§ The positive wedge is then the set

$$K = \{x | x \geq 0\} .$$

Definition 2. A vector $x \geq 0$ will be called *nonnegative*, and a vector $x \leq 0$ *nonpositive*. In the case when $x \geq 0$, $x \neq 0$, we shall write $x > 0$. The "inequality" $x < 0$ has an analogous sense.

Definition 3. If the positive wedge K is solid, then it is convenient to have a separate symbol for $x \in \text{Int } K$. Thus we shall write $x > 0$ for $x \in \text{Int } K$ and $x < 0$ for $-x \in \text{Int } K$. We shall call a vector $x > 0$ *positive*, and a vector $x < 0$ *negative*. The inequality $x > y$ means that $x-y > 0$.

The relation $\leq$ possesses the two characteristics of a *quasiorder*:

(1) $x \leq x$ (*reflexivity*);

(2) if $x \leq y$ and $y \leq z$, then $x \leq z$ (*transitivity*).

---

§Note the special meaning we have given to the signs $\geq$ and $\leq$.

We can introduce additional structure by stipulating:

(3) if $x \le y$, then $\alpha x \le \alpha y$ for any $\alpha \ge 0$ (*positive homogeneity*);

(4) if $x_1 \le y_1$ and $x_2 \le y_2$, then $x_1 + x_2 \le y_1 + y_2$ (*additivity*).

1. If a quasiorder relation which satisfies conditions (3) and (4) is given in E, then the inequality $x \ge 0$ defines a wedge which generates that relation.

2. If the wedge K is a cone, then

(5) $x \le y$ and $y \le x$ imply $x = y$ (*antisymmetry*).

Definition 4. If the positive wedge is a cone, then the quasiorder defined by it becomes an *order* by stipulation (5). The space E in this case is called *ordered*. It is clear that a quasiorder or order relation need not hold between all pairs of vectors in E. When this fact needs emphasis, we shall speak of a partial quasiorder or a partial order.

3. If an order relation is given in E which satisfies conditions (3) and (4), then the inequality $x \ge 0$ defines a cone which generates that relation.

4. If the positive wedge K is solid, then

(6) for any $x \in E$ there exists a $y > x$.

Conversely, if the quasiorder satisfies (6), then the positive wedge is solid.

5. If the positive wedge K is closed, then

(7) $x_k \ge 0$ and $\lim_{k\to\infty} x_k = x$ imply $x \ge 0$.

Conversely, if the quasiorder satisfies condition (7), then the positive wedge is closed.

6. If an order relation satisfying (3) and (4) has been defined on E and any pair $x,y \in E$ is comparable (i.e., $\ge$ holds between any x and y), then $\dim E = 1$.

Definition 5. A *Kantorovich space* is a space E with an order relation $\le$ which satisfies axioms (1) - (7).

From the propositions formulated above it follows that a Kantorovich space is an ordered space in which the order relation is defined by a solid closed cone.

The following propositions in this section hold in a Kantorovich space.

If for two vectors x,y of a Kantorovich space there exists a vector z such that $z \geq x$, $z \geq y$, and for any w satisfying $w \geq x$, $w \geq y$ the inequality $w \geq z$ holds, then we shall write $z = \sup\{x,y\}$ and use $\inf\{x,y\}$ analogously.-- For an arbitrary set $F \subset E$ the notions sup F and inf F are introduced in the same way as for a set of two elements.

7. If the element $z = \sup\{x,y\}$ exists, then it is uniquely determined.

8. If $z = \sup\{x,y\}$ exists, then

$$z + u = \sup\{x + u, y + u\} \quad (u \in E) .$$

9. If $\sup\{x,y\}$ and $\inf\{x,y\}$ exist, then

$$\sup\{x,y\} + \inf\{x,y\} = x + y .$$

10. If $\sup\{x,y\}$ exists for any x,y, then $\inf\{x,y\} = -\sup\{-x,-y\}$ also exists for any x,y.

Definition 6. A closed solid cone K for which $\sup\{x,y\}$ exists for any $x,y \in K$ is called a *minihedral cone*.

11. The positive hyperoctant in R^n is a minihedral cone.

12. The cone K_p mentioned before VII,(52) is minihedral only for dim E = 2.

13. A solid closed cone is minihedral iff for any $x,y \in K$ there exists $z \in K$ such that

$$(K + x) \cap (K + y) = K + z .$$

For this case $z = \sup\{x,y\}$.

14. A closed solid cone K is minihedral iff it is a finite cone of order n.

Hence

15. A Kantorovich space E with a minihedral positive cone is monotonically isomorphic with the arithmetic space R^n in which the order relation is generated by the positive hyperoctant, i.e., there exists an isomorphism $E \simeq R^n$ which preserves order (*Judin's Theorem*).

16. If E is a Kantorovich space with a minihedral positive cone, then for any $x \in E$ there exists a unique decomposition

$$x = x_+ - x_- \quad (x_- \geq 0,\ x_+ \geq 0)$$

such that if $x = x' - x''$ with $x' \geq 0$, $x'' \geq 0$, then $x' \geq x_+$, $x'' \geq x_-$.

Definition 7. A decomposition with the indicated properties is called *minimal*. The vectors x_+ and x_- are called the *positive* and *negative* parts of the vector x, and the vector

$$x_+ + x_- ,$$

called the *absolute value* of the vector x, is denoted by $|x|$.

17. $|x| = \sup\{x,-x\}$.

18. $|-x| = |x|$.

19. $|x + y| \leq |x| + |y|$.

Here follow some remarks about passage to the limit in a Kantorovich space.

20. If the nondecreasing sequence of vectors $\{x_k\}_1^\infty$ is bounded from above, i.e., if there exists an x such that $x_k \leq x$ $(k = 1,2,\ldots)$, then it converges, and $\lim_{k\to\infty} x_k = \sup_k x_k$.

21. If the series $\Sigma_{k=1}^\infty y_k$ with nonnegative terms converges, then it converges unconditionally.

22. If for the series

$$\Sigma_{k=1}^\infty x_k \qquad (*)$$

there exists a convergent series $\Sigma_{k=1}^\infty y_k$ with nonnegative terms such that

$$-y_k \leq x_k \leq y_k \quad (k = 1,2,3,\dots) \;,$$

then the series (*) converges unconditionally.

23. If the positive cone is minihedral and the series $\Sigma_{k=1}^{\infty}|x_k|$ converges, then the series $\Sigma_{k=1}^{\infty}x_k$ converges unconditionally.

In conclusion let us observe that

24. In a Kantorovich space with a minihedral positive cone sup F exists for any set F which is bounded above.

2. Theory of Linear Inequalities

Let the closed wedge K define a quasiorder in the space E. The conjugate wedge K' (VII, D11) defines the <u>conjugate quasiorder</u> in E'. We shall assume that in the conjugate space the quasiorder is always introduced by this method.

Let now E and E_1 be quasiordered spaces with positive wedges K and K_1, and let $h \in \text{Hom}\,(E,E_1)$. Define

$$\text{Im}^+ h = hK, \; \text{Ker}^+ h = h^{-1} K_1 \quad \text{(cf. I,D51)}.$$

25. If K = E, then $\text{Im}^+ h = \text{Im}\, h$ and $\text{Ker}^+ h' = \text{Ker}\, h'$ (cf. I,D69).

Dually,

26. If $K_1 = 0$, then $\text{Ker}^+ h = \text{Ker}\, h$ and $\text{Im}^+ h' = \text{Im}\, h'$.

Henceforth in this section the spaces E and E_1 are assumed to be Kantorovich spaces. Then the spaces E' and E_1' will also be Kantorovich spaces (cf. VII,(42)).

27. $\dim \text{Im}^+ h = \text{rk}\, h$.

The following theorem, where $(\cdot)'$ denotes the conjugate wedge of the wedge $(\cdot)$, generalizes the first and the fourth identity in the fundamental set of identities I,(197).

28. $\text{Ker}^+ h' = (\text{Im}^+ h)'$, $\text{Im}^+ h' = (\text{Ker}^+ h)'$.

(For $\mathrm{Im}^+ h' \supset (\mathrm{Ker}^+ h)'$, the contrapositive assertion may be proved: apply the separation theorem VII,(87) to the closed wedge $\mathrm{Im}^+ h'$ and use the canonical isomorphism between E" and E.)

The following theorems (29) - (31) are analogs of the third Fredholm theorem (I,202). Theorem (28) implies

29. The inhomogeneous equation $hx = y$ has a nonnegative solution iff $g(y) \geqq 0$ for any solution g of the homogeneous inequality $h'g \geq 0$.

30. The inhomogeneous inequality $hx \geq y$ has a nonnegative solution iff $g(y) \leqq 0$ for any nonnegative solution of the homogeneous equation $h'g = 0$.

Theorem (30) can be derived from (29) by considering the space $\tilde{E} = E \oplus R^1$ and defining the homomorphism $\tilde{h}$ by the formula

$$\tilde{h}(x,\xi) = hx + \xi y \quad (x \in E,\ \xi \in R^1).$$

Thus $\tilde{h}'g = (h'g,\ g(y))$. Consider the equation $\tilde{h}'g = (0,1)$. One can establish theorems (31, (33), ans (34) analogously.

31. The inequality $hx \geq y$ has a nonnegative solution iff $g(y) \leqq 0$ for any nonnegative solution of the homogeneous inequality $h'g \leq 0$.

The next propositions (32) - (34) are analogous to the first Fredholm theorem I,(97).

32. $hx = y$ has a nonnegative solution for any $y \in E_1$ iff $h'g \geq 0$ has only the trivial solution $g = 0$.

In particular, it is necessary that h' be a monomorphism.

33. $hx \geq y$ has a solution for any $y \in E_1$ iff the only nonnegative solution of $h'g = 0$ is the trivial solution.

In particular, it is sufficient that h' be a monomorphism.

34. $hx \geq y$ has a nonnegative solution for each $y \in E_1$ iff the only nonnegative solution of $h'g \leq 0$ is the trivial solution.

The following three propositions about homogeneous equations and inequalities can be regarded as analogs of the second Fredholm theorem.

35. $hx = 0$ has a nontrivial nonnegative solution (i.e., a solution $x > 0$) iff $h'g > 0$ has no solution.

36. $hx > 0$ has a solution iff $h'g = 0$ has no positive solution.

37. $hx \geq 0$ has a nontrivial nonnegative solution iff $h'g < 0$ has no nonnegative solution.

38. The set $F \subset E_1$ of those y for which the equations in (29) are solvable is closed.

As in the case of equations (cf. I,(211)-(213)) one can pass from the language of homomorphisms to the language of functionals. In doing so, E_1 becomes the m-dimensional real arithmetic space R^m. We shall take as the positive cone in R^m (previously K_1) the positive hyperoctant (clearly a finite cone).

A translation into the language of functionals transforms propositions (29) - (37) into (39) - (47), where $f_k \in E'$ $(k = 1,2,\ldots,m)$. The results (40) and (41) can be established directly (cf. Nr. 11 of the "Special Literature" to this chapter).

39. The system of equations

$$f_k(x) = \alpha_k \quad (k=1,2,\ldots,m)$$

has a nonnegative solution iff the inequality

$$\Sigma_{k=1}^{m} \gamma_k f_k \geq 0$$

implies

$$\Sigma_{k=1}^{m} \gamma_k \alpha_k \geqq 0 .$$

40. The system of inequalities

$$f_k(x) \geqq \alpha_k \quad (k = 1,2,\ldots,m)$$

as a solution iff the equality

$$\Sigma_{k=1}^{m} \gamma_k f_k = 0$$

or $\gamma_k \geqq 0$ implies

$$\Sigma_{k=1}^{m} \gamma_k \alpha_k \leqq 0 \ .$$

1. The system of inequalities

$$f_k(x) \geqq \alpha_k \qquad (k = 1,2,\ldots,m)$$

as a nonnegative solution iff the inequality

$$\Sigma_{k=1}^{m} \gamma_k f_k \leq 0$$

or $\gamma_k \geqq 0$ implies

$$\Sigma_{k=1}^{m} \gamma_k \alpha_k \leqq 0 \ .$$

2. The system of equations

$$f_k(x) = \alpha_k \qquad (k = 1,2,\ldots,m)$$

as a solution for any right-hand side iff the inequality

$$\Sigma_{k=1}^{m} \gamma_k f_k \geq 0$$

mplies $\gamma_k = 0$ $(k = 1,2,\ldots,m)$.

3. The system of inequalities

$$f_k(x) \geqq \alpha_k \qquad (k = 1,2,\ldots,m)$$

as a solution for any right-hand side iff the equation

$$\Sigma_{k=1}^{m} \gamma_k f_k = 0$$

or $\gamma_k \geqq 0$ implies $\gamma_k = 0$ $(k = 1,2,\ldots,m)$.

4. The system of inequalities

$$f_k(x) \geqq \alpha_k \qquad (k = 1,2,\ldots,m)$$

as a nonnegative solution for any right-hand side iff the
nequality

$$\Sigma_{k=1}^{m} \gamma_k f_k \leq 0$$

for $\gamma_k \geqq 0$ implies $\gamma_k = 0$ $(k = 1,2,\ldots,m)$.

45. The system of equations

$$f_k(x) = 0 \quad (k = 1,2,\ldots,m)$$

has a nontrivial nonnegative solution iff the inequality

$$\Sigma_{k=1}^{m} \gamma_k f_k > 0$$

is not satisfied for any γ_k.

46. The system of inequalities

$$f_k(x) \geqq 0 \quad (k = 1,2,\ldots,m)$$

has an f-nontrivial solution iff the equation

$$\Sigma_{k=1}^{m} \gamma_k f_k = 0$$

is not satisfied for any $\gamma_k > 0$ (a solution is called f-nontrivial if all $f_k(x)$ are not simultaneously equal to zero).

47. The system of inequalities

$$f_k(x) \geqq 0 \quad (k = 1,2,\ldots,m)$$

has a nontrivial nonnegative solution iff

$$\Sigma_{k=1}^{m} \gamma_k f_k < 0$$

is not satisfied for any $\gamma_k \geqq 0$.

The language of functionals is suitable when one has to individualize certain equations of a system. Let us consider some examples of this kind.

48. Let the system

$$f_k(x) = \alpha_k \quad (k = 1,2,\ldots,m-1)$$

$$f_m(x) \geqq \alpha_m$$

e consistent. All solutions of the system

$$f_k(x) = \alpha_k \quad (k = 1,2,\ldots,m-1)$$

atisfy the inequality

$$f_m(x) \geqq \alpha_m$$

ff the functional f_m is a linear combination of the unctionals $f_1, f_2, \ldots, f_{m-1}$.

9. Let the system

$$f_k(x) \geqq \alpha_k \quad (k = 1,2,\ldots,m) \qquad (*)$$

e consistent and $\mathrm{rk}\{f_1, f_2, \ldots, f_m\} = r \geqq 1$ (cf. I,D12). hen a subsystem $f_{k_1}, f_{k_2}, \ldots, f_{k_r}$ can be chosen from the unctionals $\{f_k\}_1^m$ such that each solution of the system of quations

$$f_{k_s}(x) = \alpha_{k_s} \quad (s = 1,2,\ldots,r)$$

s a solution of (*) (<u>principle of bounding solutions</u>).

0. Let the system

$$f_k(x) \geqq \alpha_k \quad (k = 1,2,\ldots,m-1) \qquad (**)$$

e consistent. If the inequality

$$f_m(x) \geqq \alpha_m \qquad (***)$$

s a consequence of this system but of none of its subystems, then

$$\mathrm{rk}\{f_1, f_2, \ldots, f_{m-1}\} = m-1 .$$

1. The inequality (***) follows from the consistent system **) iff there exist numbers $\gamma_k \geqq 0$ $(k = 1,2,\ldots,m-1)$ such hat

$$f_m = \Sigma_{k=1}^{m-1} \gamma_k f_k \quad \text{and} \quad \Sigma_{k=1}^{m-1} \gamma_k \alpha_k \geqq \alpha_m .$$

52. The system of inequalities

$$f_k(x) \geqq \alpha_k \quad (k = 1,2,\ldots,m)$$

is consistent iff each of its subsystems of n+1 inequalities is consistent. (n has its usual meaning.)

Helly's theorem VII,(73) follows from theorem (52) also.

Definition 8. A system

$$f_k(x) \geqq \alpha_k \quad (k = 1,2,\ldots,m) \qquad (*)$$

is called *irreducibly inconsistent* if it is inconsistent but any of its proper subsystems is consistent.

53. The system (*) is irreducibly inconsistent iff any m-1 of the functionals $f_1, f_2, \ldots, f_m$ are linearly independent and there exist numbers $\gamma_k > 0$ such that

$$\Sigma_{k=1}^{m} \gamma_k f_k = 0 \quad \text{and} \quad \Sigma_{k=1}^{m} \gamma_k \alpha_k > 0 \ .$$

If the spaces E and E_1 are Euclidean and the positive cones $K \subset E$ and $K_1 \subset E_1$ are self-dual, then the propositions (29) - (37) assume the forms (54) - (59).

54. The inhomogeneous equation $hx = y$ has a solution $x \geq 0$, iff $(y,z) \geqq 0$ for any solution z of the homogeneous inequality $h'z \geq 0$. There is a solution for each y iff $h'z \geq 0$ has only the trivial solution $z = 0$.

55. The inhomogeneous inequality $hx \geq y$ has a solution iff $(y,z) \leqq 0$ for any solution $z \geq 0$ of the homogeneous equation $h'z = 0$. There is a solution for each y iff $h'z \geq 0$ has only the trivial nonnegative solution $z = 0$.

56. The inhomogeneous inequality $hx \geq y$ has a solution $x \geq 0$ iff $(y,z) < 0$ for any solution $z \geq 0$ of the inhomogeneous inequality $h'z \leq 0$. There is a solution for each y iff the only nonnegative solution of $h'z \leq 0$ is the trivial one.

57. The homogeneous equation $hx = 0$ has a solution $x \succ 0$

iff $h'z > 0$ has no solution.

58. The inequality $hx \succ 0$ has a solution iff the homogeneous equation $h'z = 0$ has no positive solution.

59. The homogeneous inequality $hx \geq 0$ has a solution $x \succ 0$ iff the inequality $h'z < 0$ has no nonnegative solution.

In particular,

60. If $E = E_1$ and h is an antisymmetric operator, then $hx \leq 0$ has a solution $x \succ 0$.

We shall now return to our original language of homomorphisms to study the structure of the set of solutions of the inequality $hx \geq y$.

61. The set V of solutions of $hx \geq y$ is convex.

62. The set Q of solutions of the homogeneous inequality $hx \geq 0$ is a wedge.

63. The set Q is the vector sum $Q = \text{Ker } h + K$, where K is a certain cone.

64. If the set Q is represented in the form of a vector sum of a subspace and a cone $Q = L + K$, then $L = \text{Ker } h$ and $K \cap L = 0$.

65. In any representation (64) the decomposition of $x \in Q$ in the form $x = x_1 + x_2$ ($x_1 \in K$, $x_2 \in L$) is unique.

66. The set V of solutions of $hx \geq y$ is bounded iff $hx \geq 0$ has only the trivial solution.

72. If the positive cone of the space E_1 is finite (VII,D8), then let W denote a polyhedron, K is a finite cone, and the general solution of $hx \geq y$ has the form

$$x = \Sigma_{k=1}^{p} \alpha_k x_{0k} + \Sigma_{k=1}^{q} \beta_k x_{1k} + \Sigma_{k=1}^{r} \gamma_k x_{2k} ,$$

where $\alpha_k \geqq 0$, $\Sigma_{k=1}^{p} \alpha_k = 1$, $\beta_k \geqq 0$, and the γ_k are arbitrary real numbers; $\{x_{0k}\}_1^p$ is a basis of the polyhedron W, $\{x_{1k}\}_1^q$ is a basis of the cone K, and $\{x_{2k}\}_1^r$ is a basis of the subspace L (*Motzkin's Theorem*).

3. Linear and Convex Extremum Problems

Definition 9. A real functional $p(x)$ defined on a convex set $W \subset E$ is called *convex* if for any $x_1, x_2 \in W$

$$p((1-\theta)x_1 - \theta x_2) \leqq (1-\theta)p(x_1) + \theta(p(x_2) \quad (0 < \theta < 1).$$

The functional p is called *strictly convex* if in this relation equality holds only for $x_1 = x_2$.

Any linear functional is convex (but not strictly convex).

73. If p_o and p_1 are convex functionals in W then the functional $\alpha p_o + \beta p_1$ $(\alpha, \beta \geqq 0)$ is also convex.

74. If $\{p_\nu\}$ is a set of convex functionals in W and if for each $x \in W$ the set of numbers $\{p_\nu(x)\}$ is bounded, then the functional

$$p(x) = \sup_\nu p_\nu(x) \quad (x \in W)$$

is convex.

75. A convex functional is continuous on Int W.

76. Let the functional p be continuous in W and differentiable in Int W (cf. VI, Sect. 4) and let the functional $f_{x,y} \in E'$ be defined by the equality

$$f_{x,y} = p'(x) - p'(y) \quad (x,y \in \text{Int } W) .$$

The functional p is convex iff

$$f_{x,y}(x - y) \geqq 0 \quad (x,y \in \text{Int } W)^{\S}$$

77. Let the functional p be continuous in W and twice differentiable in Int W, and let the homomorphism $h_x \in$ Hom (E,E') and the functional $g_{x,y} \in E'$ be defined by the equations

§In the notation of VI,(104) x,y would be replaced by x_o, y_o throughout. The homomorphism $p'(x_o) - p'(y_o)$, here a linear functional, maps the vector $x_o - y_o$ into R^1.

$$h_x = p''(x),\ g_{x,y} = h_{xy} \quad (x,y \in \text{Int } W) .$$

The functional p is convex iff

$$g_{x,y}(y) \geq 0 \quad (x,y \in \text{Int } W).^{\S}$$

From now on the domain of definition W of the convex functional p is assumed to be closed and p continuous on W.

78. All points at which a strictly convex functional takes a local minimum are extreme points of its domain of definition W.

79. Each point of local minimization of a convex functional p is a point of global minimization (i.e., p(x) assumes its least value at that point).

80. The set V_p of points at which the functional p takes its global minimum is closed and convex.

81. If p is a strictly convex functional, then the set V_p contains no more than one point.

82. A linear functional $f \neq 0$ has on a closed convex set W no local extrema other than global extrema, and they are attained only on the boundary of the set W.

Connected with this result, we have for the set V_f defined in analogy with V_p:

83. The extreme points of the set V_f appear among the extreme points of the set W.

Consequently: if the set W is strictly convex (see VII,

---

§ In the present case p"(x) or h_x is a particular case of the homomorphism F"(x), defined in the paragraph preceding VI,(115); there the "floating" x stands again for x_o as in VI,(104) and does not denote the vector argument of h_x or F"(x). For fixed x, h_x is actually a bilinear functional on Int W ⊂ E. Fixing one argument vector y ∈ Int W, we obtain the functional $h_x(y,.) \equiv h_x y \underset{\text{def}}{=} g_{x,y}$. Thus $h_x(y,y) = g_{x,y}(y) \geq 0$ must be shown. Kantorovič and Akilov (= Nr.32 in the list of General Textbooks) have an independently readable account of Fréchet derivatives in Ch. 17 of their book.

D23), then V_f consists of no more than one point.
Analogous assertions are obviously true for the set of points in W on which f attains its global maximum.
We shall look more closely at the special case of a quadratic functional

$$q(x) = (Sx,x) - 2(x,y) \quad (x \in E)$$

in a Euclidean space E. Here S is a self-adjoint operator, and $y \in E$.

84. The quadratic functional q is convex (strictly convex) iff the operator S is nonnegative (positive) (cf. III,D38).

85. If S is a nonnegative operator and y = 0, then

$$\min q(x) = 0, \quad V_q = \text{Ker } S .$$

(V_q is defined in analogy with V_p.)

86. If S is a positive operator, then the functional q takes its minimum at the unique point which is the solution of the equation

$$Sx = y . \tag{*}$$

87. If S is a nonnegative operator and equation (*) has no solution, then the functional q is not bounded from below.

88. If S is a nonnegative operator and equation (*) is solvable, then the functional q is bounded from below and the set V_q coincides with the set of solutions of equation (*), which is an affine manifold (VII,D15).

And so in all cases $V_q = \{x | Sx = y\}$.

Now let E and E_1 be Kantorovich spaces, $f \in E'$, $h \in \text{Hom}(E,E_1)$, and $y \in E_1$. We shall study the problem of finding the minimum

$$\min_x f(x) \quad (x \in E) \tag{I}$$

under the restrictions

$$hx \geq y , \tag{II}$$

$$x \geq 0 . \tag{III}$$

Definition 10. This problem, in which f, h, and y are given, is called the *problem of linear optimization* or the *problem of linear programming*. We shall denote it by <f,h,y>.

Definition 11. A vector x is a *feasible* vector for the problem <f,h,y> if it satisfies conditions (II), (III). If the problem has a feasible vector, it is called feasible. If for a feasible problem the functional f is bounded from below on the set of feasible vectors, the problem is called bounded.

89. If the problem <f,h,y> is bounded, then $\min_x f(x)$ exists.

Definition 12. The number $\min_x f(x)$ is the *value* of the problem, and the vector on which this value is taken is called an *optimal* vector.

90. The set of optimal vectors of the problem <f,h,y> is convex.

The following is the dual problem to <f,h,y>: find the maximum

$$\max_g g(y) \qquad (g \in E_1') \tag{I'}$$

under the restrictions

$$h'g \le f , \tag{II'}$$

$$g \ge 0 . \tag{III'}$$

Definitions 10 and 11 with y, h_1' and f given, carry over in a natural way to the dual problem which we denote by <y,h',f>.

91. Let $x \in E$ and $g \in E'$ be feasible vectors for the problems <f,h,y> and <y,h',f> respectively. Then $f(x) \geqq g(y)$.

92. If equality holds in (91), then x and g are optimal vectors of the corresponding problems (optimality criterion).

93. If one of the two mutually dual problems <f,y,h> and <y,h',f> is not feasible, then the other has no optimal vectors (i.e., it is either infeasible or feasible but not bounded).

94. If both mutually dual linear optimization problems are feasible, then they both have optimal vectors and the value of either problem is the same (theorem of duality).

For the proof it is sufficient to establish the existence of nonnegative solutions $x \geq 0$, $g \geq 0$ to the systems of inequalities

$$hx \geq y \ ,$$
$$-h'g \geq -f \ ,$$
$$g(y) - f(x) \geq 0 \ ,$$

which can be represented in the form

$$\tilde{h}\begin{pmatrix} x \\ g \\ R^1 \end{pmatrix} \geq \begin{pmatrix} y \\ -f \\ 0 \end{pmatrix} .$$

Here $\tilde{h}$ denotes the homomorphism, defined as indicated above, from $E \times E_1' \times R^1$ into $E_1 \times E' \times R^1$, where the order in R^1 is the usual one and the Cartesian product is ordered componentwise.

Assertions (93) and (94) can be unified in the following existence theorem:

95. A linear optimization problem has an optimal vector iff it and its dual problem are feasible.

Using (94) we can also establish the following proposition.

96. Let $\check{x}$ be an optimal vector for the problem $<f,h,y<$, and define the sets G and Z by

$$G = \{g \mid g(h\check{x} - y) = 0,\ g \geq 0\} \ ,$$
$$Z = \{z \mid g(z) = 0,\ g \in G\} \ .$$

If $\hat{g}$ is an optimal vector for the dual problem $<y,h',f>$, then $\hat{g}(z) = 0$ for all $z \in Z$.

A more complete assertion, called the equilibrium theorem, is possible:

97. The feasible vectors x_o and g_o of the problems $<f,h,y>$ and $<y,h',f>$ are optimal iff

$$g_o(z) = 0 \quad \text{for all } z \in Z ,$$
$$h(x_o) = 0 \quad \text{for all } h \in H ,$$

where

$$Z = \{z | g(z) = 0, \ g \in G\}, \ G = \{g | g(hx_o - y), \ g \geq 0\}$$

and

$$H = \{k | k(x) = 0, \ x \in X\}, \ X = \{x | (f - h'g)(x) = 0, \ x \geq 0\}.$$

Definition 13. The *Lagrange functional* (or *function*) of the pair of mutually dual problems <f,h,y> and <y,h',f> is defined on $E \times E_1'$ for $x \geq 0$, $g \geq 0$ by the equation

$$\varphi(x,g) = f(x) + g(y-hx) = g(y) + (f-h'g)(x) \quad (x \in E, \ g \in E_1').$$

If the sign of inequality in (II) or (II') is replaced by an equality sign, then $\varphi(x,y)$ becomes the Lagrange function for one of the two resulting problems of constrained extremum[§] (this is well known from analysis).

Definition 14. The pair (x_o, g_o) is a *saddle point* of the functional $\varphi(x,y)$ if for all $x \geq 0$, $g \geq 0$

$$\varphi(x_o,g) \leqq \varphi(x_o,g_o) \leq \varphi(x,g_o) .$$

A saddle point of the function $\varphi(x,g)$ obviously is one of its stationary points in the sense of differential calculus.

The importance of the Lagrange functional for the linear optimization problem is demonstrated by propositions (98) and (99). The first of these is established immediately, and the second can be derived from the theorem of duality (94) and the equilibrium theorem (97), or established with the aid of the separation theorem for convex sets, VII,(103).

8. If (x_o, g_o) is a saddle point of the Lagrange functional

---

§ The functional g plays the role of the Lagrange multiplier for the problem <f,h,y>, and the vector x plays that role for the problem <y,h',g>.

$\varphi(x,g)$, then x_o and g_o are optimal vectors for the problems $\langle f,h,y\rangle$ and $\langle y,h',f\rangle$. Thus $\varphi(x_o,g_o)$ is the common value of these problems.

Conversely,

99. If x_o and g_o are optimal vectors for the problems $\langle f,h,y\rangle$ and $\langle y,h',f\rangle$, then $\varphi(x_o,g_o)$ is the common value of these problems and (x_o,g_o) is a saddle point of the Lagrange functional.

The results (98), (99) can also be presented in the minimax form:

100. If the problems $\langle f,h,y\rangle$ and $\langle y,h',g\rangle$ are feasible, then

$$\max_{g\geq 0}\min_{x\geq 0}\varphi(x,g) = \min_{x\geq 0}\max_{g\geq 0}\varphi(x,g) \;;$$

Conversely, this equation implies the feasibility of both problems, and the common value of either side coincides with the common value of either problem.

Now we shall turn to convex extremum problems.

Definition 15. A mapping $F:E \to E_1$ is *concave* if for any x_1, $x_2 \in E$ and $0 < \theta < 1$

$$F((1-\theta)x_1 + \theta x_2) \geq (1-\theta)Fx_1 + \theta Fx_2 \,.$$

101. A concave mapping is continuous.

102. If the mapping F is concave, then the set of solutions of the inequality $Fx \geq 0$ is closed and convex.

Definition 16. Let p be a convex functional on E and $F:E \to E_1$ a concave mapping. The problem $\langle p,F\rangle$ of finding

$$\min_x p(x) \,,$$

under the restrictions

$$Fx \geq 0, \; x \geq 0,$$

is known as the *problem of convex optimization* or *convex programming*.

The terminology introduced above for the problem

<f,h,y> naturally carries over to the problem <p,F>.

Definition 17. The *Lagrange functional* (or *function*) for the problem <p,f> is defined by

$$\varphi(x,g) = p(x) - g(Fx) \qquad (x \in E,\ g \in E_1')$$

for $x \geq 0$, $g \geq 0$.

103. If (x_o, g_o) is a saddle point of the Lagrange functional $\varphi(x,g)$ of the problem <p,F>, then x_o is an optimal vector of this problem and $\varphi(x_o, g_o)$ is its value.

Conversely,

104. If $Fx > 0$ for some $x \geq 0$, then from the optimality of the vector x_o follows the existence of a functional $g_o \in E_1'$, $g_o \geq 0$ such that (x_o, g_o) is a saddle point of the Lagrange functional $\varphi(x,g)$ of the problem <p,F>.

For the proof one can define the convex sets

$$\tilde{W}_1 = \left\{ \begin{bmatrix} y \\ \xi \end{bmatrix} \middle| \ \exists x \geq 0 \text{ such that } \begin{bmatrix} y \\ \xi \end{bmatrix} \leq \begin{bmatrix} Fx \\ -p(x) \end{bmatrix} \right\}$$

$$\tilde{W} = \left\{ \begin{bmatrix} y \\ \xi \end{bmatrix} \middle| \begin{bmatrix} y \\ \xi \end{bmatrix} > \begin{bmatrix} 0 \\ -p(x_o) \end{bmatrix} \right\}$$

in the space $\tilde{E} = E \times R^1$ and use again the separation theorem VII,(103).

Propositions (103) and (104) together form the *Kuhn-Tucker Theorem.* This theorem and proposition (100) are modifications of the following *Theorem of von Neumann* about *saddle points.*

105. Let $h \in \text{Hom}(E, E_1)$, and U and V be compact convex sets in E and E_1. Then

$$\min_{x \in U} \max_{g \in V} g(hx) = \max_{g \in V} \min_{x \in U} g(hx)$$

We note the related theorem:

106. Let U and V be compact convex sets in E and E_1. Then

$$\min_{x\in U} \max_{f\in V} |f(x)| = \max_{f\in V} \min_{x\in U} |f(x)| .$$

In connection with convex and linear optimization let us still consider one problem of Chebyshev approximation.

Definition 18. The *Chebyshev point* of the system of equations

$$f_x(x) = \alpha_k \quad (f_k \in E', \ k = 1,2,\dots,m) \qquad (*)$$

is the name of a vector $\check{x} \in E$ for which

$$\max_k |f_k(\check{x}) - \alpha_k| = \min_{x\in E} \max_k |f_k(x) - \alpha_k| .$$

If the system (*) has a solution, then its set of Chebyshev points coincides with its set of solutions. But regardless of the existence of solutions, a Chebyshev point exists for any system (*).

Because the functional

$$p(x) = \max_k |f_k(x) - \alpha_k| \quad (x \in E) \qquad (**)$$

is convex, the search for a Chebyshev point of the system (*) is a problem of convex optimization, but it can be reduced to a problem of linear optimization:

107. Let $\tilde{E} = E \times R^1$, and let the functional $\tilde{f} \in \tilde{E}^1$ be defined by $\tilde{f}(\tilde{x}) = \xi$ where $\tilde{x} = \binom{x}{\xi}$, $x \in E$, $\xi \in R^1$. The set of Chebyshev points of the system of equations (*) coincides with the set of vectors $x \in E$ for which the vector $\tilde{x}$ is a solution of the minimization problem for the linear functional $\tilde{f}(\tilde{x})$ under the linear constraints

$$|f_k(x) - \alpha_k| \leq \xi \quad (k = 1,2,\dots,m) .$$

108. The Chebyshev point of the system of equations (*) with arbitrary α_k is unique iff $m \geq n$ and every n functionals out of $\{f_k\}_1^m$ are linearly independent (*Haar's Criterion*).

09. Let $\check{x}$ be the Chebyshev point of the system (*) and let

$$P = \max_k |f_k(\check{x}) - \alpha_k| .$$

f $m > n$ and Haar's criterion is satisfied, then for at least +1 values of the index k the equality $|f_k(\check{x}) - \alpha_k| = P$ holds.

Note that a reduction analogous to (107), transforming the convex optimization problem to a linear optimization problem, is always possible when the functional is piecewise linear. In particular, the search for a Chebyshev point satisfying additional linear relations in the form of equalities (Markov's problem) or inequalities is such a problem.

In closing this section we shall dwell briefly on the problem of minimizing an "arbitrary" functional for "any" constraints, but now we mean minimum in the local rather than in the global sense. First let us note a general proposition which follows from the theorem of separation.

10. Let $K_o, K_1, \ldots, K_m$ be open solid wedges and $K_{m+1} \neq 0$ closed wedge. Then $\bigcap_{j=o}^{m+1} K_j = 0$ iff there exist functionals $f_j \in K_j'$ not all zero which satisfy the condition $\sum_{j=o}^{m+1} f_j = 0$.

The local minimum problem consists of finding the local minima of the functional $\Phi(x)$ under the constraints

$$\varphi_j(x) \geqq 0 \quad (j = 1,2,\ldots,m) \; , \qquad (*)$$
$$\psi_k(x) = 0 \quad (k = 1,2,\ldots,r) \; , \qquad (**)$$

where the functionals $\Phi(x)$, $\varphi_j(x)$, $\psi_k(x)$ are assumed to be defined and continuous in all of E.

We shall assume r=1 and concern ourselves only with necessary conditions for minima.

Definition 19. A vector $y_o \neq 0$ is called a *forbidden variation* at the point x_o if for all y from some neighborhood of the point y_o and all $\varepsilon > 0$ from some neighborhood of zero

$$\Phi(x_o + \varepsilon y) < \Phi(x_o) .$$

A vector $y_o \neq 0$ is called an *admissible variation* for the constraint $\varphi(x) \geqq 0$ at the point x_o if for all y from some neighborhood of the point y_o and all $\varepsilon > 0$ from some neighborhood of zero

$$\varphi(x_o + \varepsilon y) \geqq 0 .$$

A vector $y_o \neq 0$ is called an *admissible variation* for the constraint $\psi(x) = 0$ if for any neighborhood U of the point y_o and any $\varepsilon_o > 0$ a number ε $(0 < \varepsilon < \varepsilon_o)$ and a vector $y \in U$ can be found such that

$$\psi(x_o + \varepsilon y) = 0 .$$

The sets of variations defined above are denoted by K_Φ, K_φ, and K_ψ respectively. Adjoining to each of them the point $y_o = 0$ we obtain sets $\dot{K}_\Phi$, $\dot{K}_\varphi$, $\dot{K}_\psi$. These sets are in general not wedges, but:

111. If $y \in \dot{K}_\Phi$ (or $y \in \dot{K}_\varphi$, or $y \in \dot{K}_\psi$), then $\alpha y \in \dot{K}_\Phi$ (respectively $\alpha y \in \dot{K}_\varphi$ and $\alpha y \in \dot{K}_\psi$) for $\alpha \geqq 0$.

112. If the sets K_Φ, K_φ, K_ψ are convex and $K_\Phi \neq \emptyset$, $K_\varphi \neq \emptyset$, then $\dot{K}_\Phi$ and $\dot{K}_\varphi$ are open solid wedges, and $\dot{K}_\psi$ is a closed wedge.

113. Let the functional $\Phi(x)$ have a local minimum at a point x_o satisfying the constraints (*), (**), i.e., $\Phi(x_o) \leqq \Phi(x)$ for any point x satisfying (*), (**), and belonging to some neighborhood of the point x_o. If the sets of variations K_Φ, K_{φ_j}, K_ψ $(j = 1,2,\ldots,m)$ are convex, then there exist functionals $f_o \in \dot{K}'_\Phi$, $f_j \in \dot{K}'_{\varphi_j}$, $f_{m+1} \in \dot{K}'_\psi$ not all zero which satisfy the condition

$$\Sigma_{j=0}^{m+1} f_j(x) = 0$$

(*Euler-Lagrange Equation*).

4. Extremum Problems in the Space of Operators

In this section we shall consider several different extremum problems in the normed space $\mathbb{M}(E)$, where E is a complex Euclidean space (IV,D7) and the norm in $\mathbb{M}(E)$ is the corresponding operator norm or any unitarily invariant norm (VII,D29). Therefore we shall denote the operator norm by $|\cdot|$ as in VII, Sect. 7.

We begin with a problem about normal operators (III,D47). Let A_o be a normal operator with spectrum $\sigma(A_o) = \{\lambda_k\}_1^n$ and orthonormal characteristic basis $\Delta = \{e_k\}_1^n$, and let $\{\mu_k\}_1^n$ be a given system of points of the complex plane. Let $\mathbb{B}$ denote the set of normal operators B with spectrum $\sigma(B) = \{\mu_k\}_1^n$. We wish to compute the least and greatest values of the Hilbert norm $\|A_o - B\|$ for $B \in \mathbb{B}$ (IV,D60).

The solution of this problem is contained in the following propositions (114) - (118).

14. If $(u_{jk})_1^n$ is the matrix of the unitary operator U in he basis Δ, then

$$\|A_o - B\|^2 = \|A_o - UB_oU^*\|^2 =$$

$$\mathrm{tr}(A_oA_o^* + B_oB_o^*) - \Sigma_{j,k=1}^n (\lambda_j\bar{\mu}_k + \bar{\lambda}_j\mu_k)|\mu_{jk}|^2 .$$

15. The matrix $(|\mu_{jk}|^2)$ is doubly stochastic (VII,D26).
e denote by $\mathbb{W}_o$ the set of operators which are defined in he canonical basis Δ_o of the arithmetic space R^n by all ossible matrices $(|\mu_{jk}|^2)$ $(U \in \mathbb{U}(E))$.

16. The set $\mathbb{W}_o$ contains all operators which are permutators elative to Δ_o (VII,D25).

17. The expression $\|A_o - B\|$ takes its least and its great- st values for $B \in \mathbb{B}$ if B has Δ as its characteristic basis.

Therefore

118. $$\min_{B\in\mathbf{B}} \|A_o - B\| = \min(\Sigma_{k=1}^{n} |\lambda_k - \mu_{j_k}|^2)^{\frac{1}{2}} ,$$

$$\max_{B\in\mathbf{B}} \|A_o - B\| = \max(\Sigma_{k=1}^{n} |\lambda_k - \mu_{j_k}|^2)^{\frac{1}{2}} ,$$

where minimum and maximum on the right are taken over all possible permutations $(j_1,j_2,\ldots,j_n)$ of the indices $1,2,\ldots,n$.

119. If A_o is a self-adjoint operator and the real numbers $\{\lambda_k\}_1^n$, $\{\mu_k\}_1^n$ are enumerated in increasing order, then

$$\min_{B\in\mathbf{B}} \|A_o - B\| = (\Sigma_{k=1}^{n} |\lambda_k - \mu_k|^2)^{\frac{1}{2}}$$

$$\max_{B\in\mathbf{B}} \|A_o - B\| = (\Sigma_{k=1}^{n} |\lambda_k - \mu_{n-k+1}|)^{\frac{1}{2}} .$$

Definition 20. An operator $A \in M(E)$ is said to *orthogonalize* the normalized basis Δ of the space E if $A\Delta$ is an orthonormal basis. The set of operators that orthogonalizes the basis Δ will be denoted by $\mathbf{O}_\Delta$.
If $A \in \mathbf{O}_\Delta$, $U \in \mathbf{U}(E)$, then $UA \in \mathbf{O}$. Also, if $A_1,A_2 \in \mathbf{O}_\Delta$ and U_1,U_2 are the right phase multipliers in the polar representation of the operators A_1, A_2 (III,D45), then $U_1^{-1}A_1 = U_2^{-1}A_2$.

120. If $\Delta = \{\mu_k\}_1^n$ is a normal basis, then

$$\min_{A\in\mathbf{O}_\Delta} \Sigma_{k=1}^{n} \|\mu_k - Au_k\|^2 = \mathrm{tr}(S^{\frac{1}{2}} - I)^2 ,$$

where S is the operator generated by the Gram matrix of the system Δ in any orthonormal basis. The minimum is achieved for $A = U^{-1}B$ where $B \in \mathbf{O}_\Delta$ and U is the right phase multiplier in the polar decomposition of the operator B (a theorem of M. G. *Kreĭn*).

121. If $S \in \mathbf{S}(E)$ and $\omega_1 \geq \omega_2 \geq \cdots \omega_n$ are the moduli of the eigenvalues of the (self-adjoint) operator S, then

$$\min_{\substack{X \in S(E) \\ \mathrm{rk}\, X=m}} |S - X| = \omega_{m+1}$$

The minimum is achieved for X = PS, where P is the orthoprojector onto the linear hull of those eigenvectors of the operator S which correspond to its m absolutely largest eigenvalues.

122. The value of the minimum in (121) and the point where it is assumed do not change under the wider variation:

$$X \in \mathbb{M}(E), \quad \mathrm{rk}\, X = m .$$

123. For each $A \in \mathbb{M}(E)$

$$\min_{\substack{X \in M(E) \\ \mathrm{rk}\, X=m}} |A - X| = s_{m+1}(A) :$$

The minimum is achieved when X is the segment of the Schmidt decomposition of A (III,(250)), given by

$$\Sigma_{k=1}^{m} s_k(A) \; (\cdot, e_k') e_k .$$

124. Let $A \in \mathbb{M}(E)$ and let $\mathbb{P}_m \subset \mathbb{M}(E)$ be the set of orthoprojectors of rank m. Then

$$\min_{X \in \mathbb{P}_m} \max_{\substack{y=Ax \\ \|x\| \leq 1}} \|y - Xy\| = s_{m+1}(A)$$

and

$$\max_{\substack{y=Ax \\ \|x\| \leq 1}} \|y - Py\| = s_{m+1}(A) ,$$

where $P = \Sigma_{k=1}^{m} (\cdot, e_k) e_k$.

Thus the linear hull of the system of vectors $\{e_k\}_1^m$ is the m-dimensional subspace which deviates least from the A-image of the unit ball of the space E.

The following proposition is the generalization of (123) to arbitrary unitarily invariant norms.

125. If $\tilde{p}(x) = \tilde{p}(\xi_1, \ldots, \xi_n)$ is a symmetric seminorm (VII,

D31) in R^n and $A \in \mathbb{M}(E)$, then

$$\min_{\substack{X \in \mathbb{M}(E) \\ \operatorname{rk} X = m}} \tilde{p}(s(A - X)) = \tilde{p}(0,0,\ldots,0,s_{m+1}(A),\cdots,s_n(A)) .$$

The minimum is achieved on the segment of the Schmidt decomposition

$$X = \Sigma_{k=1}^{m} s_k(A)\ (\cdot,e_k')e_k$$

In particular:

126. For each $A \in \mathbb{M}(E)$

$$\min_{\substack{X \in \mathbb{M}(E) \\ \operatorname{rk} X = m}} \operatorname{tr}[(A^* - X^*)(A - X)] = \Sigma_{k=m+1}^{n} s_k^2(A) .$$

The minimum is achieved for $X = \Sigma_{k=1}^{m} s_k(A)\ (.,e_k')e_k$.

In closing we mention the extremal properties of the Cartesian and polar decompositions of the operator $A \in \mathbb{M}(E)$ (III,D35,45).

127. If $A = S + iT$ $(S,T \in \mathbb{S}(E))$, then for any unitarily invariant norm

$$\min_{X \in \mathbb{S}(E)} \|A - X\| = \|A - S\| .$$

128. If $A = UR$, where U is a unitary operator and R is a nonnegative self-adjoint operator, then for any unitarily invariant norm

$$\min_{X \in \mathbb{U}(E)} \|A - X\| = \|A - U\|, \quad \max_{X \in \mathbb{U}(E)} \|A - X\| = \|A + U\|.$$

Theorem (128) can be reduced to the inequality

129. $\|R - I\| \leq \|R - X\| \leq \|R + I\| \quad (X \in \mathbb{U}(E))$.

This inequality follows from theorem VII,(189) and the inequality

130. $\Sigma_{k=1}^{m} s_k(R-I) \leq \Sigma_{k=1}^{m} s_k(R-X) \leq \Sigma_{k=1}^{m} s_k(R+I)$

$(m = 1,2,\ldots,n)$.

For the proof of the right-hand inequality in (130) it is sufficient to establish

131. $s_k(R - X) \leq s_k(R + I) \quad (k = 1,2,\dots,n)$

by the use of theorem III,(263).

For the proof of the left-hand inequality in (130) one can apply Wielandt's theorem VII,(151) to the operators $S = \tilde{R} - \tilde{X}$ and $T = \tilde{R} - \tilde{I}$ where the symbol $\tilde{}$ has the same meaning as in theorem VII,(152).

5. Monotone Operators

Definition 21. Let E be a Kantorovich space with positive cone K. A mapping $F : E \to E$ is *monotone* if $Fx_1 \geq Fx_2$ for $x_1 \geq x_2$ $(x_1, x_2 \in E)$.

132. The operator $A \in \mathbb{M}(E)$ is monotone iff $Ax \geq 0$ for all $x \geq 0$ $(x \in E)$.

This last condition means that the cone K is invariant relative to the operator A

$$AK \subset K .$$

133. The set $\mathbb{K}$ of monotone operators is a closed solid cone in $\mathbb{M}(E)$.

Therefore the space $\mathbb{M}(E)$ ordered by the cone $\mathbb{K}$ is a Kantorovich space. The monotonicity of the operator A considered as an element of this space is taken to mean that A is nonnegative.[§] Thus the relations $A \geq 0$ $(A > 0)$ will be understood in this section to refer to the order in $\mathbb{M}(E)$. In the space $\mathbb{M}(E')$ order is introduced analogously by the conjugate cone.

The aim of this section is the study of the spectral properties of operators nonnegative in $\mathbb{M}(E)$.

134. $A > 0$ iff $Ax > 0$ for all $x \succ 0$.

---

§ Nonnegativity so defined does not coincide with nonnegativity defined in III,D38 for self-adjoint operators.

135. The inequalities $A \geq 0$ and $A' \geq 0$ are equivalent. Analogously, the inequalities $A > 0$ and $A' > 0$ are equivalent. (A' is the linear operator in $\mathbb{M}(E')$ defined in II,D3.)

136. The cone $\mathbb{K} \subset \mathbb{M}(E)$ is finite iff the cone $K \subset E$ is finite.

137. If the cone K is minihedral and $A \geq 0$, then $|Ax| \leq A|x|$ for any $x \in E$ (see D7).

138. If $A \geq 0$, then $R_\lambda(A) \leq 0$ for $\lambda > \rho(A)$.

139. If the operator $A \geq 0$ has $\rho = \rho(A)$ as a positive eigenvalue of order r, then in the Laurent expansion of the resolvent

$$R_\lambda = \frac{c_{-r}}{(\lambda-\rho)^r} + \frac{c_{-r+1}}{(\lambda-\rho)^{r-1}} + \cdots$$

the coefficient c_{-r} is nonpositive.

140. Under the conditions of theorem (139) there corresponds to the eigenvalue $\rho = \rho(A)$ at least one eigenvector $x_o > 0$ of the operator A.

Theorem (140) represents a preliminary version of the theorem of Frobenius (143). In its final formulation the inclusion of $\rho(A)$ in the spectrum follows from the nonnegativity of the operator A and is not assumed beforehand. In the case when the circumference $|\lambda| = \rho(A)$ contains an eigenvalue λ for which arg λ is a rational multiple of 2π, the theorem of Frobenius follows from (141); in the general case it follows from (141) and (142).

141. If $A \geq 0$ and for some natural number p

$$A^p y = \mu y \quad (\mu > 0,\ y \succ 0),$$

then for

$$x = \rho^{p-1}y + \rho^{p-2}Ay + \cdots + \rho A^{p-2}y + A^{p-1}y,\ \rho = \sqrt[p]{\mu} > 0,$$

we have $x \succ 0$ and $Ax = \rho x$.

142. If λ ($|\lambda| = \rho(A)$) is an eigenvalue of the operator $A \geq 0$ having the largest real part among all eigenvalues with modulus $\rho(A)$, then for a sufficiently small $\varepsilon > 0$ the numbers $\lambda + \varepsilon\lambda^2$ and $\bar{\lambda} + \varepsilon\bar{\lambda}^2$ are eigenvalues of the operator $A + \varepsilon A^2$ of largest modulus.

143. For any nonnegative operator A the spectral radius $\rho(A)$ is an eigenvalue. To this eigenvalue corresponds at least one nonnegative eigenvector x_o (*Theorem of Frobenius*).

The first part of Frobenius' theorem can also be proved by applying to the decomposition of the resolvent

$$R_\lambda = -\Sigma_{k=0}^{\infty} \frac{A^k}{\lambda^{k+1}} \quad (|\lambda| > \rho(A))$$

Pringsheim's Theorem: the radius of convergence of a power series with nonnegative coefficients is a singular point of the sum of the series.

In the next two problems of this section we assume the cone K minihedral. It is then possible to make the following interesting inference on the absolutely largest eigenvalues of A.

144. If the cone $K \subset E$ is minihedral and the operator A nonnegative, then each of the eigenvalues of modulus $\rho = \rho(A)$ has the form $\rho\omega$ where ω is some p-th root of unity (p is a natural number). Therefore all points $\rho\omega^k$ ($k = 1,2,\ldots$) belong to the spectrum $\sigma(A)$.

Describing K by the system of inequalities $f_k(x) \geq 0$ ($f_k \in E'$, $k = 1,2,\ldots,n$) and complexifying the space E, we can with the aid of VII,(124) reduce (144) to

145. If in the hypothesis of (144) $\rho(A) = 1$, $Ax_o = x_o$ ($x_o \geq 0$) and $A^{(c)}x = e^{i\alpha}x$, then

$$e^{i\alpha} f_j^{(c)}(x) = \Sigma_{k=1}^{n} \alpha_{jk} f_k^{(c)}(x) \quad (j = 1,2,\ldots,n) .$$

and $(\alpha_{jk})_{j,k=1}^{n}$ is a stochastic matrix.

If $A > 0$, then the theorem of Frobenius admits of a

series of more precise results which we shall consider step by step; the final result is formulated in (157).

146. If $A > 0$, then $\rho(A) > 0$.

Thus a positive operator cannot be nilpotent.

147. An operator $A > 0$ cannot have an eigenvector in the boundary ∂K of the cone K.

148. The order (II,D26) of the eigenvalue $\rho(A)$ of the operator $A > 0$ is one.

149. The multiplicity of the eigenvalue $\rho(A)$ of the operator $A > 0$ is one.

150. The eigenvector x_o of the operator $A > 0$ belonging to the eigenvalue $\rho(A)$ is (up to scalar multiples) the unique nonnegative eigenvector of the operator A. The vector x_o is positive.

The last assertion can be established on the basis of the following lemma.

151. If $A \geq 0$ and $Ax = \mu x$ $(\mu > 0,\ x > 0)$, then for any $y > 0$ the sequence $\{\mu^{-m}A^m y\}_1^\infty$ has no limit point in ∂K.

Proposition (151) in its turn follows from the following simple remark.

152. $y \geq \alpha x$ for any $x > 0$ and $y > 0$ if $\alpha > 0$ is sufficiently small.

153. If $A > 0$, then the eigenvalue $\rho(A)$ is larger than the modulus of any other eigenvalue.

For the proof of (153) one can utilize lemmas (154) - (156) and also (151).

154. If $\lambda = re^{i\alpha}$ $(0 < r \leqq \rho,\ 0 < \alpha < 2\pi)$ is some eigenvector of the operator $A > 0$, and $x_1 + ix_2$ $(x_1, x_2 \in E)$ is the corresponding eigenvector, then the real linear hull of the vectors x_1, x_2 intersects the cone K only in its vertex.

155. For any $x > 0$ we have $x + \alpha_1 x_1 + \alpha_2 x_2 \in \partial K$ for some real α_1, α_2.

156. If in (153) $p\alpha = 2q\pi$ (p,q natural numbers), then $r < \rho(A)$.

Finally we obtain:

157. Any operator $A > 0$ has a unique nonnegative eigenvector x_o; this vector is positive, and the corresponding eigenvalue ρ is simple and larger than the modulus of any other eigenvalue of the operator A (*Perron's Theorem*).

The following proposition is a generalization of (146).

158. If $A \geq 0$, $x > 0$, and for some $\alpha > 0$

$$Ax \geq \alpha x ,$$

then $\rho(A) \geqq \alpha$.

In making this proposition more precise we are led to the following theorem which is an analog of the extreme property of the largest eigenvalue of a self-adjoint operator.

159. The spectral radius of the operator $A \geq 0$ equals

$$\rho(A) = \max_{\alpha \in \mathcal{A}} \alpha$$

where $\mathcal{A} = \{\alpha | Ax \geq \alpha x \text{ for some } x > 0\}$. It also equals

$$\rho(A) = \min_{\beta \in \mathcal{B}} \beta$$

where $\mathcal{B} = \{\beta | Ax \leq \beta x \text{ for some } x > 0\}$.

Therefore

160. For any $x > 0$ the inequalities $\alpha x \leq Ax$ and $\beta x \geq Ax$ imply the inequalities $\alpha \leqq \rho(A)$ and $\beta \geqq \rho(A)$.

From (159) one can obtain the following two formulae for $\rho(A)$:

161. $$\rho(A) = \inf_{\substack{g \geq o \\ g \in E'}} \sup_{\substack{x \geq o \\ x \in E}} g(Ax)/g(x) ,$$

$$\rho(A) = \sup_{\substack{x \geq o \\ x \in E}} \inf_{\substack{g \geq o \\ g \in E'}} g(Ax)/g(x) .$$

Thus

162. For the functional $\varphi(x,g) = g(Ax)/g(x)$ in $E \times E'$ one has the relation

$$\inf_{g \in K'} \sup_{x \in K} \varphi(x,g) = \sup_{x \in K} \inf_{g \in K'} \varphi(x,g)$$

(cf. (105)).

163. The lower bound in the first relation and the upper bound in the second relation of (161) are assumed on the corresponding eigenvectors of the operators A and A'.

Using (159) we can easily obtain also the following theorem.

164. The spectral radius $\rho(A)$ for $A \geq 0$ is a monotone function of A, i.e. $\rho(A_2) \geqq \rho(A_1)$ if $A_2 \geq A_1 \geq 0$.

165. If $A \geq 0$ and $R_\lambda(A) \leq 0$, then $\lambda > \rho(A)$ (cf. (138)).

Let us quote one application of theorems (138) and (165) to the study of stable operators. For this we introduce

Definition 22. An operator $B \in \mathbb{M}(E)$ is *strictly stable* if its spectrum $\sigma(B)$ lies in the open left half-plane.

Definition 23. An operator $B \in \mathbb{M}(E)$ is *translation-positive* if for some $\mu > 0$

$$B + \mu I > 0 .$$

166. A translation-positive operator B is strictly stable iff $B^{-1} \leq 0$.

Definition 24. An operator $A \geq 0$ is *primitive* if $A^p > 0$ for some natural number p.

167. All assertions of Perron's theorem (157) remain true when positivity of the operator A is replaced by primitivity.

This generalization is final in the following sense:

168. If the operator $A \geq 0$ possesses a positive simple eigenvalue which is larger than the moduli of the remaining eigenvalues, and if the corresponding eigenvectors of the operators A and A' are positive, then the operator A is primitive.

In closing this section let us note a proposition which itself represents a development of Perron's theorem. With this aim let us consider the exterior power $\overset{r}{\wedge}E$ with the positive cone $\overset{r}{\wedge}K$ for a minihedral cone K.

Definition 25. The operator $A \in \mathbb{M}(E)$ is *totally positive* iff $\overset{r}{\wedge} A > 0$ $(r = 1,2,\ldots,n)$.

From V,(89) and Perron's theorem follows

169. All eigenvalues of a totally positive operator are positive and simple (*Theorem of Gantmakher-Krein*).

6. Order Relations in the Space of Operators

Definition 26. Let $K \subset E$ be a closed solid cone. K induces an order in the space of operators $\mathbb{M}(E)$. This order is determined by the cone $\mathbb{K} = \mathbb{K}(K)$ of monotone operators. It is called the *operator order*.

Compared with arbitrary orders in the space $\mathbb{M}(E)$, operator orders occupy a place analogous to operator norms among arbitrary norms in $\mathbb{M}(E)$. The statements of the problems and results of this section are analogous to the exposition in IV, Sect. 10 (see also V, Sect. 6).

Definition 27. A cone $\mathbb{K} \subset \mathbb{M}(E)$ has the *ring property* (is a *ring cone*), if $AB \in \mathbb{K}$ for $A,B \in \mathbb{K}$.

170. The cone $\mathbb{K}(K)$ is a ring and contains the identity operator.

171. The cone Int $\mathbb{K}(K)$ is a ring, but it does not contain the identity operator.

Definition 28. The order in $\mathbb{M}(E)$ is a *ring order* if $A \geq 0$, $B \geq 0$ imply $AB \geq 0$. According to (170) any operator order is a ring order; note also $I \geq 0$.

We shall study the structure of the dual cone $\mathbb{K}' = [\mathbb{K}(K)]'$. This will lead us to one of the characterizations of operator orders.

Let $f \in E'$, $x \in E$. Consider the linear functional defined in $\mathbb{M}(E)$ by the formula

$$\varphi(A) = f(Ax) .$$

A functional of this form can be interpreted as a tensor product and may thus be denoted by $f \otimes x$.§ At the same time this product can be considered as an operator in E. The choice of the interpretation of the tensor product $f \otimes x$ will be clear each time from the context.

172. If $x \in K$ and $f \in K'$, then $f \otimes x \in \mathbb{K}'$. Conversely

173. If $f \otimes x \in \mathbb{K}'$, then $x \in K$, $f \in K'$ (or $-x \in K$, $-f \in K'$).

Denote by $\mathbb{F}'$ the set of all functionals of the form $f \otimes x$ ($x \in K$, $f \in K'$). By (172) $\mathbb{F}' \subset \mathbb{K}'$, but $\mathbb{F}' \neq \mathbb{K}'$, since the set $\mathbb{F}'$ is not a wedge. However

174. The cone $\mathbb{K}'$ is the closed convex hull of the set $\mathbb{F}'$:

$$\mathbb{K}' = \overline{\mathrm{Co}}\, \mathbb{F}' .$$

175. If $x \in K$ and $f \in K'$ are extremal vectors of the cones K and K', then the functional $f \otimes x$ is an extremal vector of the cone $\mathbb{K}'$.

Conversely:

176. If $\Psi = f \otimes x$ is an extremal vector of the cone $\mathbb{K}'$, then x and f are extremal vectors of the cones K and K'.

Theorems (172) - (176) with the obvious changes in formulation carry over to the operator interpretation of the tensor product $f \otimes x$.

---

§With A running over $\mathbb{M}(E)$, $\varphi(A)$ maps the fixed pair (f,x) into C^1. Since φ, for varying (f,x), is clearly bilinear, it is a bilinear functional on $E'\times E$, whereas $f \otimes x$ is a bilinear functional on $E''\times E'$ or [c.i.] on $E\times E'$. Thus $\dim(E'\times E) = \dim(E\times E') = n^2$ shows that $f(Ax) \simeq f\otimes x$.--On the other hand, $f\otimes x$ maps any pair (y,g), ($y\in E, g\in E'$) onto $y(f)g(x)$. Fix y and let g run over E'. Then $f\otimes x$ becomes $\Phi(y,g)$, a linear functional on E', and hence assigns to y a certain vector from E, denoted by x in I(169). Thus $f\otimes x$ can be interpreted as an endomorphism $E\to E$.

177. Let $\mathbb{Q}$ be a closed solid cone contained in $M(E)$. Then $\mathbb{Q}$ is representable in the form $\mathbb{Q} = \mathbb{K}(K)$ iff the following conditions are satisfied:

1) If

$$K_1 = \{x \mid x \in E,\ f \otimes x \in \mathbb{Q} \quad \text{for some } f \in E'\},$$
$$K_2 = \{f \mid f \in E',\ f \otimes x \in \mathbb{Q}' \text{ for some } x \in E\},$$

then K_1 is a closed solid cone and $K_2 = K_1'$.

2) $\mathbb{Q}'$ is the closed convex hull of the set of all functionals of the form $f \otimes x$ which belong to $\mathbb{Q}$.

In this way we have obtained a characterization of operator orders in terms of the pair of cones $\mathbb{Q}$, $\mathbb{Q}'$. A characterization in terms of the cone $\mathbb{Q}$ alone will result from the following theorems.

178. If K_1, K_2 are closed solid cones in E, then the cones $\mathbb{K}_1 = \mathbb{K}(K_1)$ and $\mathbb{K}_2 = \mathbb{K}(K_2)$ are either incomparable (i.e. neither is a proper part of the other) or they coincide. In the latter case $K_2 = K_1$ or $K_2 = -K_1$.

Further:

179. If $\mathbb{Q}$ is a closed solid ring cone in $\mathbb{M}(E)$, then there exist $f \in E'$ and $x \in E$ ($f \neq 0$, $x \neq 0$) such that $f \otimes x \in \mathbb{Q}$.

Thus

180. The set

$$K = \{x \mid x \in E,\ f \otimes x \in \mathbb{Q} \quad \text{for some } f \in E'\}$$ is a

closed cone in E.

181. $\mathbb{Q} \in \mathbb{K}(K)$.

Consequently the cone K is solid.

Theorems (178) and (181) imply:

182. A closed solid ring cone $\mathbb{Q} \subset \mathbb{M}(E)$ defines in $\mathbb{M}(E)$ an operator order iff it is maximal with respect to inclusion.

Now it is easy to establish the existence of ring orders in $\mathbb{M}(E)$ which are not operator orders, although they satisfy the condition $I \geq 0$. For example:

183. If K_1 and $K_2 \neq K_1$ are closed solid cones in E and the cone $K_1 \cap K_2$ is also solid, then the intersection

$\mathbb{K}(K_1) \cap \mathbb{K}(K_2)$ is a closed solid ring cone which contains the identity, but the order it defines is not an operator order.

7. The Ordered Space of Self-adjoint Operators

We shall consider nonnegative (positive) self-adjoint operators <u>in the sense of III,D41</u> from the point of view of the theory of ordered spaces.

184. The set $\mathbb{S}_+(E)$ of nonnegative operators is a closed solid cone in $\mathbb{S}(E)$.

Thus the space $\mathbb{S}(E)$ with the order relation defined by the cone $\mathbb{S}_+(E)$ is a Kantorovich space.

Definition 29. The order relation defined in $\mathbb{S}(E)$ by $\mathbb{S}_+(E)$ is called the *spectral order*.

185. The set of positive operators coincides with Int $\mathbb{S}_+(E)$.

186. For $n \geqq 2$ the cone $\mathbb{S}_+(E)$ is not finite.

187. For $n \geqq 2$ the cone $\mathbb{S}_+(E)$ is not a ring cone.

Thus the spectral order is not an operator order.

188. The cone $\mathbb{S}_+(E)$ is unitarily invariant: if $S \in \mathbb{S}_+(E)$, then

$$USU^{-1} \in \mathbb{S}_+(E) \quad (U \in \mathbb{U}(E)).$$

189. The general form of a nonnegative linear functional in the space $\mathbb{S}(E)$ is determined by the formula $f(S) = \mathrm{tr}(CS)$, where $C \in \mathbb{S}_+(E)$.

Thus

190. The cone $\mathbb{S}_+(E)$ is selfconjugate (under the natural identification of the spaces $[\mathbb{S}(E)]'$ and $\mathbb{S}(E)$).

191. Let A be a self-adjoint operator. Then the set of operators of the form $X = \varphi(A)$, where φ is any real function, is a subspace in $\mathbb{S}(E)$, and its intersection with the cone $\mathbb{S}_+(E)$ is a finite cone (cf. remark preceding III,(148)).

.92. If $S_1 \geq 0$, $S_2 \geq 0$, and $S_1 \nabla S_2$, then $S_1S_2 \geq 0$.
The condition $S_1 \nabla S_2$ cannot be omitted, i.e., from $S_1 \geq 0$, $S_2 \geq 0$ does not follow $S_1S_2 \geq 0$. The <u>Schur product</u> of matrices (VI, after D31) remedies this "deficiency".

.93. If the operators S_1, S_2 are nonnegative and if their ıatrices in some orthogonal basis are (S_1) and (S_2), then :he self-adjoint operator S defined in this same basis by :he matrix $(S) = (S_1) \circ (S_2)$ is nonnegative. If, in addi- :ion, $S_1 > 0$, $S_2 > 0$, then $S > 0$ also.

.94. If $S_1 \leq S_2$, $S \geq 0$ and $S \nabla S_1$, $S \nabla S_2$ (or at least ; $\nabla (S_1 - S_2)$), then $S_1S \leq S_2S$.

Definition 30. We now introduce the *positive and negative parts of a self-adjoint operator* by the formulas

$$S_+ = p_+(S), \quad \text{where } p_+(\lambda) = \max\{0,\lambda\} ,$$
$$S_- = p_-(S), \quad \text{where } p_-(\lambda) = \min\{0,\lambda\} ,$$
$$\lambda \in \sigma(S) .$$

Note that the right and left operator moduli (III,D44) of a self-adjoint operator coincide and equal

$$|S| \underset{\text{def}}{=} p_+(S) + p_-(S), \quad \text{while}$$
$$S = p_+(S) - p_-(S) .$$

.95. The operator $|S|$ satisfies the relations $|S| \nabla S$, $-|S| \leq S \leq |S|$ and is the least operator which satisfies :hese relations for a given S (cf.(17)).

.96. The operator S_+ satisfies the relations $S_+ \nabla S$ and $S_+ \geq S$ and is the least operator satisfying these relations :or a given S.

The analogous characterization applies to the operator S_-. These propositions are particular cases of the following theorem:

l97. If $S_1 \nabla S_2$, then in the class of self-adjoint operators commuting with S_1 and S_2

$$\sup\{S_1,S_2\} = \frac{1}{2}(S_1 + S_2 + |S_1 - S_2|) ,$$

$$\inf\{S_1,S_2\} = \frac{1}{2}(S_1 + S_2 - |S_1 - S_2|) .$$

(For sup and inf see the remark preceding (7)).

We now consider the spectral order relation for orthoprojectors.

198. For orthoprojectors the inequality $P \leq Q$ is equivalent with $\operatorname{Im} P \subset \operatorname{Im} Q$.

Thus the relation $P \leq Q$ implies $P \nabla Q$.

199. If the orthoprojectors P_1 and P_2 commute, then in the class of all orthoprojectors which commute with P_1 and P_2

$$\sup\{P_1,P_2\} = P_1 + P_2 - P_1P_2, \ \inf\{P_1,P_2\} = P_1P_2 .$$

200. The orthoprojector $P = P[\operatorname{Im} S]$ depends monotonically upon S for $S \geq 0$:

$$P[\operatorname{Im} S_1] \leq P[\operatorname{Im} S_2] \quad (0 \leq S_1 \leq S_2) .$$

The next theorems refer to the existence of suprema and infima of infinite sets in $\mathbb{S}(E)$. Let $\mathbb{H}$ be any set in $\mathbb{S}(E)$.

201. If $T = \sup \mathbb{H}$, then

$$(Tx,x) = \sup_{S\in\mathbb{H}} (Sx,x) \quad (x \in E) .$$

Conversely,

202. If

$$(Tx,x) = \sup_{S\in\mathbb{H}} (Sx,x) \quad (x \in E) ,$$

then $T = \sup \mathbb{H}$.

203. If

$$\sup_{S\in\mathbb{H}} (Sx,x) < \infty \quad (x \in E) \qquad (*)$$

nd if for any $S_1, S_2 \in \mathbb{H}$ there is a majorant S $S \geq S_1$, $S \geq S_2$) in $\mathbb{H}$, then the set $\mathbb{H}$ possesses a least pper bound.

The necessity of condition (*) is obvious.

04. If $T = \sup \mathbb{H}$, then for any $\varepsilon > 0$ there exists an perator $S_\varepsilon \in \mathbb{H}$ such that for all $S \in \mathbb{H}$ which satisfy the ondition $S \geq S_\varepsilon$, the inequality $\|T - S\| \leqq \varepsilon$ holds.

Assertions for greatest lower bounds analogous to (201) - (204) are true.

We consider now some special problems on the construction of suprema.

Let $\mathbb{A} = \{S \mid S \in \mathfrak{S}(E),\ S \leqq A,\ \operatorname{Ker} S \supset L\}$, where L is a subspace and $A \geq 0$.

Denote by Q the orthoprojector onto the orthogonal complement to the subspace $A^{\frac{1}{2}}L$.

05. $A^{\frac{1}{2}}QA^{\frac{1}{2}} \in \mathbb{A}$.

06. For any $S \in \mathbb{A}$

$$(Sx,x) \leqq \|A^{\frac{1}{2}}x - A^{\frac{1}{2}}y\| \quad (x \in E,\ y \in L)\ .$$

07. For any $S \in \mathbb{A}$

$$S \leq A^{\frac{1}{2}}QA^{\frac{1}{2}}\ .$$

Thus $\sup \mathbb{A} = A^{\frac{1}{2}}QA^{\frac{1}{2}}$, and the least upper bound is attained.

Note also that

08. $\operatorname{Im} A^{\frac{1}{2}}QA^{\frac{1}{2}} = L^{\perp} \cap \operatorname{Im} A^{\frac{1}{2}}$.

Finally we mention an iteration process for extracting the square root of an operator.

09. If $0 \leq S \leq I$, then the sequence

$$S_{k+1} = S_k + \frac{1}{2}(S - S_k^2) \quad (k = 0,1,2,\ldots;\ S_o = 0)$$

as the limit $\lim_{k\to\infty} S_k = \sup_k S_k = S^{\frac{1}{2}}$ (cf. (20)).

8. Positive Operators and Inequalities for Eigenvalues

We shall devote this section to various problems connected with the spectral order in the space $\mathcal{S}(E)$. Some of these are related to the Fischer-Courant theory (cf. III, Sect. 8).

Let $S > 0$. Put $\lambda_1 = \lambda_1(S)$, $\lambda_n = \lambda_n(S)$.

210. $$\min_{\|x\|=1} [(Sx,x)\,(S^{-1}x,x)] = 1 .$$

211. $$\max_{\|x\|=1} [(Sx,x)\,(S^{-1}x,x)] = (\lambda_1+\lambda_n)^2/4\lambda_1\lambda_n$$

This is called the *Kantorovich Equality*.

212. $$\max_{\|x\|=1} [\|Sx\|^2 - (Sx,x)^2] = (\lambda_1 - \lambda_n)^2/4$$

213. $$\min_{\|x\|=1} \frac{(Sx,x)^2}{\|Sx\|^2} = 4\lambda_1\lambda_n/(\lambda_1+\lambda_n)^2 .$$

Observe also the following inequality.

214. If $(S) = (s_{jk})_{j,k=1}^{n}$ is the matrix of the operator S in some orthonormal basis and if

$$d_m = \det(s_{jk})_{j,k=1}^{m} \quad (m = 1,2,\ldots,n;\ d_o = 1) ,$$

then $$\Sigma_{k=1}^{n} (\lambda_k(S))^{\frac{1}{2}} \geq \Sigma_{k=1}^{n} (d_k/d_{k-1})^{\frac{1}{2}}$$

where equality holds only when (S) is a diagonal matrix (*M. G. Krein's Inequality*).

This theorem can be deduced easily from III,(265) if one makes use of the following two lemmas.

215. If $\{u_k\}_1^n$ is a basis and G_j is the Gram matrix of the system of vectors $\Gamma_j = \{u_k\}_{k=1}^{j}$, then the distance ρ_j of the vector u_j from the linear hull of the system Γ_{j-1} ($\mathrm{sp}\,\Gamma_o \underset{\mathrm{def}}{=} 0$) equals $\rho_j = (\det G_j/\det G_{j-1})^{\frac{1}{2}}$ ($j = 1,2,\ldots,n$, $\det G_o = 1$).

216. If the orthonormal system $\{e_k\}_1^n$ is obtained from the system $\{u_k\}_1^n$ by the Gram-Schmidt-Sonin process, then

$(e_j, u_j) = \rho_j$ $(j = 1,2,\ldots,n)$.

Note now the generalized Schwarz inequality (cf. IV,(9)).

217. If $T \geq 0$, then

$$|(Tx,y)|^2 \leq (Tx,x)(Ty,y) \quad (x,y \in E) .$$

Hence,

218. If $(Tx,x) = 0$, then $Tx = 0$.

Let $T > 0$. Then the (bilinear) functional $[x,y] = (Tx,y)$ is a new inner product in E. Thus we may introduce

Definition 31. The unitary space with the inner product $[x,y] = (Tx,y)$ is denoted by $E[T]$ and is called a *Friedrichs space.*

219. The formula $A = T^{-1}S$ establishes a bijection between $\mathbb{S}(E)$ and $\mathbb{S}(E[T])$, and the cone $\mathbb{S}_+(E)$ goes into the cone $\mathbb{S}_+(E[T])$.

The spectrum of the operator $T^{-1}S$ coincides with the set of values λ for which

$$Sx - \lambda Tx = 0 \qquad (*)$$

has a solution $x \neq 0$, and these solutions coincide with the corresponding eigenvectors.

Definition 32. $S - \lambda T$, an operator-valued function of λ, is called a *linear pencil of operators*, and the values λ and the vectors x satisfying (*) are its eigenvalues and eigenvectors.

220. The eigenvalues of the pencil of operators $S - \lambda T$ $(T > 0)$ are real, and the eigenvectors which belong to distinct eigenvalues are T-orthogonal: $(Tu,v) = 0$.

221. There is a T-orthogonal basis of eigenvectors of the pencil of operators $S - \lambda T$ $(T > 0)$.

One of the consequences of (221) is

222. If S is self-adjoint and T a positive operator in E, then $A = TS$ is an operator of scalar type with real spectrum.

223. The eigenvalues of the pencil $S - \lambda T$ $(T > 0)$ are con-

tained in the interval (α,β) iff $S - \alpha T > 0$ and $\beta T - S > 0$.

In particular:

224. If $S > 0$, $T > 0$, then the eigenvalues of the pencil $S - \lambda T$ are positive.

Finally we shall discuss inequalities for the eigenvalues considered as functionals on $\mathcal{S}(E)$.

225. The spectra of the operators S and $T \geq S$ satisfy the inequalities

$$\lambda_k(S) \leqq \lambda_k(T) \quad (k = 1,2,\ldots,n) .$$

226. The number of equality signs in (225) does not exceed $\operatorname{def}(T - S)$. In particular,

$$\lambda_k(S) = \lambda_k(T) \quad (k = 1,2,\ldots,n) \text{ and } T \geq S$$

imply $T = S$.

The first and last eigenvalues (but not, in general, the intermediate ones) have the following properties:

227. The functional $\lambda_1(S)$ is convex, and the functional $\lambda_n(S)$ is concave: for $0 < \theta < 1$

$$\lambda_1((1-\theta)S_1 + \theta S_2) \leqq (1-\theta)\lambda_1(S_1) + \theta\lambda_1(S_2) ,$$

$$\lambda_n((1-\theta)S_1 + \theta S_2) \geqq (1-\theta)\lambda_n(S_1) + \theta\lambda_n(S_2) .$$

We can estimate the perturbation of the spectrum of a self-adjoint operator by the norm or the rank of the perturbing operator.

228. $|\lambda_k(T) - \lambda_k(S)| \leqq \|T - S\|$.

229. If $S \leq T$ and $\operatorname{rk}(T - S) \leqq r$, then

$$\lambda_k(T) \leqq \lambda_{k-r}(S) \quad (k = r+1,\ldots,n) .$$

In particular:

230. If $S \leq T$ and $\operatorname{rk}(T - S) = 1$, then

$$\lambda_k(T) \leqq \lambda_{k-1}(S) \quad (k = 2,\ldots,n) .$$

In this case the spectra of the operators S and T

intertwine.

$$\lambda_1(T) \geqq \lambda_1(S) \geqq \lambda_2(T) \geqq \cdots \geqq \lambda_n(S)$$

(cf. III,(231)).

Theorems (228) and (230) do not admit a converse, but

231. If

$$|\lambda_k(T) - \lambda_k(S)| \leqq \varepsilon \quad (k = 1,2,\ldots,n; \ \varepsilon > 0) ,$$

then there exists a unitary operator U such that

$$\|UTU^{-1} - S\| \leqq \varepsilon.$$

And if $S \leq T$, then the operator U can be chosen so that $S \leq UTU^{-1}$.

232. If $S \leq T$ and

$$\lambda_k(T) \leqq \lambda_{k-1}(S) \quad (k = 2,\ldots,n) ,$$

where at least one of the inequalities is strict, then there exists a unitary operator U such that $S \leq UTU^{-1}$ and $\mathrm{rk}(UTU^{-1} - S) = 1$.

For the proof one can restrict oneself to the case of a real space E and seek an operator U in the form of a rotation.

9. Monotone and Convex Functions of Self-adjoint Operators

Let $\varphi(\lambda)$ be a real-valued scalar function defined on the finite or infinite interval (α,β).

Definition 33. A function $\varphi(S)$ of the operator $S \in \mathbf{S}(E)$ is *monotone* in the interval $(\alpha I,\beta I) = \{S|\ \alpha I \leq S \leq \beta I\}$, if $S \leq T$ and $S,T \in (\alpha I,\beta I)$ imply $\varphi(S) \leq \varphi(T)$.

233. The function $\varphi(S) = -S^{-1}$ is monotone in the intervals $S > 0$ and $S < 0$.

234. The function $\varphi(S) = -R_\lambda(S)$ is monotone in the intervals $S > \lambda I$ and $S < \lambda I$.

235. The linear fractional function

$$\varphi(S) = (\gamma_{11}S + \gamma_{12}I)(\gamma_{21}S + \gamma_{22}I)^{-1}$$

where all γ_{jk} are real and $\gamma_{11}\gamma_{22} - \gamma_{12}\gamma_{21} > 0$, is monotone in the intervals $S > -(\gamma_{22}/\gamma_{21})I$ and $S < -(\gamma_{22}/\gamma_{21})I$.

236. The rational function

$$\varphi(S) = \beta_o I + \beta_1 S + \Sigma_{k=1}^{m} \gamma_k (S - \alpha_k I)^{-1} ,$$

where $\beta_o \geqq 0$, $\beta_1 \geqq 0$, $\gamma_k \leqq 0$, and $\mathrm{Im}\, \alpha_k = 0$, is monotone in each interval which does not contain a "pole" $\alpha_k I$ $(k = 1,2,\ldots,m)$.

237. The function $\varphi(S) = S^{\frac{1}{2}}$ is monotone on the "semi-axis" $S \geq 0$.

This is a consequence of the following proposition:

238. If $0 \leq S \leq T$, then the real part of the operator

$$(T^{\frac{1}{2}} + S^{\frac{1}{2}} + \mu I)(S^{\frac{1}{2}} - T^{\frac{1}{2}} + \mu I)$$

for any $\mu > 0$ is positive.

239. If $S_1, S_2 \in \mathbf{S}(E)$ and $S_1 > 0$, then the operator $S_1 + S_2$ is invertible (cf. (224)).

A general criterion for monotonicity can be obtained with the aid of Taylor's formula for functions of an operator (cr. VI,(144)).

240. If $\varphi(\lambda) \in C_1(\alpha,\beta)^{\S}$, then the function $\varphi(S)$ is monotone iff

$$\left.\frac{d\varphi(S + t\Delta S)}{dt}\right|_{t=0} \geq 0$$

for all $S \in (\alpha I, \beta I)$ and $\Delta S \geq 0$.

---

§ $C_n(\alpha,\beta)$ is the class of scalar functions which are n-times continuously differentiable in the interval (α,β).

241. If $\varphi(\lambda) \in C_1(\alpha,\beta)$, then the function $\varphi(S)$ is monotone iff

$$\varphi^{[1]}(S) \circ dS \geq 0 \qquad \text{(notation in VI, after D31)}$$

for all $S \in (\alpha I, \beta I)$ and $dS \geq 0$.

This condition is equivalent to the inequality $\varphi^{[1]}(S) \geq 0$.

242. If $\varphi(\lambda) \in C_1(\alpha,\beta)$, then the function $\varphi(S)$ is monotone iff the quadratic form

$$\Sigma_{j,k=1}^{n}[(\varphi(\lambda_j) - \varphi(\lambda_k))/(\lambda_j-\lambda_k)]\xi_j\xi_k$$

is nonnegative for any $\lambda_1 \geqq \lambda_2 \geqq \cdots \geqq \lambda_n$ from the interval (α,β).

Below we shall point out still another equivalent condition for monotonicity. For its deduction the differentiability of the function $\varphi(\lambda)$ need not be assumed. Moreover, one can consider a function $\varphi(\lambda)$ defined on any set of points $E \subset (\alpha,\beta)$ containing no fewer than 2n distinct points, and the function $\varphi(S)$ need be defined only on the set $\{S \mid \sigma(S) \subset E\}$. As a preliminary let us note the following simple lemma.

243. If the function $\varphi(S)$ is monotone in $\mathbf{S}(E)$, then it is monotone in $\mathbf{S}(E_1)$ where $\dim E_1 < \dim E$.

244. If the function $\varphi(S)$ is monotone, then the quadratic form

$$\Sigma_{j,k=1}^{n} [(\varphi(\lambda_j) - \varphi(\mu_k))/(\lambda_j - \mu_k)]\xi_j\xi_k$$

is nonnegative for any $\mu_1 > \lambda_1 > \mu_2 > \ldots > \mu_n > \lambda_n$ from E.

For the proof one can utilize (232), after which one can compute the matrix of the operator $\varphi(U^{-1}SU) - \varphi(S)$ in a characteristic basis of the operator S.

Theorem (244) permits one to find a series of unexpected properties of monotone functions, of which the most important are (248) and (249). In theorems (245) - (249)

dim E > 1 is assumed. By (243), one can take dim E = 2 for the proofs.

245. Any function $\varphi(S)$, which is monotone in the interval $(\alpha I, \beta I)$ and not a constant, is strictly increasing, i.e., $\varphi(S) < \varphi(T)$ for $S < T$.

Therefore, the function $\varphi(\lambda)$ has an inverse. However, the inverse function $\varphi^{-1}(S)$ is in general not monotone:

246. If the function $\varphi(S)$ and its inverse are both monotone, then $\varphi(S) = (\gamma_{11}S + \gamma_{12}I)(\gamma_{21}S + \gamma_{22}I)^{-1}$.

In particular,

247. The unique function $\varphi(S)$ monotone on the whole axis $(-\infty I, \infty I)$ and possessing a monotone inverse is the linear function $\varphi(S) = \beta_o I + \beta_1 S$.

248. If $\varphi(S)$ is monotone in the interval $(\alpha I, \beta I)$, then $\varphi(\lambda)$ is continuous in the interval (α, β).

Stronger than that:

249. If $\varphi(S)$ is monotone in the interval $(\alpha I, \beta I)$, then $\varphi(\lambda)$ is continuously differentiable in the interval (α, β).

Hence the criterion of (242) can be strengthened:

250. The function $\varphi(S)$ is monotone in the interval $(\alpha I, \beta I)$ iff it is continuously differentiable and the quadratic form

$$\Sigma_{j,k=1}^{n} [(\varphi(\lambda_j) - \varphi(\lambda_k))/(\lambda_j - \lambda_k)] \xi_j \xi_k$$

is nonnegative for any $\lambda_1 \geqq \lambda_2 \geqq \cdots \geqq \lambda_n$ from (α, β).

Our presentation of the theorems about monotone functions of operators embodies the basis of the <u>theory of Loewner</u>. The most remarkable results of this theory refer to properties of smoothness and analyticity. One can show that under the conditions of (249) the function $\varphi(\lambda)$ possesses not only first but subsequent continuous derivatives up to order n-1. And if the operator-function $\varphi(S)$ is monotone in the interval $(\alpha I, \beta I)$ for a space E of any dimension, then the function $\varphi(\lambda)$ is analytic in the interval (α,β), continuable into the upper half plane, and has there a nonnegative imaginary part (i.e., it belongs to the class of functions, described in and after III,(147). The last property is characteristic of functions $\varphi(\lambda)$, which generate functions $\varphi(S)$ that are monotone for all n.

Let us now pass to convex functions of operators.

Definition 34. A function $\varphi(S)$ is *convex* in the interval $(\alpha I,\beta I)$ if for any operators S,T from the interval $(\alpha I,\beta I)$

$$\varphi((1-\theta)S + \theta T) \le (1-\theta)\,\varphi(S) + \theta\varphi(T) \qquad (0 < \theta < 1) .$$

51. The function $\varphi(S) = S^2$ is convex on the axis $(-\infty I,\infty I)$.

52. The function $\varphi(S) = |S|$ is convex in the interval $\alpha I,\beta I)$ if (and for dim $E > 1$ only if) the interval (α,β) oes not contain zero.

53. If $\varphi(\lambda) \in C_2(\alpha,\beta)$, then the function $\varphi(S)$ is convex iff

$$\left.\frac{d^2\varphi(S + t\Delta S)}{dt^2}\right|_{t=0} \ge 0$$

or all $S \in (\alpha I,\beta I)$ and $\Delta S \in \mathfrak{S}(E)$ (cf. (240)).

54. If $\varphi(\lambda) \in C_2(\alpha,\beta)$, then the function $\varphi(S)$ is convex iff

$$\varphi^{[2]}(S) \circ dS^{(2)} \ge 0$$

or all $S \in (\alpha I,\beta I)$ and $dS \in \mathfrak{S}(E)$ (cf. (241)).

55. If $\varphi(\lambda) \in C_2(\alpha,\beta)$, then the function $\varphi(S)$ is convex iff he quadratic form

$$\Sigma_{j,k=1}^{n}\ \varphi^{[2]}(\lambda_j,\lambda_o,\lambda_k)\xi_j\xi_k$$

s nonnegative for any $\lambda_o,\lambda_1,\ldots,\lambda_n$ from the interval (α,β) cf. (242)).

56. If $\varphi(\lambda) \in C_2(\alpha,\beta)$, then the function $\varphi(S)$ is convex iff he function $\psi_\mu(S) = \varphi^{[1]}(\mu,S)$ is monotone for any $\mu \in (\alpha,\beta)$.

additive quasiorder, following D3
admissible variation D19
antisymmetric quasiorder, following (2)

bounded problem D11

Chebyshev point D18
comparable vector pair (6)
concave mapping D15
convex function D9
convex operator function D34
convex optimization (programming) D16

duality theorem (94)

equilibrium theorem (97)
Euler-Lagrange equation (113)
existence theorem for optimization problem (95)

feasible problem D11
f-nontrivial (46)
forbidden variation D19
Friedrichs space D31
Frobenius' Theorem (143)

Gantmakher-Kreĭn Theorem (169)

Haar's Criterion (108)

irreducibly inconsistent D8

Judin's Theorem (15)

Kantorovich space D5
Kreĭn's inequality (214)
Kuhn-Tucker Theorem (103),(104)

Lagrange function D13
linear optimization D10
linear pencil of operators D32
linear programming D10
Loewner's theory, following (250)

Markov's problem, following (109)
minihedral cone D6
minimal decomposition D7
monotone mapping D21
monotone operator (132)
monotone operator function D33
Motzkin's Theorem (72)

v. Neumann's Theorem on saddle points (105)
nonnegative operator, following (133)
nonnegative (nonpositive) vector D2

operator order D26
optimality criterion (93)
optimal vector D12
ordered space D4
orthogonalizing operator D20

partial (quasi-) order D4
Perron's Theorem (157)
piecewise linear functional, following (109)
positive homogeneous quasiorder, following D3
positive (negative) part of self-adjoint operator D30
positive (negative) part of vector D7
positive (negative) vector D3
positive wedge D1

primitive operator D24
principle of bounding solutions (49)
Pringsheim's Theorem, following (143)

quasiorder D3
quasiordered space D1

ring cone D27
ring order D28

saddle point of functional D14

spectral order D29
strictly convex function D9
strictly stable operator D22
sup F (inf F) in Kantorovich space, following D5

totally positive operator D25
translation-positive operator D23

value of problem D12

CHAPTER IX

EXTENSION OF OPERATORS

1. Linear Operators Defined on a Subspace of a Linear Space.

In this chapter we shall study homomorphisms $A \in \mathrm{Hom}(D_A, E)$, where D_A is a subspace of the complex space E. When $D_A = E$, the homomorphism A is an operator in the sense of Ch. II. This term will also be retained for $D_A \neq E$ (although the names "partial operator" or "pre-operator" might be more natural). Unlike the case in Ch. II, the vector Ax exists now only for $x \in D_A$; the relations

$$A(x_1 + x_2) = Ax_1 + Ax_2, \quad A(\alpha x) = \alpha Ax$$

are required to hold for all x_1, x_2, $x \in D_A$ and any complex α.

Definition 1. The subspace D_A is called the *domain of definition* and will be denoted by Dom A. The codimension of D_A, sometimes called the *defect number* of the operator A, will be denoted by

$$\delta(A) \underset{\text{def}}{=} n - \dim A \quad .$$

Im A is known as the *range* of the operator A. All terms introduced in connection with homomorphisms in Ch. I retain their sense for operators defined on a subspace.

Definition 2. The operators A_1 and A_2 are considered *equal* if

$$\text{Dom } A_1 = \text{Dom } A_2 \quad \text{and} \quad A_1x = A_2x$$

for all permissible x.

Definition 3. An operator A (in agreement with I, D56) will again be called an *extension* of the operator A (to Dom $\tilde{A}$), and the operator A the *restriction* of the operator $\tilde{A}$ (to Dom A), if

$$\text{Dom } A \subset \text{Dom } \tilde{A} \quad \text{and} \quad \tilde{A}x = Ax \qquad (x \in \text{Dom } A).$$

We write $A \subset \tilde{A}$. An extension A is *complete* if

$$\text{Dom } \tilde{A} = E.$$

. The set of all complete extensions of an operator A is convex.

First consider arithmetic operations on operators: The definitions D4 and D5, given below for the sum (difference) and the product of operators, agree with the operations on homomorphisms introduced in I,D52 and D55. The operators 0 and I retain their original meaning of II,D1 and D9 so that Dom 0 = E and Dom I = E.

Definition 4. The *sum* and the *difference* $A = A_1 \pm A_2$ of the operators A_1 and A_2 are defined by

$$Ax = A_1x \pm A_2x$$

on the subspace

$$\text{Dom } A = \text{Dom } A_1 \cap \text{Dom } A_2 .$$

. $(A_1 + A_2) + A_3 = A_1 + (A_2 + A_3)$.

. $A_1 + A_2 = A_2 + A_1$.

. $A + 0 = A$.

. $A_1 - A_2 = A_1 + (-A_2)$.

. $A - A \subset 0$.

7. $(A_1 - A_2) + A_2 \subset A_1$.

Definition 5. The *product* $A = A_2A_1$ of the operators A_1 and A_2 is defined by

$$Ax = A_2(A_1x)$$

on the subspace

$$D_A = \{x \mid x \in \text{Dom } A_1,\ A_1x \in \text{Dom } A_2\} .$$

Definition 6. If $A_1A_2 = A_2A_1$, then A_1 and A_2 are called *commuting operators;* as before, this is written $A_1 \nabla A_2$. The m-fold product of A with itself is denoted by A^m.

8. $(A_1A_2)A_3 = A_1(A_2A_3)$.

9. $(A_1 + A_2)A_3 = A_1A_3 + A_2A_3$.

However,

10. $A_3(A_1 + A_2) \supset A_3A_1 + A_3A_2$.

11. $IA = AI = A$.

12. $\dim(\text{Dom } A_1A_2) = \dim(\text{Dom } A_1 \cap \text{Im } A_2) + \text{def } A_2$ (cf. I, (133)-(136)).

Here of course def A_2 is the relativized concept, that is the set $\{x \in \text{Dom } A_2 \mid A_2x = 0\}$.

13. If A_2^{-1} exists, then
$\dim(\text{Dom } A_1A_2) = \dim(\text{Dom } A_1) + \dim(\text{Dom } A_2) - \dim(\text{Dom } A_1 + \text{Dom } A_2^{-1})$.

Again, A_2^{-1} is understood to be the <u>unique inverse map</u> from Im A_2 to Dom A_2.

Concerning the powers of an operator we note only that

14. For any natural number m

$$\text{Dom } A^m \supset \text{Dom } A^{m+1} .$$

If equality holds in (14) for some m_o, then it holds for all $m > m_o$.

15. $m_o + \delta(A) \le n$.

Definition 7. The subspace $L \subset \text{Dom } A$ is *invariant relativ to* A if $Ax \in L$ for all $x \in L$.

16. The subspace

$$L_A = \bigcap_{k=1}^{n-1} \text{Dom } A^k = \text{Dom } A^{m_o}$$

is invariant relative to A.

17. The subspace L_A is not contained in any other invariant subspace of the operator A.

18. The subspace L_A contains any invariant subspace of the operator A.

Thus L_A is the largest invariant subspace of the operator A.

Definition 8. An operator A is *simple* if it has no invariant subspace $\neq 0$.

19. An operator A is simple iff

$$\bigcap_{k=1}^{n-1} \text{Dom } A^k = 0.$$

In particular,

20. If Ker A = 0 and Dom A ∩ Im A = 0, then the operator A is simple.

21. Any restriction of a simple operator is a simple operator.

The ideas of eigenvectors and characteristic subspaces to be used in this chapter are defined as in Ch. II.

22. An operator is simple iff it has no eigenvectors.

23. If the operator A is simple, then the dimension of each characteristic subspace of any complete extension $\tilde{A}$ of the operator A is no greater than $\delta(A)$.

Conversely,

24. If the dimension of each characteristic subspace of any complete extension $\tilde{A}$ of the operator A does not exceed $\delta(A)$, then the operator A is simple.

For the proof it is sufficient to use the following proposition:

25. Let L be any complement to the subspace Dom A. For that complete extension $\tilde{A} \supset A$ which is defined by $\tilde{A}x = 0$ $(x \in L)$

we have

$$\text{Ker } \tilde{A} = \text{Ker } A \oplus L .$$

Definition 9. We shall call the operator A the *direct sum* of the operators A_1 and A_2 and denote it by

$$A = A_1 \oplus A_2 ,$$

if the subspaces Dom A_1 and Dom A_2 are mutually independent (I,D29a) and

$$\text{Dom } A = \text{Dom } A_1 \oplus \text{Dom } A_2, \quad Ax = A_1x_1 + A_2x_2,$$

where $x = x_1 + x_2$, $x_1 \in \text{Dom } A_1$, $x_2 \in \text{Dom } A_2$.

2. Linear Operators Defined on a Subspace of a Unitary Space.

In the following sections of this chapter, E denotes a unitary space.

Definition 10. An operator S is *hermitian* if

$$(Sx,y) = (x,Sy) \qquad (x,y \in \text{Dom } S).$$

When Dom S = E, a hermitian operator S is *self-adjoint.*

Definition 11. An operator V is *isometric* if

$$\|Vx\| = \|x\| \qquad (x \in \text{Dom } V).$$

When Dom V = E, an isometric operator V is *unitary.*

26. Any restriction of a self-adjoint operator is a hermitian operator.

27. Any restriction of a unitary operator (III,D42) is an isometric operator.

Conversely,

28. Any isometric operator admits a unitary extension.

The question of a converse of theorem (26) is postponed until Section 4. The following propositions are generalizations of the corresponding theorems in Ch. III, Sections 5 and 6.

29. An operator S is hermitian iff it satisfies the condition

$$\operatorname{Im}(Sx,x) = 0 \qquad (x \in \operatorname{Dom} S).$$

30. If the operator S is hermitian, then the operator αS is hermitian for any real α.

31. The sum of hermitian operators is a hermitian operator.

32. If a hermitian operator S is invertible, then the operator S^{-1} is also hermitian.

33. The eigenvalues of a hermitian operator are real.

34. The eigenvectors of a hermitian operator corresponding to distinct eigenvalues are orthogonal.

35. An operator V is isometric iff it satisfies the condition

$$(Vx,Vy) = (x,y) \qquad (x,y \in \operatorname{Dom} V).$$

36. If the operator V is isometric, then for $|\alpha| = 1$ the operator αV is also isometric.

37. The product of isometric operators is an isometric operator.

38. The operator which is inverse to an isometric operator exists and is isometric.

39. The eigenvalues of an isometric operator have modulus one.

40. The eigenvectors of an isometric operator corresponding to distinct eigenvalues are orthogonal.

41. For any operator A there exists a unique *adjoint operator* A^* which is defined on *all* of E and satisfies the conditions

(1) $(Ax,y) = (x,A^*y) \qquad (x \in \operatorname{Dom} A,\ y \in E).$

(2) $\operatorname{Im} A^* \subset \operatorname{Dom} A.$

42. $A \subset A^{**}$.

43. If $\tilde{A}$ is the complete extension of the operator A defined by the equality $\tilde{A}x = 0$ $(x \in (\operatorname{Dom} A)^{\perp})$, then $(\tilde{A})^* = A^*$.

Conversely,

44. If B is a complete extension of the operator A and $B^* = A^*$, then $B = \tilde{A}$.

45. $\operatorname{Im} A \oplus \operatorname{Ker} A^* = E$ (notation in III,D35; cf. III,115).

In the following problems P_A denotes the orthoprojector onto Dom A.

46. If the operators A and B are connected by the relation

$$(Ax,y) = (x,By) \quad (x \in \text{Dom } A,\ y \in \text{Dom } B),$$

then $P_A B \subset A^*$.

Conversely,

47. If $P_A B \subset A^*$, then the operators A and B are connected by the relation

$$(Ax,y) = (x,By) \quad (x \in \text{Dom } A,\ y \in \text{Dom } B).$$

48. The operator A is hermitian iff $P_A A \subset A^*$.

49. The operator B is the adjoint of some hermitian operator iff it has the form $B = PS$, where S is a self-adjoint operator and P is an orthoprojector.

50. Let S, $\tilde{S}$ be hermitian operators and let Dom $\tilde{S} \supset$ Dom S. In order that $\tilde{S} \supset S$, it is necessary and sufficient that $P_S(\tilde{S})^* \subset S$.

51. If $D_B \supset D_A$ and $P_A B^* = A^*$, then $B \supset A$.

Conversely,

52. If $B \supset A$, then $P_A B^* = A^*$.

53. Let S be a hermitian and $\tilde{S}$ a self-adjoint operator. $\tilde{S}$ is an extension of S iff $P_S \tilde{S} = S^*$.

54. The set of all self-adjoint extensions of a given hermitian operator is convex (cf. (1)).

55. No matter what the operator A may be, the operators A^*A and AA^* are self-adjoint (the first defined on Dom A, the second on E) and nonnegative (III,D41).

56. The operator A^*A (on Dom A) is positive iff Ker $A = 0$.

57. The operator AA^* (on E) is positive iff Im $A = E$.

58. The operator V is isometric iff $V^*V \subset I$.

59. If V is an isometric operator, then VV^* is the orthoprojector onto the range of V.

60. An operator B is the adjoint of some isometric operator iff it has the form $B = PU$ where U is a unitary operator and P is an orthoprojector.

61. If V is an isometric operator, then $V^{-1} \subset V^*$.

Conversely,

62. If the operator V is invertible and $V^{-1} \subset V^*$, then V is an isometric operator.

63. Let V and V* be isometric operators and let $\mathrm{Dom}\,\tilde{V} \supset \mathrm{Dom}\,V$. Then $\tilde{V} \supset V$, iff $V^{-1} \subset (N)^*$.

64. Let V be an isometric and $\tilde{V}$ a unitary operator. $\tilde{V}$ is an extension of V iff $V^*\tilde{V} = P_V$.

Definition 12. The operator B defined on all of E is a *partial isometry* (or a *partially isometric operator*), if it is an extension of some isometric operator V and $Bx = 0$ for $x \in (\mathrm{Dom}\,V)^{1}$.

For example, the adjoint of any isometric operator is a partial isometry.

65. If B is a partial isometry, then B* is also a partial isometry.

66. If B is a partial isometry, then $B^*B = P$ and $BB^* = Q$, where P and Q are the orthoprojectors onto Dom B and Im B, respectively.

67. If the operator B is defined on all of E and if B^*B (or BB^*) is an orthoprojector, then B is a partial isometry.

68. An operator B ($\mathrm{Dom}\,B = E$) is a partial isometry iff its operator moduli are orthoprojectors (cf. III,D47).

69. $L_A^{\perp} = \Sigma_{k=o}^{n-1} (A^*)^k (\mathrm{Dom}\,A)^{\perp}$ (cf. (16) and recall I,D28).

70. If $B \supset A$, then

$$L_A^{\perp} = \Sigma_{k=o}^{n-1} (B^*)^k (\mathrm{Dom}\,A)^{\perp}.$$

71. If S is a hermitian operator and $\tilde{S}$ is any of its self-

adjoint extensions, then

$$L_S^{\perp} = \Sigma_{k=o}^{n-1} \tilde{S}^k (\text{Dom } S)^{\perp} .$$

72. If V is an isometric operator and U is any of its unitary extensions, then

$$L_V^{\perp} = \Sigma_{k=o}^{n-1} U^{-k} (\text{Dom } V)^{\perp} .$$

From (69) and (72) one can obtain criteria for the simplicity (D8) of an operator. Let us first introduce the concept of a generating subspace.

Definition 13. A subspace L is *generating* for an operator T (Dom T = E) if $\Sigma_{k=o}^{n-1} T^k L = E$ (cf. Ch. II, Sect. 3).

73. The operator A is simple iff the subspace $(\text{Dom } A)^{\perp}$ is generating for A*.

74. The symmetric operator S (VI,D14) is simple only if the subspace $(\text{Dom } S)^{\perp}$ is generating for <u>any</u> self-adjoint extension $\tilde{S} \supset S$. If $(\text{Dom } S)^{\perp}$ is generating for <u>some</u> self-adjoint extension $\tilde{S} \supset S$, then S is simple.

75. A necessary condition that an isometric operator V be simple is that the subspace $(\text{Dom } V)^{\perp}$ be generating for <u>any</u> unitary extension $\tilde{V} \supset V^{-1}$, and a sufficient condition is that it be generating for <u>some</u> unitary extension $\tilde{V} \supset V^{-1}$.

3. Generalized Inverse of an Operator

Let A be an arbitrary operator defined on all of the unitary space E. The operator A^{-1} need not exist, but when we contract A to A_o with domain of definition $\text{Dom } A_o = (\text{Ker } A)^{\perp}$, we obtain an invertible operator $A_o \subset A$. However, this operator A_o^{-1} will in general not be defined on all of E.

Definition 14. The extension $A^{(-1)}$ of the operator A_o^{-1} to all of E defined by the formula

$$A^{(-1)} = A_o^{-1} P ,$$

where P is the orthoprojector onto Im A, is known as the *generalized inverse operator* to A.

76. If the operator A^{-1} exists, then $A^{(-1)} = A^{-1}$.

77. Im $A^{(-1)} = (\text{Ker } A)^{\perp}$, Ker $A^{(-1)} = (\text{Im } A)^{\perp}$.

78. $AA^{(-1)} = P$.

Denoting the orthoprojector onto $(\text{Ker } A)^{\perp}$ by Q, we find

79. $A^{(-1)}A = Q$,

80. $A^{(-1)}P = QA^{(-1)} = A^{(-1)}$.

81. The operator $A^{(-1)}$ satisfies the following system of independent equations:

$$AXA = A, \quad XAX = X, \quad AX = X^*A^*, \quad XA = A^*X^*.$$

82. The system of equations in (81) has no solution other than $A^{(-1)}$.

83. $[A^{(-1)}]^{(-1)} = A$.

84. $[A^*]^{(-1)} = [A^{(-1)}]^*$.

85. If the operator B has an inverse and the subspace Im A is invariant relative to B*B, then $(BA)^{(-1)} = A^{(-1)}B^{-1}$.

86. If the operator B has an inverse and the subspace Ker A is invariant relative to BB*, then $(AB)^{(-1)} = B^{-1}A^{(-1)}$.

87. $A^{(-1)} = A^*$ iff A is a partial isometry.

The usefulness of the generalized inverse for the solution of the inhomogeneous equation

$$Ax = y \tag{*}$$

shows itself in the following propositions.

88. The equation (*) is solvable iff $AA^{(-1)}y = y$.

89. The general solution of the homogeneous equation

$$Ax = 0$$

is given by the formula $x = [I - A^{(-1)}A]z \quad (z \in E)$.

90. If the equation (*) is solvable, then its general solution has the form

$$\check{x} = A^{(-1)}y + [I - A^{(-1)}A]z \quad (z \in E).$$

91. For any y, the vectors $\check{x}$ defined by the preceding formula are the solutions to (*) in the sense of least-square approximations, i.e.

$$\|A\check{x} - y\| = \min_{x \in E}\|Ax - y\|$$

(Cf. here the problem of Chebyshev approximation (VIII,D18) and also IV,(127).)

Passing from vector- to operator-equations we can obtain the following proposition which fully characterizes the extremal properties of the generalized inverse operator. (In (92)-(94), $\|.\|$ is the Hilbert norm IV,D60.)

92. $$\|AA^{(-1)} - I\| = \min_{X \in \mathbb{M}(E)} \|AX - I\|.$$

93. The functional $p(X) = \|AX - I\|$ has a unique point of minimization in $\mathbb{M}(E)$. Here and in (94), the minimal operator X satisfies Im X $\perp$ Ker A.

94. For any $B \in \mathbb{M}(E)$

$$\|A(A^{(-1)}B) - B\| = \min_{X \in \mathbb{M}(E)} \|AX - B\|,$$

and the minimum is achieved only for $X = A^{(-1)}B$.

The concept of the generalized inverse operator can be extended to the case Dom $A \neq E$ by defining $A^{(-1)}$ by the same formalism as in D14. If this is done, theorems (88)-(94) remain in force.

4. Theory of Extensions of Hermitian and Isometric Operators

Definition 15. The *direct sum* of the operators A_1 and A_2 in a unitary space E will be called *orthogonal* if $\mathrm{Dom}\, A_1 \perp \mathrm{Dom}\, A_2$. The orthogonal sum of operators is written $A_1 \boxplus A_2$.

95. The formula $\tilde{A} = A \boxplus B$, where B runs through the set of all operators with domain of definition $(\mathrm{Dom}\, A)^\perp$, establishes a bijective correspondence between the complete extensions $\tilde{A}$ of the operator A and the operators B.

96. If V is an isometric operator, then the formula $\tilde{V} = V \boxplus B$, where B runs through the set of all isometric operators for which

$$\mathrm{Dom}\, B = (\mathrm{Dom}\, V)^\perp, \quad \mathrm{Im}\, B = (\mathrm{Im}\, V)^\perp,$$

establishes a bijective correspondence between the unitary extensions V of the operator $\tilde{V}$ and the operators B.

97. If S is a hermitian operator, then the formula

$$\tilde{S} = S \boxplus (S^*P + A),$$

where P is the orthoprojector onto $(\mathrm{Dom}\, S)^\perp$ and A runs through the set of all self-adjoint operators in the subspace $(\mathrm{Dom}\, S)^\perp$, establishes a bijective relation between the self-adjoint extensions $\tilde{S}$ of the operator S and the operators A.

Note that (96) immediately yields a description of all isometric extensions of the operator A.

98. If V is an isometric operator, then the formula $\tilde{V} = V \boxplus B$, where B runs through the set of all isometric operators for which

$$\mathrm{Dom}\, B \subset (\mathrm{Dom}\, V)^\perp, \quad \mathrm{Im}\, B \subset (\mathrm{Im}\, V)^\perp,$$

establishes a bijective relation between the isometric

extensions $\tilde{V}$ of an operator V and the operators B.

Unlike (98), proposition (97) does not immediately yield a description of all hermitian extensions of an operator S. The latter can be obtained with the aid of the Cayley transformation which for self-adjoint and unitary operators has already been defined earlier in III,D45.

Definition 16. The *Cayley transform* V_ω *of a hermitian operator* S is the operator

$$V_\omega = (S - \bar{\omega}I)(S - \omega I)^{-1},$$

where the imaginary part of ω is $\neq 0$.

99. The operator V_ω is isometric and

$$\text{Dom } V_\omega = \text{Im}(S - \omega I), \quad \text{Im } V_\omega = \text{Im}(S - \bar{\omega}I).$$

100. The number one is not an eigenvalue of the operator V_ω.

101. The operator S is expressed in terms of its Cayley transform V_ω by the formula

$$S = (\omega V_\omega - \bar{\omega}I)(V_\omega - I)^{-1}.$$

Definition 17. If V is an isometric operator for which one is not an eigenvalue, then for Im $\omega \neq 0$ the operator

$$S_\omega = (\omega V - \bar{\omega}I)(V - I)^{-1}$$

exists and is called the *Cayley transform* of the isometric operator V.

102. The operator S_ω is hermitian, and

$$\text{Dom } S_\omega = \text{Im}(V - I), \quad \text{Im } S_\omega = \text{Im}(\omega V - \bar{\omega}I).$$

103. The operator V is expressed in terms of its Cayley transform by the formula

$$V = (S_\omega - \bar{\omega}I)(S_\omega - \omega I)^{-1}.$$

Note that the formulas for the Cayley transforms introduced above, contrary to the corresponding formulas in III, Section 6, do not in general permit an interchange of the factors.

104. A hermitian operator is simple (D8) iff its Cayley transform is simple.

105. If $\tilde{S}$ is a hermitian (self-adjoint) extension of a hermitian operator S, then the Cayley transform of the operator $\tilde{S}$ is an isometric (unitary) extension of the Cayley transform of the operator S.

106. If $\tilde{V}$ is an isometric (unitary) extension of an isometric operator V and if one is not an eigenvalue of the operator $\tilde{V}$, then the Cayley transform of the operator $\tilde{V}$ is a hermitian (self-adjoint) extension of the Cayley transform of the operator V.

107. If S is a hermitian operator and does not have λ as an eigenvalue (this is so, in particular, if $\operatorname{Im} \lambda \neq 0$), then $\operatorname{rk}(S - \lambda I) + \delta(S) = n$ (notation in D1).

Definition 18. The subspace

$$N_\lambda = (\operatorname{Im}(S - \lambda I))^\perp,$$

where $\lambda \notin \sigma(S)$, is called the *defect subspace of the operator* S *at the point* λ.

Proposition (107) implies the following proposition (108), which is the *theorem of the invariance of dimension for defect subspaces*.

108. The number $\operatorname{rk}(S - \lambda I)$ does not depend upon λ for $\lambda \notin \sigma(S)$.

109. The defect subspace coincides with the set of solutions to the equation

$$S^*x = \bar{\lambda}Px;$$

here P is the orthoprojector onto Dom S.

110. The hermitian operator S is self-adjoint iff $\sigma(S)$ is zero.

111. If V_λ is the Cayley transform of the hermitian opera-

tor S, then

$$(\mathrm{Dom}\ V_\lambda)^\perp = N_\lambda, \qquad (\mathrm{Im}\ V_\lambda)^\perp = N_{\bar\lambda} \quad (\mathrm{Im}\ \lambda \neq 0).$$

It follows from (111) that the defect number is left invariant by the Cayley transform.

Now the following description of hermitian extensions of a (hermitian) operator S follows from (98):

112. Let V_λ be the Cayley transform of an operator S. Put $\tilde V_\lambda = V_\lambda \boxplus B$, where B is an isometric operator such that $\mathrm{Dom}\ B \subset N_\lambda$, $\mathrm{Im}\ B \subset N_{\bar\lambda}$, and one is not an eigenvalue of the operator $\tilde V_\lambda$. Then the formula

$$\tilde S = (\lambda\tilde V_\lambda - \bar\lambda I)(\tilde V_\lambda - I)^{-1}$$

establishes a bijective correspondence between hermitian extensions S of the operator S and the operators B.

The further constructions have the aim of replacing the condition $1 \notin \sigma(\tilde V_\lambda)$ by a more "transparent" restriction on the operator B. The final result is contained in (119).

113. $N_\lambda \cap \mathrm{Dom}\ S = 0$.

114. If Q_λ is the orthoprojector onto N_λ, then

$$V_\lambda Q_\lambda^\perp x = Q_{\bar\lambda}^\perp x \qquad (x \in (\mathrm{Dom}\ S)^\perp).$$

115. The formulas

$$\left.\begin{aligned} y_\lambda &= Q_\lambda x, \\ Wy &= Q_{\bar\lambda} x \end{aligned}\right. \qquad (x \in (\mathrm{Dom}\ S)^\perp)$$

define an isometric operator $W = W_\lambda$ with domain of definition $\mathrm{Dom}\ W_\lambda = N_\lambda$ and range $\mathrm{Im}\ W_\lambda = N_{\bar\lambda}$.

116. Let $\tilde V_\lambda$ be an isometric extension of the Cayley transform V_λ of the operator S. If there exists a vector

$$y \in \mathrm{Dom}\ W_\lambda \cap \mathrm{Dom}\ \tilde V_\lambda \qquad (y \neq 0)$$

such that $\tilde V_\lambda y = W_\lambda y$, then one is an eigenvalue of the operator $\tilde{V}_\lambda$.

This proposition is easy to invert by first noting:

117. If $\tilde{V}_\lambda x = x$, then $x \in (\mathrm{Dom}\ S)^\perp$.

118. If one is an eigenvalue of the operator $\tilde{V}_\lambda$, then there exists a vector

$$y \in \mathrm{Dom}\ W_\lambda \cap \mathrm{Dom}\ \tilde{V}_\lambda \quad (y \neq 0) \quad \text{such that } \tilde{V}_\lambda y = W_\lambda y.$$

Now (112) can be presented in the following form:

119. Let B be the set of all isometric operators B with Dom $B \subset N_\lambda$ and Im $B \subset N_{\bar{\lambda}}$, which satisfy the condition $(B - W_\lambda)y \neq 0$ $(y \neq 0)$. Then the formula

$$\tilde{S} = (\lambda\tilde{V}_\lambda - \bar{\lambda}I)(\tilde{V}_\lambda - I)^{-1},$$

where $\tilde{V}_\lambda = V_\lambda \boxplus B$ establishes a bijective correspondence between the set of hermitian extensions $\tilde{S}$ of the operator S and the set B.

120. The extension $\tilde{S}$ in (119) is self-adjoint iff Dom $B = N_\lambda$.

The result obtained can also be expressed by the *parametric formulas of von Neumann:*

121. If the extension $\tilde{S}$ corresponds to the operator $B \in$ B, then any $x \in \mathrm{Dom}\ \tilde{S}$ can be uniquely represented in the form

$$x = u + v - Bv \qquad (u \in \mathrm{Dom}\ S,\ v \in \mathrm{Dom}\ B)$$

and

$$\tilde{S}x = Su + \bar{\lambda}v - \lambda Bv.$$

5. Self-Adjoint Extensions under Preservation of the Norm.

In agreement with IV,D21, the norm of an operator A (considered as a homomorphism) is defined by

$$\|A\| = \max_{x \in \mathrm{Dom}\ A} \frac{\|Ax\|}{\|x\|}.$$

We also have, as in Ch. IV,

122. $$\|A\| = \max_{\substack{x \in \text{Dom } A \\ y \in E}} \frac{|(Ax,y)|}{\|x\| \cdot \|y\|}.$$

123. $\|A^*\| = \|A\|$.

124. If $B \supset A$, then $\|B\| \geqq \|A\|$.

But "continuation by zero" preserves the norm of an operator:

125. The operator $\tilde{A} = A \boxplus C$, where $\text{Dom } C = (\text{Dom } A)^\perp$ and $C = 0$, is an extension of the operator A to the whole space which preserves the norm: $\|\tilde{A}\| = \|A\|$ (cf. IV,(72)).

126. If the operator S is hermitian and $\text{Im } S \subset \text{Dom } S$, then its extension by formula (125) is self-adjoint. Therefore $\tilde{S} = S^*$.

In constructing the theory of norm-preserving self-adjoint extensions for an arbitrary hermitian operator we shall assume without loss of generality that

$$\|S\| = 1.$$

Representing the operator S in the form $S = S_1 + S_2$ where $S_1 = PS$, $S_2 = P^\perp S$ (P is the orthoprojector onto Dom S), let us extend S_1 and S_2 separately.

127. The operator $A_1 = S_1^*$ is a complete extension of the operator S_1 with norm $\|A_1\| \leqq 1$.

We now introduce the operator $I - A_1^* \geq 0$, put $L = \text{Ker}(I - A_1^*A_1)$, form the factor space E/L, and convert it into a unitary space by the scalar product

$$<[x],[y]> = (\{I - A_1^*A_1\}x,y).$$

134. If $x,y \in \text{Dom } S$ and $x \equiv y \pmod{L}$, then $S_2x = S_2y$.

Therefore S_2 can be considered as a homomorphism from Dom S/L to $(\text{Dom } S)^\perp$.

135. Let A_2 be the extension of our homomorphism

$$S_2 \in \text{Hom}(\text{Dom } S/L, (\text{Dom } S)^{\perp})$$

to all of E/L. We define this extension by $A_2 \underset{\text{def}}{=} S_2 Q$, where Q is the orthoprojector from E/L onto Dom S/L. Then for any $[x] \in E/L$

$$(A_2[x], A_2[x]) \leqq (Q[x], Q[x]).$$

136. $\|A_2\| \leqq 1$.

It remains to transform in a natural way the homomorphism

$$A_2 \in \text{Hom}(E/L, (\text{Dom } S)^{\perp})$$

into an operator $A_2 \in \mathbb{M}(E)$.

With this aim put $A_2 x = A_2[x]$ $(x \in E)$.

137. The operator A_2 is a complete extension of the operator S_2 and satisfies the inequality $\|A_2 x\|^2 \leqq \|x\|^2 - \|A_1 x\|^2$.

Two further steps need to be made.

138. The operator $A = A_1 + A_2$ is a complete extension of the operator S, and $\|A\| \leqq 1$.

139. The operator A^* is an extension of the operator S.

140. The operator $\tilde{S} = \frac{1}{2}(A + A^*)$ is a self-adjoint extension of the operator S, and $\|\tilde{S}\| = 1$.

And so we obtain the following fundamental result:

141. Any hermitian operator admits a self-adjoint extension which preserves the norm.

Let $\mathbb{W}$ denote the set of all self-adjoint extensions of the operator S which preserve the norm.

142. The set $\mathbb{W}$ is convex.

Therefore,

143. If a norm-preserving self-adjoint extension is not unique, then the set of all such extensions is infinite.

We now wish to investigate the structure of the set $\mathbb{W}$.

144. Let $\tilde{S} \in \mathbb{W}$. The operator $T \in \mathbb{W}$, iff T can be represented in form $T = \tilde{S} + C$, where C is a self-adjoint operator satisfying the conditions

$$\operatorname{Ker} C \supset \operatorname{Dom} S, \ -(I + \tilde{S}) \leq C \leq (I - \tilde{S}).^{\S}$$

Now it is natural to make use of the results VIII,(204)-(207).

145. Let $T \in \mathbb{W}$ and let Q_1 and Q_2 be the orthoprojectors onto the orthogonal complements of the subspaces $(I + \tilde{S})^{\frac{1}{2}}\operatorname{Dom} S$ and $(I - \tilde{S})^{\frac{1}{2}}\operatorname{Dom} S$, respectively. Then the operators

$$S' = \tilde{S} - (I + \tilde{S})^{\frac{1}{2}} Q_1 (I + \tilde{S})^{\frac{1}{2}}$$

and

$$S'' = \tilde{S} + (I - \tilde{S})^{\frac{1}{2}} Q_2 (I - \tilde{S})^{\frac{1}{2}}$$

belong to $\mathbb{W}$.

146. A self-adjoint extension T of an operator S belongs to $\mathbb{W}$ iff it satisfies the inequality

$$S' \leq T \leq S''.$$

Hence it follows that the operators S' and S'' constructed in (145) do not depend upon the particular choice of the extension $\tilde{S} \in \mathbb{W}$. The following proposition permits an improvement of the result (146).

147. Any self-adjoint operator T which satisfies the inequality in (146) is an extension of the operator S.

Thus the final description of all norm-preserving self-adjoint extensions of an operator S takes the following form:

148. The set $\mathbb{W}$ is the closed interval $[S', S'']$ in the ordered space $\mathbb{S}(E)$ (*M. G. Krein's Theorem*).

149. The operator $\tilde{S}$ is the unique norm-preserving self-adjoint extension of the operator S, iff

---

§ The order is the spectral order of VIII,D29.

$$(I + \tilde{S})^{\frac{1}{2}} Q_1 (I + \tilde{S})^{\frac{1}{2}} = 0 \quad \text{and} \quad (I - \tilde{S})^{\frac{1}{2}} Q_2 (I - \tilde{S})^{\frac{1}{2}} = 0.$$

The defect of this uniqueness criterion is its formulation in terms of $\tilde{S}$ instead of S. We shall, however, use proposition (149) as an intermediate result on the way to a criterion in terms of S itself.

150. If in (145) we replace $\tilde{S}$ by $\tilde{S}_o = \frac{1}{2}(S' + S'')$, then

$$(I + \tilde{S}_o)^{\frac{1}{2}} Q_1 (I + \tilde{S}_o)^{\frac{1}{2}} = (I - \tilde{S}_o)^{\frac{1}{2}} Q_2 (I - \tilde{S}_o)^{\frac{1}{2}}.$$

151. Let $x_o \in \text{Dom } S$ and $\|x_o\| = 1$, $\|Sx_o\| = 1$. Then for any $T \in \mathbb{W}$ we have the representation $x_o = x_1 + x_2$, where $Tx_1 = x_1$, $Tx_2 = -x_2$.

152. If the operator $\tilde{S}_o$ is defined as in (150), and the vector x_o as in (151), then for any $y \neq 0$ from the range of the operator $(I - \tilde{S}_o)^{\frac{1}{2}} Q_2 (I - \tilde{S}_o)^{\frac{1}{2}}$ we have $(Sx_o, y) = 0$.

153. If for all vectors x_o satisfying the conditions of (151) for some $y \in (\text{Dom } S)^{\perp}$, $y \neq 0$, the relation $(Sx_o, y) = 0$ is fulfilled, then a norm-preserving self-adjoint extension is not unique.

For the proof of (153) it is convenient first to consider the case $\delta(S) = 1$.

The final criterion for uniqueness has the form:

154. The operator S admits a unique norm-preserving self-adjoint extension iff for any $y \in (\text{Dom } S)^{\perp}$, $y \neq 0$, there exists a vector x satisfying the conditions $x \in \text{Dom } S$, $\|x\| = 1$, $\|Sx\| = 1$, such that $(Sx, y) \neq 0$.

6. The Spectra of Self-Adjoint and Unitary Extensions.

In this section S denotes a hermitian operator with $\delta(S) = m$.

Definition 19. An interval (α,β) $(-\infty < \alpha < \beta < \infty)$ is called a *spectral lacuna* (or *gap*) *of a hermitian operator* S if

$$\|Sx - \frac{\alpha+\beta}{2} x\| \leqq \frac{\beta-\alpha}{2}\|x\| \qquad (x \in \text{Dom } S).$$

155. A spectral lacuna does not contain any eigenvalues.

For $m = 0$ the converse is also true:

156. If the interval (α,β) contains no eigenvalues of a self-adjoint operator, then (α,β) is a spectral lacuna of it.

For $m > 0$ one can only assert:

157. If the real point λ_o is not an eigenvalue of the operator S, then some neighborhood of this point is a spectral lacuna of the operator S.

Definition 20. If for all $x \in \text{Dom } S$

$$a(x,x) \leqq (Sx,x) \leqq b(x,x),$$

then the intervals $(-\infty,a)$ and (b,∞) are semi-infinite spectral lacunae of the operator S.

Propositions (155)-(156) carry over (in obvious form) to the case of semi-infinite lacunae.

The next propositions are based on an estimate of the dimension of the intersection $\text{Dom } S \cap F$ under the assumption

$$\dim F > \delta(S),$$

for a suitably chosen subspace F.

158. The sum of the multiplicities of the eigenvalues of any self-adjoint extension $\tilde{S} \supset S$, which lie in the spectral lacuna (a,b) of the operator S, does not exceed m.

159. The sum of the multiplicities of the eigenvalues of the self-adjoint extension $\tilde{S} \supset S$, which lie in a semi-infinite spectral lacuna of the operator S, does not exceed m.

160. The sum of the multiplicities of the eigenvalues of any

self-adjoint extension $\tilde{S} \supset S$, which, in modulus, are greater than $\|S\|$, does not exceed m.

161. If ν is the maximal dimension for a subspace $L \subset \text{Dom } S$, for which

$$\left\|Sx - \frac{\alpha+\beta}{2} x\right\| \leqq \frac{\beta-\alpha}{2} \|x\| \qquad (x \neq 0),$$

then the sum of the multiplicities of the eigenvalues of any self-adjoint extension $\tilde{S} \supset S$ which lie in the interval (α,β) does not exceed $\nu+m$.

162. Under the conditions of (161) the sum of the multiplicities of the eigenvalues of any self-adjoint extension $\tilde{S} \supset S$ lying in the interval (α,β) is no less than ν.

163. If ν is the maximal dimension of a subspace $L \subset E$ for which $(Sx,x) < a(x,x)$ $(x \neq 0)$, then the sum of the multiplicities of the eigenvalues of any self-adjoint extension $\tilde{S} \supset S$ lying to the left of the point a does not exceed $\nu+m$.

164. Under the conditions of (163) the sum of the multiplicities of the eigenvalues of any self-adjoint extension $\tilde{S} \supset S$, which lie to the left of the point a, is no less than ν.

165. If ν is the maximal dimension of the subspaces $L \subset E$ on which $\|Sx\| > \rho\|x\|$ $(x \neq 0)$, then the sum of the multiplicities of the eigenvalues of any self-adjoint extension of modulus greater than ρ does not exceed $\nu+m$.

166. Under the conditions of (165) the sum of the multiplicities of the eigenvalues of any self-adjoint extension of modulus greater than ρ is no less than ν.

167. If some interval (α,β) contains r eigenvalues of some self-adjoint extension $\tilde{S} \supset S$, then this interval contains no fewer than r-m eigenvalues of any other self-adjoint extension of the operator S (eigenvalues are counted acc. to multiplicity).

168. For m = 1 the spectra of two arbitrary self-adjoint extensions of an operator S <u>intertwine</u> (term in III,(231)).

169. For strict intertwining of the spectra in (168) it is necessary and sufficient that the operator S be simple (D8).

We now go on to some problems on the construction of self-adjoint extensions with given spectral properties.

170. Let Ker S = 0 and $(\text{Dom } S)^{\perp} \cap \text{Im } S = 0$. Then

$$\text{Dom } S \oplus (\text{Im } S)^{\perp} = E.$$

171. Under the conditions of (170) the extension $\tilde{S} \supset S$, defined by $\tilde{S}x = 0 \quad (x \in (\text{Im} S)^{\perp})$, is self-adjoint.

Definition 21. Set $L_{\lambda} \underset{\text{def}}{=} (\text{Dom } S)^{\perp} \cap N_{\lambda}^{\perp}$ (notation in D18).

172. If the real number λ is not an eigenvalue of the operator S and $L_{\lambda} = 0$, then there exists a self-adjoint extension $\tilde{S} \supset S$ for which the number λ is an eigenvalue of multiplicity m.

Conversely,

173. If the real number λ is not an eigenvalue of the operator S, and if there exists a self-adjoint extension $\tilde{S} \supset S$ for which the number λ is an eigenvalue of multiplicity m, then $L_{\lambda} = 0$.

174. If the number λ is an r-fold eigenvalue of the operator S, and if $L_{\lambda} = 0$, then there exists a self-adjoint extension $\tilde{S} \supset S$ for which the number λ is an eigenvalue of multiplicity r+m.

Conversely,

175. If the number λ is an r-fold eigenvalue of the operator S, and if there exists a self-adjoint extension $\tilde{S} \supset S$ for which the number λ is an eigenvalue of multiplicity r+m, then $L_{\lambda} = 0$.

176. If the real number λ is not an eigenvalue of the operator S, and if $\dim L_{\lambda} = s < m$, then there exists a self-adjoint extension $\tilde{S} \supset S$ for which the number λ is an (m-s)-

fold eigenvalue.

Conversely,

177. If the real number λ is not an eigenvalue of the operator S, and if the maximal multiplicity of the eigenvalue λ in the spectra $\sigma(\tilde{S})$ of all possible self-adjoint extensions $\tilde{S} \supset S$ equals $m-s > 0$, then $\dim L_\lambda = s$.

178. If the number λ is an eigenvalue of the operator S of multiplicity $r < m$, and if $\dim L_\lambda = s < m+r$, then there exists a self-adjoint extension $\tilde{S} \supset S$ for which the number λ is an eigenvalue of multiplicity $m+r-s$.

Conversely,

179. If the number λ is an eigenvalue of the operator S of multiplicity $r < m$, and if the maximal multiplicity of the eigenvalue λ in the spectra $\delta(S)$ of all possible self-adjoint extensions $\tilde{S} \supset S$ equals $m+r-s > 0$, then $\dim L_\lambda = s$.

180. If the real number λ satisfies the inequality

$$\lambda(x,x) < (Sx,x) \qquad (x \in \text{Dom } S, \quad x \neq 0),$$

then the subspaces Dom S and N_λ are mutually independent (I,D29a).

181. Under the hypothesis of (180) Dom S $\oplus N_\lambda = E$.

182. The operator $\tilde{S} \supset S$, which under the hypothesis of (180) is defined by the equality $\tilde{S}x = \lambda x \quad (x \in N_\lambda)$, is a self-adjoint extension of S.

We thus have established, if (159) is also taken into account:

183. If the point λ belongs to one of the semi-infinite lacunae $(-\infty,a)$ or (b,∞) of the operator S, then there exists a self-adjoint extension $\tilde{S} \supset S$ for which the number λ is an eigenvalue of multiplicity m. Any such extension has no other eigenvalue in the corresponding lacuna.

For what follows it is convenient to specialize (183):

184. If $S \geq 0$ and $\lambda < 0$, then there exists a self-adjoint

extension $\tilde{S} \supset S$ for which the number λ is an eigenvalue of multiplicity m. Any such extension has no other negative eigenvalues.

185. If $\lambda > \|S\|$, then there exists a self-adjoint extension $\tilde{S} \supset S$ for which the number λ is an eigenvalue of multiplicity m. Any such extension has no other eigenvalue which in modulus exceeds $\|S\|$.

Note that in passing to the limit $\lambda \to \|S\|$ we can obtain once more (141) from (185), but the analogous passage to the limit $\lambda \to 0$ in (184) is not always possible, and a self-adjoint extension which preserves the lower bound[§] of the operator may not exist.

A development of the method described in (180)-(182) makes it possible to obtain the following generalization of theorem (184).

186. If $S \geq 0$, then for any given negative numbers λ_k and positive integers r_k $(k = 1,\ldots,p)$,

$$\Sigma_{k=1}^{p} r_k = m,$$

there exists a self-adjoint extension $\tilde{S} \supset S$ for which the numbers λ_k are eigenvalues of multiplicity r_k $(k = 1,\ldots,p)$. Any such extension has no other negative eigenvalues.

187. The extension $\tilde{S}$ in (186) is unique iff m = 1.

The result (186) can be reformulated in an obvious way for the case of an arbitrary semi-infinite gap. In (191) this result will be transferred to the domain $\lambda > \|S\|$ with the aid of theorems (188) and (189) about linear fractional transformations.

188. If the operator S is hermitian and $\|S\| < 1$, then the operator

[§] = $\inf(Sx,x)$ for $\|x\| = 1$, $x \in \mathrm{Dom}\, S$.

$$T = (I + S)(I - S)^{-1}$$

with the domain of definition Dom T = Im(I - S) is also hermitian and $T > 0$. Thus $\delta(T) = \delta(S)$.

Conversely,

189. If the operator T is hermitian and $T > 0$, then the operator

$$S = (T - I)(T + I)^{-1}$$

with domain of definition Dom S = Im(T + I) is also hermitian, and $\|S\| < 1$. Thus $\delta(S) = \delta(T)$.

190. If the spectrum of the self-adjoint extension $\tilde{T}$ of the operator $T > 0$ does not contain the point $\lambda = -1$, then the operator

$$\tilde{S} = (\tilde{T} - I)(\tilde{T} + I)^{-1}$$

is a self-adjoint extension of the operator

$$S = (T - I)(T + I)^{-1}.$$

191. For arbitrarily assigned real λ_k such that $|\lambda_k| > \|S\|$, and arbitrarily natural numbers r_k $(k = 1,\ldots,p)$,

$$\Sigma_{k=1}^{p} r_k = m,$$

there exists a self-adjoint extension $\tilde{S} \supset S$, for which the numbers λ_k are eigenvalues of multiplicities r_k. Any such extension has no other eigenvalues which in modulus exceed $\|S\|$.

192. The extension $\tilde{S}$ in (191) is unique iff $m = 1$.

193. If the real sequences $\{\lambda_k'\}_{k=1}^{n}$ and $\{\lambda_k''\}_{k=1}^{n}$ intertwine, then there exists a hermitian operator S with $\delta(S) = 1 = m$ and self-adjoint extensions $\tilde{S}_1 \supset S$ and $\tilde{S}_2 \supset S$ such that

$$\sigma(\tilde{S}_1) = \{\lambda_k'\}_{k=1}^{n}, \quad \sigma(\tilde{S}_2) = \{\lambda_k''\}_{k=1}^{n}.$$

The converse spectral problem to (193) has an essentially unique solution:

194. The operator S in (193) is uniquely determined by the two given spectra (within unitary similarity).

In the remaining part of this section V denotes an isometric operator with $\delta(V) = m$.

Definition 22. An arc[§] of the unit circle: $(e^{i\alpha}, e^{i\beta})$, where $\alpha<\beta<\alpha+2\pi$, is called a *spectral gap* (or *gap*) *of the isometric operator* V if

$$\|Vx - \exp\{i\tfrac{\alpha+\beta}{2}\}x\| \geqq 2\sin\tfrac{\beta-\alpha}{4}\|x\| \qquad (x \in \mathrm{Dom}\, V).$$

195. If the union of two arcs covers the unit circle, then at least one of them is not a spectral gap.

196. A spectral lacuna contains no eigenvalues.

For $m = 0$ the converse is also true:

197. If an arc of the unit circle contains no eigenvalues of a unitary operator, then the arc is a spectral lacuna.

For $m > 0$ one can only assert:

198. If the point λ_o $(|\lambda_o| = 1)$ is not an eigenvalue of the operator V, then some arc of the unit circle which contains this point is a spectral gap of the operator V.

199. For any unitary extension $\tilde{V} \supset V$ the sum of the multiplicities of the eigenvalues lying in a spectral lacuna of the operator V does not exceed m.

200. If ν is the maximum of the dimensions of the subspaces $L \subset \mathrm{Dom}\, V$ on which

$$\|Vx - (\exp\{i\tfrac{\alpha+\beta}{2}\})x\| < 2\sin\tfrac{\beta-\alpha}{4}\|x\| \qquad (x \neq 0),$$

then the sum of the multiplicities of the eigenvalues of any unitary extension $\tilde{V} \supset V$ lying on the arc $(e^{i\alpha}, e^{i\beta})$ is not smaller than ν.

[§]Any arc is considered to be open.

201. Under the conditions of (200) it is true that for any unitary extension $\tilde{V} \supset V$ the sum of the multiplicities of the eigenvalues lying on the arc $(e^{i\alpha}, e^{i\beta})$ is no less than ν.

202. If an arc of the unit circle contains r eigenvalues of some unitary extension $\tilde{V} \supset V$, then this arc contains at least r-m eigenvalues of any other unitary extension of the operator V (eigenvalues are counted acc. to multiplicity).

203. For m = 1 the spectra of two arbitrary unitary extensions of the operator V intertwine. ("Intertwining" refers to the arguments of the eigenvalues.)

204. The spectra in (203) are strictly intertwining iff the operator V is simple.

Let us now consider some problems concerning the construction of unitary extensions with given spectral properties.

205. If $(e^{-i\pi/2}, e^{i\pi/2})$ is a spectral lacuna of the isometric operator V, then its Cayley transform

$$S = i(V + I)(V - I)^{-1}$$

is a hermitian operator with norm $\|S\| \leqq 1$, and $\delta(S) = \delta(V)$.

206. If the hermitian operator S satisfies the inequality $\|S\| \leqq 1$, then the arc $(e^{-i\pi/2}, e^{i\pi/2})$ is a spectral lacuna of its Cayley transform

$$V = (S + iI)(S - iI)^{-1}$$

and

$$\delta(V) = \delta(S).$$

207. For any given numbers λ_k $(|\lambda_k| = 1)$ which lie in a spectral lacuna of the isometric operator V, and for any natural numbers r_k $(k = 1,\ldots,p;\ \Sigma_{k=1}^{p} r_k = m)$ there exists a unitary extension $\tilde{V} \supset V$ for which the numbers λ_k are eigenvalues with multiplicities r_k. Any such extension has no other eigenvalues in this lacuna.

From (207) follows in particular:

208. If the number λ_o ($|\lambda_o| = 1$) is not an eigenvalue of the operator V, then there exists a unitary extension $\tilde{V} \supset V$ for which the number λ_o is an eigenvalue of multiplicity m.

209. Whatever the distinct numbers λ_k ($|\lambda_k| = 1$; $k = 1,\dots,m$) may be, unless λ_k is an eigenvalue of V there exists a unitary extension $\tilde{V} \supset V$ for which the numbers λ_k are simple eigenvalues.

210. The extension $\tilde{V}$ in (207) is unique iff $m = 1$.

By passage to the limit one obtains from (207):

211. Among the unitary extensions of an isometric operator V with the spectral lacuna $(e^{i\alpha}, e^{i\beta})$ there exists an extension which preserves this spectral lacuna.

212. If the sequences of points $\{\lambda'_k\}_{k=1}^n$ and $\{\lambda''_k\}_{k=1}^n$ on the unit circle intertwine, then there exists an isometric operator V with defect number $m = 1$ and unitary extensions $\tilde{V}_1 \supset V$, $\tilde{V}_2 \supset V$ such that

$$\sigma(\tilde{V}_1) = \{\lambda'_k\}_{k=1}^n, \quad \sigma(V_2) = \{\lambda''_k\}_{k=1}^n .$$

213. The two given spectra determine the operator V in (212) uniquely within unitary similarity.

7. Quasi-selfadjoint and Quasi-unitary Extensions.

Definition 23. A complete extension $\tilde{S}$ of a hermitian operator S is called *quasi-selfadjoint* if $S \subset (\tilde{S})^*$. The *rank of non-selfadjointness* of any operator A with Dom A = E is defined as $\mathrm{rk}(A - A^*)$.

214. The rate of non-selfadjointness of any quasi-selfadjoint extension $\tilde{S}$ of a hermitian operator S does not exceed its defect number m.

215. Any operator A with rank of non-selfadjointness m is a quasi-selfadjoint extension of some hermitian operator S

with defect number m.

216. Let A be a linear operator defined on all of E. Among the subspaces $L \subset E$, which are invariant relative to A and A^* and on which $Ax = A^*x$, there is a largest subspace.

We shall denote it by L_A.

217. The restriction $A|L_A$ is a self-adjoint operator.

Definition 24. That restriction is called the *self-adjoint component* of the operator A. When $L_A = 0$, the operator A is called *simple non-selfadjoint*. The restriction $A|L_A^{\perp}$ is the maximal simple non-selfadjoint part of the operator A.

218. $L_A = L_{A^*}$.

219. $L_A \subset \mathrm{Ker}(A - A^*)$.

220. $L_A^{\perp} = \Sigma_{k=0}^{n-1} A^k\{\mathrm{Im}(A-A^*)\}$.

221. The subspace $L_A^{\perp}$ is invariant relative to A, and the restriction $A|L_A^{\perp}$ is a simple non-selfadjoint operator.

222. The spectra of two distinct quasi-selfadjoint extensions of a simple hermitian operator S with $\delta(S) = 1$ do not have a common point (cf. (168),(169)).

Definition 25. An extension $\tilde{V}$ of an isometric operator V to all the space E is called *quasi-unitary* if $(\tilde{V})^*V \subset I$. For any operator B defined on all of E, the number $\mathrm{rk}(I - B^*B)$ is called the *non-unitary rank* of B.

223. The non-unitary rank of any quasi-unitary extension $\tilde{V}$ of an isometric operator V does not exceed its defect number m.

224. Any operator B with non-unitary rank m is a quasi-unitary extension of some isometric operator V with defect number m.

225. Let B be an arbitrary linear operator defined on all of E. Among the subspaces $M \subset E$ which are invariant relative

to B and B^* and on which $B^*Bx = x$ there is a largest. Denote it by M_B.

226. The restriction $B|M_B$ is a unitary operator.

Definition 26. The restriction $B|M_B$ is called the *unitary component* of the operator B. When $M_B = 0$, the operator B is called *simple non-unitary*. The restriction $B|M_B$ is the maximal simple non-unitary part of the operator B.

227. $M_B = M_{B^*}$.

228. $M_B \subset \mathrm{Ker}(I - BB^*) \cap \mathrm{Ker}(I - B^*B)$.

229. $M_B^{\perp} = \Sigma_{k=0}^{n-1} B^k\{\mathrm{Im}(I - BB^*) \cap \mathrm{Im}(I - B^*B)\}$.

230. The subspace $M_B^{\perp}$ is invariant relative to B, and the restriction $B|M_B^{\perp}$ is a simple non-unitary operator.

231. The spectra of two distinct quasi-unitary extensions of a simple isometric operator V with defect number $\delta(V) = 1$ do not have common points.

232. If $\tilde{S}$ is a quasi-selfadjoint extension of a hermitian operator S and if the nonreal number ω does not belong to the spectrum $\sigma(S)$, then the operator

$$\tilde{V} = (\tilde{S} - \bar{\omega}I)(\tilde{S} - \omega I)^{-1}$$

is a quasi-unitary extension of the isometric operator

$$V = (S - \bar{\omega}I)(S - \omega I)^{-1}$$

and $1 \notin \sigma(\tilde{V})$.

233. If $\tilde{V}$ is a quasi-unitary extension of the isometric operator V and if $1 \notin \sigma(\tilde{V})$, then for any nonreal ω the operator

$$\tilde{S} = (\omega\tilde{V} - \bar{\omega}I)(\tilde{V} - I)^{-1}$$

is a quasi-selfadjoint extension of the hermitian operator

$$S = (\omega V - \bar{\omega}I)(V - I)^{-1}$$

and $\omega \notin \sigma(S)$.

234. If S is a hermitian operator, if the numbers α, β, γ, δ are real, and if $-\delta/\gamma$ is not an eigenvalue of the operator S, then the operator

$$T = (\alpha S + \beta I)(\gamma S + \delta I)^{-1}$$

is also hermitian and $\delta(T) = \delta(S)$.

235. If under the hypotheses of (234) $\tilde{S}$ is a quasi-selfadjoint extension of the operator S and if $-\delta/\gamma \notin \sigma(\tilde{S})$, then the operator

$$\tilde{T} = (\alpha\tilde{S} + \beta I)(\gamma\tilde{S} + \delta I)^{-1}$$

is a quasi-selfadjoint extension of the operator T.

236. If V is an isometric operator, if the number $\bar{\alpha}^{-1}$ is not an eigenvalue of V, and if $|\gamma| = 1$, then the operator

$$W = \gamma(V - \alpha I)(I - \bar{\alpha}V)^{-1}$$

is also isometric and $\delta(W) = \delta(V)$.

237. If under the hypotheses of (236) $\tilde{V}$ is a quasi-unitary extension of the operator V and if $\bar{\alpha}^{-1} \notin \sigma(V)$, then the operator

$$\tilde{W} = \gamma(\tilde{V} - \alpha I)(I - \bar{\alpha}\tilde{V})^{-1}$$

is a quasi-unitary extension of the operator W.

In theorems (238)-(244) below, V is a simple isometric operator with defect number $\delta(V) = 1$, and U_0 is a fixed unitary extension of it.

238. For any unit vector $u \in (\text{Dom } V)^{\perp}$, $|u| = 1$, the formula $U_\varphi u = e^{i\varphi}U_0 u$ establishes a bijective correspondence between the interval $0 \leqq \varphi < 2\pi$ and the set of unitary extensions $U_\varphi \supset V$.

239. Let $R_\lambda[U_\varphi]$ be the resolvent of the extension $U_\varphi \supset V$ and

$$w_\varphi(\lambda) = \lambda(R_\lambda[U_\varphi]u,u)/\{1 + \lambda(R_\lambda[U_\varphi]u,u)\}.$$

Then $w_\varphi(\lambda) = e^{i\varphi}w_0(\lambda)$.

240. The set of roots of the function $w_o(\lambda)$ coincides with the spectrum of the quasi-unitary extension $VP \supset V$, where P is the orthoprojector on Dom V.

241. If the operator $\tilde{V}$ runs through the set of unitary extensions of the operator V, then the points $(R_\lambda[\tilde{V}]u,u)$ run over the circular contour $C_\lambda(V)$ defined by the formula

$$(R_\lambda[\tilde{V}]u,u) = \frac{1}{\lambda}\,\frac{w_o(\lambda)e^{i\varphi}}{1-w_o(\lambda)e^{i\varphi}} \qquad (0 \leqq \varphi < 2\pi).$$

If $\lambda \notin \sigma(VP)$, then this circle does not degenerate to a point, and the relation between the set of unitary extensions $\tilde{V} \supset V$ and the circle $C_\lambda(V)$ is bijective.

Definition 27. The (closed) disk $W_\lambda(V)$ with the circular boundary $C_\lambda(V)$ for $\lambda \notin \sigma(VP)$ is called the *Weyl-Hamburger disc*.

The following proposition supplements theorem (241).

242. The interior of the Weyl-Hamburger disc $W_\lambda(V)$ is in a bijective correspondence with the set of quasi-unitary extensions $\tilde{V} \supset V$ for which $(\tilde{V})^*V < I$. The exterior of the disc $W_\lambda(V)$ is in a bijective correspondence with the set of quasi-unitary extensions $\tilde{V} \supset V$ for which $(V)^*\,\tilde{V} < I$.

Definition 28. The function $w_o(\lambda)$ defined in (239) is called the *characteristic function of the simple isometric operator* V.

Obviously, $w_o(\lambda)$ is a rational function. From (239) it follows that the characteristic function of the operator V is defined by this operator to within a constant multiple of modulus one. Below we shall write $w(\lambda;V)$ instead of $w_o(\lambda)$.

243. The characteristic function $w(\lambda;V)$ is regular analytic in the unit disc; it maps the unit disc onto itself, and $w(0;V) = 0$.

The following two propositions express the fundamental properties of the characteristic function.

244. If $\tilde{V}$ is a quasi-unitary extension of the operator V defined by the formula $\tilde{V}u = \alpha U_o u$, $\|u\| = 1$, $(|\alpha| \neq 1)$, then the spectrum $\sigma(\tilde{V})$ coincides with the set of roots of the equation

$$w(\lambda;V) = \alpha.$$

245. The simple isometric operators V_1 and V_2 are unitarily similar iff their characteristic functions $w(\lambda;V_1)$ and $w(\lambda;V_2)$ are the same (within a constant factor of modulus one).

In (246)-(251) below, the results (238)-(245) are transferred to hermitian operators. The proofs can be obtained with the aid of the Cayley transformation. In (246)-(250), S is a simple hermitian operator with $\delta(S) = 1$.

246. Let

$$V = (S - iI)(S + iI)^{-1},$$

let U_φ be the operator defined in (238) and S_φ its Cayley transform, viz.

$$S_\varphi = i(U_\varphi + I)(U_\varphi - I)^{-1},$$

and let $R_\mu[S]$ be the resolvent of the operator S_φ. Then the function $w_\varphi(\lambda)$ of (239) becomes for real μ

$$w_\varphi\left(\frac{\mu-i}{\mu+i}\right) = \frac{\mu-i}{\mu+i} \cdot \frac{(\{I+(\mu-i)R_\mu[S_\varphi]\}u,U_\varphi u)}{(\{I+(\mu-i)R_\mu[S_\varphi]\}u,u)} .$$

247. If the operator $\tilde{S}$ runs through the set of self-adjoint extensions of the operator S, then the points $(R_\mu[S]u,u)$ run through some circular contour $C_\mu(S)$. Unless the number μ belongs to the spectrum of the Cayley transform of the operator VP (P is the orthoprojector onto Dom V), this contour does not degenerate to a point and the correspondence

between the set of self-adjoint extensions and $C_\mu(S)$ is bijective.

The corresponding (closed) disc $W_\mu(S)$ is called the Weyl-Hamburger disc, as before. The following proposition supplements theorem (247).

248. The interior of the Weyl-Hamburger disc $W_\mu(S)$ is in bijective correspondence with the set of quasi-selfadjoint extensions $\tilde{S} \supset S$ for which $\tilde{S} - \tilde{S}^* > 0$. The exterior of the disc $W_\mu(S)$ corresponds bijectively with the set of quasi-selfadjoint extensions $\tilde{S} \supset S$ for which $\tilde{S} - \tilde{S}^* < 0$.

Definition 29. The function

$$v(\mu;S) = w_o\left(\frac{\mu-i}{\mu+i}\right)$$

is called the *characteristic function of the simple hermitian operator S.*

Again $v(\mu;S)$ is a rational function. The characteristic function $v(\mu;S)$ is determined by the operator S up to a constant factor of modulus one.

249. The characteristic function $v(\mu;S)$ is regular in the upper halfplane, maps it onto the unit disc, and $v(i;S) = 0$.

250. If the quasi-selfadjoint extension $\tilde{S} \supset S$ is defined by the formula

$$\tilde{S} = i(\tilde{V} + I)(\tilde{V} - I)^{-1},$$

where $\tilde{V} \supset V$ and $\tilde{V}u = \alpha U_o u$, $\|u\| = 1$, $(|\alpha| = 1)$, then the spectrum $\sigma(S)$ coincides with the set of roots of the equation $v(\mu;S) = \alpha$.

251. The simple hermitian operators S_1 and S_2 are unitarily similar iff their characteristic functions $v(\mu;S_1)$ and $v(\mu;S_2)$ coincide up to a constant multiple of modulus one.

The results (238)-(245) and (246)-(251) can be generalized to isometric and hermitian operators with arbitrary defect numbers. We shall sketch these generalizations in (252)-(261) only for isometric operators.

Let V be a simple isometric operator, and $\delta(V) = m$. For a unitary extension $U_o \supset V$ we introduce the vector function

$$\hat{R}_o(\lambda) = QR_\lambda(U_o) \text{ (Dom V)}^\perp,$$

where Q is the orthoprojector onto $(\text{Dom V})^\perp$. By analogy with (239) we formulate the next definition.

Definition 30. The *characteristic operator-function* $W_o(\lambda) = W(\lambda;V)$ of the operator V is given by the formula

$$W_o(\lambda) = \lambda\hat{R}_o(\lambda)(I + \lambda\hat{R}_o(x))^{-1}.$$

Clearly, $W(\lambda;V)$ is a rational operator-function of λ and $W(0,V) = 0$.

252. When $|\lambda| < 1$, the real part of the operator $\frac{1}{2}I + \lambda\hat{R}_o(\lambda)$ is nonnegative.

253. The operator $W(\lambda;V)$ is a strict compression (IV,D56) for $|\lambda| < 1$ and unitary for $|\lambda| = 1$.

254. If U_o is a unitary extension of the operator V, then for any $u \in (\text{Dom V})^\perp$, $\|u\| = 1$, the formula $U_\Phi u = e^{i\Phi}U_o u$ establishes a bijective correspondence between the set of self-adjoint operators Φ in the space $(\text{Im V})^\perp$, which satisfy the inequality $0 \leqq \Phi < 2\pi I$, and the set of unitary extensions $U_\Phi \supset V$.

255. If

$$\hat{R}_\Phi(\lambda) = QR_\lambda(U_\Phi)\,|\,(\text{Dom V})^\perp$$

and

$$W_\Phi(\lambda) = \lambda\hat{R}_\Phi(\lambda)(I + \lambda R_\Phi(\lambda))^{-1},$$

then $W_\Phi(\lambda) = \hat{U}W(\lambda;V)$ where $\hat{U}$ is a unitary operator in the space $(\text{Dom V})^\perp$, not depending upon λ.

256. The set of roots of the equation $\det W_o(\lambda) = 0$ coincides with the spectrum of the quasi-unitary extension

$VP \supset V$ where P is the orthoprojector onto $(\mathrm{Dom}\ V)$.

For any quasi-unitary extension $\tilde{V} \supset V$ let us introduce, provided $\lambda \notin \sigma(VP)$,

$$\hat{R}(\lambda;\tilde{V}) = QR_\lambda(\tilde{V}) \,|\, (\mathrm{Dom}\ V)^\perp .$$

257. If the operator $\tilde{V}$ runs through the set of quasi-unitary extensions of the operator V, then the operator $\hat{R}(\lambda;\tilde{V})$ runs through the set described by the formula

$$\hat{R}(\lambda;\tilde{V}) = 1/\lambda\ ZW_o(\lambda)(I - W_o(\lambda)Z)^{-1} ,$$

where Z runs through the space of operators on $(\mathrm{Dom}\ V)^\perp$.

258. In (257) the unitary extensions $\tilde{V} \supset V$ are in bijective correspondence with the unitary operators Z.

Further:

259. The operator $\tilde{V}$ is a compression iff Z is a compression.

The set of operators

$$\mathbb{W} = \{\hat{R}(\lambda;\tilde{V}) \,|\, Z^*Z \le I\}$$

is the operator-analog of the Weyl-Hamburger disc.

260. Let $\tilde{V}$ be the quasi-unitary extension of the operator V defined by the formula

$$\tilde{V}u = AU_o u \qquad (u \in (\mathrm{Dom}\ V)^\perp,\ \|u\| = 1) ,$$

where A is an operator in $(\mathrm{Im}\ V)^1$. Then the spectrum $\sigma(\tilde{V})$ coincides with the set of roots of the equation

$$\det\{W(\lambda;V) - \tilde{A}\} = 0 \qquad (\tilde{A} = U_o^{-1}AU_o) .$$

261. Any two simple isometric operators V_1 and V_2 are unitarily similar iff their characteristic operator-functions $W(\lambda;V_1)$ and $W(\lambda;V_2)$ coincide up to constant unitary operator-factors left and right (*Livšic's Theorem*).

8. Block-triangular Representation of Operators with Nonzero Rank of Non-selfadjointness

In this section, A is an operator defined on all of E and $m = \mathrm{rk}(A - A^*)$. For $m = 0$ Schur's triangular representation (cf. III,D52) of the operator A is a diagonal matrix, since in this case the diagonal elements of the triangular representation determine A within unitary similarity. This property is in general not conserved for $m > 0$. However, in a definite sense, it can be restored by proper generalization of the triangular representation (For $m = 1$ no generalization is required.)

262. Assume $m = 1$ and let μ be a nonzero eigenvalue of the operator $(1/i)(A - A^*)$. If $\{e_k\}_1^n$ is an arbitrary Schur basis for the operator A, then the matrix $A = (a_{jk})_{j,k=1}^n$ of the operator A has in this basis the form

$$a_{jk} = \begin{cases} \alpha_k + (i/2)|\gamma_k|^2 \operatorname{sign} \mu & (j = k), \\ i\gamma_j\gamma_k \operatorname{sign} \mu & (j > k), \\ 0 & (j < k), \end{cases}$$

where the α_k are real and not all the γ_k are zero.

263. Under the hypotheses of (262) there exists an orthonormal basis in which the matrix A has the form

$$a_{jk} = \begin{cases} \alpha_k + (i/2)\beta_k^2 \operatorname{sign} \mu & (j = k), \\ i\beta_j\beta_k \operatorname{sign} \mu & (j > k), \\ 0 & (j < k), \end{cases}$$

where the α_k are real and the β_k are nonnegative and not all zero.

Definition 31. We shall call the triangular representation in (263) *canonical*.

From (263) follows

264. If two operators have 1 as their rank of non-selfadjointness and if their spectra coincide, then they are unitarily similar.

For $m > 1$, we go from the usual to the block-triangular representation:

Denote by μ_k $(k = 1,\dots,m)$ the nonzero eigenvalues of the operator $S = (1/i)(A - A^*)$ (rk $S = m$). Let $\{e_k\}_1^n$ be some Schur basis for the operator A, $s = \{s_{jk}\}_{j,k=1}^n$ the matrix of the operator S in this basis, and let J be a diagonal matrix with the diagonal $\{\text{sign}\ \mu_k\}_1^m$. The diagonal elements of the matrix of the operator $R = (A + A^*)/2$ in this same basis are denoted by α_k $(k = 1,\dots,n)$.

Recalling I,D28, let $\tilde{E} = E + E_1$ $(\dim \tilde{E} = nm)$ and let $\{\tilde{e}_j\}_1^{mn}$ be an orthonormal basis of the space E in which $\tilde{e}_{(k-1)m+1} = e_k$ $(k = 1,\dots,n)$. Denote by $\tilde{S} \in \mathbb{M}(E)$ the extension of the operator S defined by the condition $\tilde{S}|E_1 = 0$.

265. The matrix $\tilde{S}$ of the operator S in the basis $\{e_j\}_1^{mn}$ has the form

$$\tilde{S} = (F_{jk})_{j,k=1}^n$$

where

$$F_{jk} = (t_{rs}^{(j,k)})_{r,s=1}^m,\quad t_{rs}^{(j,k)} = \begin{cases} s_{jk} & \text{for } r = 1,\ s = 1 \\ 0 & \text{for } r \neq 1 \text{ or } s \neq 1. \end{cases}$$

266. If $U = (u_{jk})_{j,k=1}^n$ is a unitary matrix (III,D43) which transforms the matrix S to diagonal form so that

$$u^*Su = \begin{bmatrix} \mu_1 & 0 \ldots 0 & 0 \ldots 0 \\ 0 & \mu_2 \ldots 0 & 0 \ldots 0 \\ \ldots & \ldots & \ldots \\ 0 & 0 \ldots \mu_m & 0 \ldots 0 \\ 0 & 0 \ldots 0 & 0 \ldots 0 \\ \ldots & \ldots & \ldots \\ 0 & 0 \ldots 0 & 0 \ldots 0 \end{bmatrix},$$

then

$$F_{jk} = F_j J F_k^*$$

where

$$F_j = \begin{bmatrix} t_{j1} & t_{j2} & \ldots & t_{jm} \\ 0 & 0 & \ldots & 0 \\ \ldots & \ldots & \ldots & \ldots \\ 0 & 0 & \ldots & 0 \end{bmatrix},$$

$t_{j\nu} = u_{j\nu}|\mu_\nu|^{\frac{1}{2}}$ $(j = 1,\ldots,n;\ \nu = 1,\ldots,m)$.

Let $\widetilde{R} \in \mathbb{M}(\widetilde{E})$ be the extension of the operator R defined by the formula $\widetilde{R}e_k = \alpha_k \widetilde{e}_{(k-1)m+\nu}$ $(k = 1,\ldots,n;\ \nu = 2, \ldots, m)$.

Definition 32. The operator $\widetilde{A} = \widetilde{R} + (i/2)\widetilde{S}$ is called the *canonical extension* of the operator A.

267. The canonical extension $\widetilde{A}$ of the operator A is the orthogonal sum of this operator and a self-adjoint operator in E_1. The matrix $\widetilde{A}$ of the operator A in the basis $\{\widetilde{e}_j\}_1^{mn}$ has the block-triangular form

$$\widetilde{A} = (\widetilde{A}_{jk})_{j,k=1}^{n},$$

where

$$A_{jk} = \begin{cases} \alpha_k E + (i/2) F_k J F_k^* & (j = k) \\ i F_j J F_k & (j > k) \\ 0 & (j < k) \end{cases}$$

and E is the identity matrix of order m.

The representation obtained is analogous to (262). From it, a block-triangular representation can be obtained which corresponds to (263):

268. In the orthonormal basis $\tilde{g}_{(k-1)m+\nu} = D_k \tilde{e}_{(k-1)m+\nu}$ $(k = 1,\ldots,n;\ \nu = 1,\ldots,m)$, where D_k is the left phase factor (III,D48) in the polar representation $F_k = D_k B_k$, the matrix A of the operator $\tilde{A}$ has the form

$$\tilde{A} = (A_{jk})_{j,k=1}^{n},$$

where

$$A_{jk} = \begin{cases} \alpha_k E + (i/2) B_k J B_k & (j = k) \\ i B_i J B_k & (j > k) \\ 0 & (j < k) . \end{cases}$$

The ranks of the blocks A_{jk} $(j \neq k)$ do not exceed one.

The block-triangular representation described in (268) is also regarded as canonical.

269. Let $\tilde{A}^{(1)}$ and $\tilde{A}^{(2)}$ be matrices of the form

$$\tilde{A}^{(p)} = (A_{jk}^{(p)})_{j,k=1}^{n} \qquad (p = 1,2),$$

$$A_{jk}^{(p)} = \begin{cases} \alpha_k^{(p)} E + (i/2) B_k^{(p)} J B_k^{(p)} & (j = k), \\ i B_j^{(p)} J B_k^{(p)} & (j > k), \\ 0 & (j < k). \end{cases}$$

where the $B_j^{(p)}$ are nonnegative hermitian matrices and

$$\operatorname{rk} B_j^{(p)} = \operatorname{rk} B_j^{(p)} J B_j^{(p)} \leqq 1 .$$

Then, if $A_{kk}^{(1)} = A_{kk}^{(2)}$ $(k = 1,\ldots,n)$, it follows that $\tilde{A}^{(1)} = \tilde{A}^{(2)}$.

We shall call operators $A_1, A_2 \in \mathbb{M}(E)$ which have the same rank of non-selfadjointness strongly unitarily similar, if their canonical extensions are unitarily similar.

270. If the diagonal blocks of the canonical block-triangular representations of the operators $A_1, A_2 \in \mathbb{M}(E)$ coincide, then these operators are strongly unitarily similar.

Thus:

271. If the diagonal blocks of the canonical block-triangular representations of the operators $A_1, A_2 \in \mathbb{M}(E)$ coincide, then their maximal simple non-selfadjoint parts are unitarily similar (another *theorem of Livšic*).

9. The Problem of Moments

Definition 33. Let $S \in \mathbb{S}(E)$, $x \in E$. The real numbers

$$s_k = (S^k x, x) \qquad (k = 0,1,\ldots)$$

are called the *moments of the operator* S *for the vector* x. The problem of finding a self-adjoint operator with simple spectrum (II,D19) which for some generating vector x has the given moments

$$s_k = (S^k x, x) \qquad (k = 0,1,\ldots,p;\ p \leqq \infty), \qquad (*)$$

represents the operator-interpretation of the exponential problem of moments. If S is a solution to the moment problem (*), then the operator USU^{-1} for $U \in \mathbb{U}(E)$ is also a solution to this moment problem. Unitarily similar solutions are not regarded as distinct. The moment problem (*) is said to be determinate if it has a unique solution, and indeterminate if it has more than one solution. The analytic apparatus for the study of the moment problem is the quadratic form in the following

Definition 34. The real hermitian-quadratic form

$$\mathbb{H}_m = \Sigma_{j,k=0}^{m-1} s_{j+k}\xi_j\bar{\xi}_k$$

is called a *Hankel form*.

272. If the moment problem (*) has a solution, then the Hankel form $H_{[(p/2)+1]}$ is nonnegative and

$$\text{rk } H_m = \min(m,n) \qquad (m = 1,\ldots,[(p/2)+1]).$$

273. If p = 2n-1 and the moment problem (*) has the solutions S_1, S_2, then their characteristic polynomials coincide.

Therefore,

274. If p = 2n-1 and the moment problem is solvable, then it is determinate.

275. Let p = 2n-2, let the moment problem (*) have a solution S, and let T be the contraction of the operator S to $sp(\{S^k x\}_o^{n-2}$. Then the set of solutions of the moment problem (*) coincides with the set of self-adjoint extensions $\tilde{T}$ of the operator T.

Therefore,

276. If p = 2n-2 and the moment problem (*) is solvable, then it is indeterminate.

277. Under the conditions of theorem (276) the moment problem (*) has a continuum of solutions and the spectra of two distinct solutions strictly intertwine.

278. Under the conditions of theorem (275) the formula

$$s_{p+1} = (\tilde{T}^{p+1}x,x)$$

establishes a one-to-one correspondence between the self-adjoint extensions $\tilde{T}$ of the operator T and the points s_{p+1} of the real axis.

In this way we can establish

279. If $p = 2n-2$ and the moment problem (*) is solvable, then the extended moment problem obtained by the adjunction to (*) of the additional equation

$$(S^{p+1}x,x) = s_{p+1} ,$$

is also solvable for any real s_{p+1}.

The key to the study of the solvability of the moment problem is given by theorem (280).

Let the given moments s_k $(k = 0,1,\dots,2n-2)$ be such that the Hankel form $\mathbb{H}_n$ is positive. Introduce in the space of polynomials Π^n the scalar product

$$(P,Q) = \Sigma_{j,k=o}^{n-1} s_{j+k}\xi_j\bar{\eta}_k \qquad (\mathcal{P} = \Sigma_{j=o}^{n-1} \xi_j\lambda^j,\ \mathcal{Q} = \Sigma_{k=o}^{n-1} \eta_k\lambda^k)$$

and consider the operator T which multiplies by λ:

$$T(\Sigma_{k=o}^{n-2} \xi_k\lambda^k) = \Sigma_{k=o}^{n-2} \xi_k\lambda^{k+1} \qquad (\text{Dom } T = \Pi^{n-1}) .$$

280. The operator T is simple and hermitian, its defect number equals unity, and the moments of the operator T for the polynomial $\mathcal{P}(\lambda) = 1$ equal s_k $(k = 0,1,\dots,2n-2)$.

It is now easy to carry out a full investigation of the moment problem (*).

281. For the solvability of the moment problem (*) it is necessary that the Hankel form $\mathbb{H}_{[(p/2)+1]}$ be nonnegative and that

$$\text{rk } \mathbb{H}_m = \min(m,n) \qquad (m = 1,\dots,[(p/2)+1]).$$

These conditions are also sufficient if $p \leqq 2n$.

282. If the moment problem (*) is solvable then it is determinate iff $p \geqq 2n-1$.

The operator-interpretation of the trigonometric moment problem can be treated in an analogous way.

This problem requires the determination of a unitary operator U with simple spectrum, which for any generating vector x has the given moments

$$c_k = (U^k x, x) \qquad (k = 0,1,\ldots,p;\ p \leqq \infty) \ . \qquad (**)$$

Together with U any operator which is unitarily similar to U is a solution of the moment problem (**). Unitarily similar solutions are not regarded as distinct. The moment problem (**) is called determinate if it has a unique solution, and indeterminate if it has more than one solution.

The analytic apparatus for (**) is again a real hermitian-quadratic form

$$\mathbb{T}_m = \Sigma_{j,k=o}^{m-1} c_{j-k}\xi_j\overline{\xi}_k \qquad (c_{-k} = \overline{c}_k) \ ,$$

called a Toeplitz form.

283. If the moment problem (**) has a solution, then the Toeplitz form $\mathbb{T}_{p+1}$ is nonnegative and

$$\mathrm{rk}\ \mathbb{T}_m = \min(m,n) \qquad (m = 1,\ldots,p+1) \ .$$

284. If p = n and the moment problem (**) is solvable, then it is determinate.

285. Let p = n-1, let the moment problem (**) have a solution U, and let V be the restriction of the operator U to $\mathrm{sp}(\{U^k x\}^{n-2})$. Then the set of solutions of the moment problem (**) coincides with the set of unitary extensions $\widetilde{V}$ of the operator V.

Therefore,

286. If p = n-1 and the moment problem (**) is solvable, then it is indeterminate.

287. Under the conditions of theorem (286) the moment problem has a continuum of solutions and the spectra of two distinct

solutions of it strictly intertwine.

288. Under the conditions of theorem (285) the formula

$$c_{p+1} = (\tilde{V}^{p+1}x,x)$$

establishes a bijective correspondence between the unitary extensions $\tilde{V}$ of the operator V and the points c_{p+1} of the circular contour

$$\det\begin{vmatrix} c_o & c_1 & \cdots & c_{n-1} & z \\ c_{-1} & c_o & \cdots & c_{n-2} & c_{n-1} \\ \cdots & \cdots & \cdots & \cdots & \cdots \\ \bar{z} & c_{1-n} & \cdots & c_{-1} & c_o \end{vmatrix} = 0 . \qquad (***)$$

Thus,

289. If $p = n-1$ and the moment problem (**) is solvable, then the extension obtained by adjoining to (**) the supplementary equation

$$(U^{p+1}x,x) = c_{p+1},$$

is also a solution for any complex c_{p+1} lying on the circle (***).

Let moments c_k $(k = 0,1,\ldots,n-1)$ be given such that the Toeplitz form $\mathbb{T}_n$ is positive. Introduce in the space T^n of trigonometric polynomials $\Sigma_{k=o}^{n-1} \xi_k e^{ik\theta}$ the scalar product

$$(P,Q) = \Sigma_{j,k=o}^{n-1} c_{j-k}\xi_j\bar{\eta}_k$$

$(P = \Sigma_{j=o}^{n-1} \xi_j e^{ij\theta},\ Q = \Sigma_{k=o}^{n-1} \eta_k e^{ik\theta})$, and consider the operator V which multiplies by $e^{i\theta}$.

290. The operator V is a simple isometry, its defect number equals unity, and the moments of V for the polynomial $P(\theta) = 1$ equal c_k $(k = 0,1,\ldots,n-1)$.

Finally we have

291. The moment problem (**) is solvable iff the Toeplitz form $\mathbb{T}_{p+1}$ is nonnegative and

$$\operatorname{rk} \mathbb{T}_m = \min(m,n) \qquad (m = 1,\ldots,p+1) .$$

292. If the moment problem (**) is solvable, then it is determinate iff $p \geqq n$.

adjoint operator (41)

bijection between self-adjoint operators and unitary extensions (254)
bijection for quasi-self-adjoint extensions (248)
bijection for unitary extensions (238),(241)
block-diagonal representation, following (264)

canonical extension of operator D32, (268)
canonical triangular representation D31
Cayley transform of herm. (isom.) operator D16 (D17)
characteristic function, fundamental properties of (244)
characteristic function of simple hermitian operator D29
characteristic function of simple isometric operator D28,D30
commuting operators D6
complete extension D3
continuation by zero, following (124)

defect number D1
defect subspace of operator at a point D18
determinate moment problem, following D33
dimensional invariance of defect subspace (108)
direct sum of operators D9
domain of definition D1

equality of operators D2
exponential problem of moments, following D33
extension-preserving spectral lacuna (211)
extension (restriction) of operator D3
extension theorem for slef-adjoint hermitian operators (141)

generalized inverse of operator D14

Hankel form D34
hermitian operator D10

intertwining spectra and existence of extensions (193)
intertwining spectra of self-adjoint extensions (168)f.
isometric operator D11

M.G. Kreĭn's Theorem on set of self-adjoint extensions (148)

L_λ defined D21
Livshitz's Theorems (261), (271)
lower bound of operator, following (185)

moment of operator for vector D33
multiplicity sum of eigenvalues in a gap (200)f.

v. Neumann's parametric formulas (120)

CHAPTER X

SOME SPECIAL CLASSES OF OPERATORS

1. Dissipative Operators and Compressions in Euclidean Space

Definition 1. An operator A in a Euclidean space E is called *dissipative* if

$$(1/2i)(A - A^*) \geq 0.$$

1. The set of dissipative operators is a solid wedge in $\mathbb{M}(E)$; its linear part is $\mathbb{S}(E)$ (VII,D5 and D7).
2. If $\tilde{S}$ is a quasi-selfadjoint extension of a hermitian operator S with defect number $\delta(S) = 1$, then at least one of the two operators S and -S is dissipative.
3. The spectrum of a dissipative operator lies in the halfplane $\mathrm{Im}\,\lambda \geqq 0$.
4. In the halfplane $\mathrm{Im}\,\lambda < 0$ the resolvent R_λ of any dissipative operator A satisfies

$$\|R_\lambda\| \leqq \frac{1}{|\mathrm{Im}\,\lambda|} \qquad (*)$$

(cf. IV,(236)).

Conversely,

5. If the resolvent R of an operator A satisfies the inequality (*) in the halfplane $\mathrm{Im}\,\lambda < 0$, then the operator A is dissipative (cf. IV,(237)).

6. If x and y are eigenvectors of a dissipative operator corresponding to eigenvalues λ and μ, and if $\|x\| = \|y\| = 1$, then

$$|(x,y)|^2 \leq \frac{4\,\mathrm{Im}\,\lambda \cdot \mathrm{Im}\,\mu}{|\lambda-\bar{\mu}|^2} .$$

7. The eigenvectors of a dissipative operator corresponding to distinct real eigenvalues are orthogonal.

8. Among the invariant subspaces L of a dissipative operator A there is a largest and it coincides with the subspace L_A defined in IX,(216).

Definition 2. When $L_A = 0$, the operator A is called a *simple dissipative operator.*

9. Each dissipative operator is the orthogonal sum (IX,D15) of a self-adjoint operator and a simple dissipative one.

10. A dissipative operator A is simple dissipative iff its spectrum lies in the open upper halfplane.

Parallel with dissipative operators we shall consider compressions (IV,D56). Relative to unitary operators, they play the same role as the dissipative operators relative to the self-adjoint ones. In Euclidean space compressions are characterized by the inequality $I-B^*B \geq 0$.

11. The set of compressions is a convex body in $\mathbb{M}(E)$; its set of extreme points is $\mathbb{U}(E)$ (VII,D17 and D23).

12. If $\tilde{V}$ is a quasi-unitary extension of an isometric operator V with defect number $\delta(V) = 1$, then at least one of the operators $\tilde{V}$ and $\tilde{V}^{-1}$ is a compression (if V is not a compression, then V^{-1} exists).

The spectral properties of compressions (cf. IV,(204)-(206)) are analogous to the spectral properties of dissipative operators.

We note an analog of theorem (6).

13. If x, y ($\|x\| = \|y\| = 1$) are eigenvectors of a compression, which correspond to eigenvalues λ and μ, then

$$|(x,y)|^2 \quad \frac{(1-|\lambda|^2)(1-|\mu|^2)}{|1-\lambda\bar{\mu}|^2} .$$

14. Among the invariant subspaces M of the compression B, on which the operator $B|M$ is unitary, there is a largest and it coincides with the subspace M_B defined in IX,(225).

Definition 3. If $M_B = 0$ we shall call the compression B *simple*.

15. Each compression is the orthogonal sum of a unitary operator and a simple compression.

16. The compression B is simple iff its spectrum lies in the open unit disc.

The Cayley transform can be used to establish a connection between dissipative operators and compressions:

17. If A is a dissipative operator, then the Cayley transform

$$B = (A - \bar{\omega}I)(A - \omega I)^{-1} \quad \text{with imaginary part of } \omega < 0$$

is a compression and $1 \notin \sigma(B)$.

Conversely,

18. If B is a compression and $1 \notin \sigma(B)$, then the Cayley transform

$$A = (\omega B - \bar{\omega}I)(B - I)^{-1} \quad \text{with imaginary part of } \omega < 0$$

is a dissipative operator.

The following two theorems correspond to linear fractional transformations of the halfplane and of the circular disc onto themselves.

19. Let A be a dissipative operator. If the numbers α, β, γ, δ are real, $\alpha\delta - \beta\gamma > 0$, and $-\delta/\gamma \notin \sigma(A)$, then the operator

$$D = (\alpha A + \beta I)(\gamma A + \delta I)^{-1}$$

is also dissipative. If the operator A is self-adjoint, then D is also self-adjoint.

20. If B is a compression and if the numbers α, γ satisfy the conditions $|\alpha| \leq 1$, $|\gamma| = 1$, $1/\bar{\alpha} \notin \sigma(B)$, then the operator

$$C = \gamma(B - \alpha I)(I - \bar{\alpha}B)^{-1}$$

is also a compression. If the operator B is unitary, then C is also unitary.

21. Let $\tilde{E} \supset E$, let $\tilde{P}$ be the orthoprojector from $\tilde{E}$ onto E, and let $\tilde{Q}$ be any orthoprojector in E. Then the operator $\tilde{P}\tilde{Q}|E$ is a nonnegative self-adjoint compression in E.

Also:

22. If $\tilde{U}$ is a unitary operator in the space $\tilde{E}$, then $\tilde{P}\tilde{U}|E$ is a compression in E.

The converses of these two obvious propositions are contained in theorems (23) and (25).

23. If the operator S is a nonnegative self-adjoint compression, then there exists a space $\tilde{E} \supset E$ and an orthoprojector $\tilde{Q} \in \mathbb{M}(E)$ such that $S = \tilde{P}\tilde{Q}|E$, where $\tilde{P}$ is the orthoprojector from $\tilde{E}$ onto E.

This theorem ensues from the following proposition which effectively contains the construction of the space $\tilde{E}$ and the operator $\tilde{Q}$.

24. In the space $\tilde{E} = E \times E$ the block-matrix

$$\begin{bmatrix} S & S^{\frac{1}{2}}(I-S)^{\frac{1}{2}} \\ S^{\frac{1}{2}}(I-S)^{\frac{1}{2}} & I-S \end{bmatrix}$$

defines an orthoprojector $\tilde{Q}$.

25. If $B \in \mathbb{M}(E)$ is a compression, then there exists a space $\tilde{E} \supset E$ and a unitary operator $\tilde{U}$ in $\tilde{E}$ such that $B = \tilde{P}\tilde{U}|E$, where P is the orthoprojector from $\tilde{E}$ onto E.

The theorem ensues from the following construction.

26. In the space $\tilde{E} = E \times E$, the block-matrix

$$\begin{bmatrix} B & (I-BB^*)^{\frac{1}{2}} \\ -(I-B^*B)^{\frac{1}{2}} & B^* \end{bmatrix}$$

defines a unitary operator $\tilde{U}$.

Proposition (25) is also contained in the more general theorem (27) which can be established with the aid of (28) and (29).

27. If $B \in \mathbb{M}(E)$ is a compression and m is a natural number, then there exists a subspace $\tilde{E} \supset E$ and a unitary operator $U \in \mathbb{U}(E)$ such that

$$B^k = \tilde{P}(\tilde{U})^k|E \qquad (k = 1,\ldots,m) .$$

Here $\tilde{P}$ is the orthoprojector from $\tilde{E}$ onto E (*Sz.-Nagy's Theorem*).

28. In the space $\tilde{E} = \underbrace{E \times E \times \ldots \times E}_{m+1}$ the block-matrix

$$\begin{bmatrix} B & S_1 & 0 & 0 & \ldots & 0 \\ 0 & 0 & -I & 0 & \ldots & 0 \\ \cdot & \cdot & \cdot & \cdot & \cdot & \cdot \\ 0 & 0 & 0 & 0 & \ldots & -I \\ -S_2 & B^* & 0 & 0 & \ldots & 0 \end{bmatrix},$$

where $S_1 = (I - BB^*)^{\frac{1}{2}}$, $S_2 = (I - B^*B)^{\frac{1}{2}}$, defines a unitary operator U.

29. The first row of the block-matrix operator U^k (k = 1, 2,...,m) has the form

$$(B^k, B^{k-1}S_1, -B^{k-2}S_1, B^{k-3}S_1, \ldots, (-1)^{k-1}S_1, \underbrace{0,\ldots,0}_{m-k}) .$$

2. Spectral Sets

Definition 4. Let T be an operator acting in the Euclidean space E. A closed set Λ of the complex plane is called a *spectral set of* T if (1) $\sigma(T) \subset \Lambda$, and (2) for any function $\varphi(\lambda)$ holomorphic in a neighborhood of Λ, which satisfies $|\varphi(\lambda)| < 1$ for $\lambda \in \Lambda$, also $\|\varphi(T)\| \leq 1$.

30. If Λ is a spectral set for the operator T and the closed set $\tilde{\Lambda}$ contains Λ, then $\tilde{\Lambda}$ is also a spectral set of the operator T.

The following proposition is analogous to the spectral mapping theorem, II,(144).

31. If Λ is a spectral set of the operator T and if the function $\psi(\lambda)$ is holomorphic in a neighborhood of the set Λ, then the set $\psi(\Lambda)$ is a spectral set for the operator $\psi(T)$.

32. The unit circumference is a spectral set for any unitary operator. The real axis is a spectral set for any self-adjoint operator.

A precise and more general form is

33. If T is a normal (in particular, a unitary or a self-adjoint) operator (III,D50), then the spectrum $\sigma(T)$ is its spectral set.

If a set Λ is spectral for any compression, then it obviously contains the unit disc $|\lambda| \leq 1$. It turns out that this remark is invertible.

34. The unit disc $|\lambda| \leq 1$ is a spectral set for any compression (*von Neumann's Theorem*).

One of the possible proofs is based on (27). Von Neumann's theorem implies

35. The halfplane $\text{Im}\ \lambda \geq 0$ is a spectral set for any dissipative operator.

Conversely, von Neumann's theorem can be deduced from (35).

Let us also note that from (35) follows

36. If λ_o is a regular point of the operator T, then the set

$$\Lambda = \{\lambda \mid |\lambda - \lambda_o|^{-1} \leq \|R_{\lambda_o}(T)\|\}$$

is a spectral set for T.

Therefore,

37. $\sigma(T)$ is the intersection of all spectral sets of any operator T.

From von Neumann's theorem and the mapping theorem (31) for spectral sets it is not difficult to obtain

38. If A is a dissipative operator, then $B = e^{iA}$ is a compression.

Now theorem (33) can be inverted.

40. If the unit circumference is a spectral set of the operator T, then T is a unitary operator.

41. If the real axis is a spectral set for the operator T, then T is a self-adjoint operator.

The following proposition contains an inverse to theorem (33).

42. If the operator T possesses a finite spectral set (in particular, if the spectrum $\sigma(T)$ is a spectral set), then the operator T is normal.

In concluding this section we shall consider the problems that arise if one tries to transfer von Neumann's theorem from Euclidean space to an arbitrary normed space. In this attempt we let the definition of the spectral set carry over to normed spaces without change.

First of all we note that (34) does not hold in arbitrary normed spaces. For example:

43. If the operator B in two-dimensional ℓ-space is defined by the formula $B\{\xi_1,\xi_2\} = \{\xi_2,\xi_1\}$, then $\|B\| = 1$, but

$$\|(2B - iI)(2I + iB)^{-1}\| > 1 ,$$

although

$$|(2\lambda - i)(2 + i\lambda)^{-1}| \leqq 1 \qquad (|\lambda| \leqq 1) .$$

Also, von Neumann's theorem holds only in Euclidean spaces:

44. If the unit disc is a spectral set for all compressions of a normed space E, then the space E is Euclidean (*Foiaş' Theorem*).

This theorem permits the simpler formulation:

45. If for all compressions B in a normed space E the linear fractional transformation

$$(B - \alpha I)(I - \bar{\alpha}B)^{-1}$$

for each α ($|\alpha| < 1$) is also a compression, then the space E is Euclidean.

The proof can be obtained if one follows (46) and (47).

46. If under the conditions of (45) the operator B is defined by the equality

$$Bx = f_o(x)x_o \qquad (x \in E),$$

where $\|f_o\|\|x_o\| \leqq 1$ ($x_o \in E$, $f_o \in E'$), then for each α ($|\alpha| \leqq 1$)

$$\|f_o(x)x_o - \alpha x\| \leqq \|x - \bar{\alpha}f_o(x)x_o\| \qquad (x \in E) .$$

47. Under the same conditions

$$\|y - \alpha x\| \leqq \|x - \bar{\alpha}y\| \qquad (|\alpha| \leqq 1) ,$$

holds for any $x,y \in E$ ($\|x\| \geqq \|y\| > 0$); and when $\|x\| = \|y\|$, then for any α

$$\|y - \alpha x\| = \|x - \bar{\alpha}y\| .$$

From (47) we easily deduce that the space E is Euclidean,

if IV,(12) is taken into account.

The results so far obtained lead naturally to the question, whether it is generally true that there is for all compressions a spectral set distinct from the whole plane.

Definition 5. We shall call the disc $|\lambda| \leqq \rho$ $(\rho > 0)$ a *Bohr disc* if for any function $\varphi(\lambda) = \Sigma_{k=0}^{\infty}\alpha_k\lambda^k$, holomorphic on this disc and satisfying $|\varphi(\lambda) \leqq 1$, the inequality $\Sigma_{k=0}^{\infty} |\alpha_k| \leq 1$ is fulfilled.

48. Each Bohr disc is a spectral set for all compressions in any normed space.

Conversely,

49. Let the operator B be defined by the formula

$$Be_k = e_{k+1} \quad (k = 0,1,\dots,n-1),\ Be_n = 0$$

relative to some basis $\{e_k\}_1^n$ of the space E. Then in the c-norm (IV,D3) relative to this basis,

$$\|B\| = 1,\ \|\varphi(B)\| = \Sigma_{k=0}^{n-1} |\alpha_k|$$

for any function

$$\varphi(\lambda) = \Sigma_{k=0}^{\infty} \alpha_k\lambda^k ,$$

holomorphic in the disc $|\lambda| \leq 1$.

Therefore:

50. If the disc $|\lambda| \leqq \rho$ is a spectral set for all compressions in all normed spaces, then it is a Bohr disc.

It follows from Foiaş' theorem that the disc $|\lambda| \leqq 1$ is not a Bohr disc. If $|\lambda| \leqq \rho_0$ is a Bohr disc, then, clearly, $|\lambda| \leqq \rho$ $(\rho > \rho_0)$ is likewise a Bohr disc. Denote by K the Bohr disc of least radius. By virtue of (48) the disc K is a spectral set for all compressions in any normed space. The radius of this disc is 3 (cf.

E. Landau, Darstellung und Begründung einiger neuerer Ergebnisse der Funktionentheorie, Berlin, 1916 (there is also a Chelsea reprint)).

3. The Abstract Cauchy Problem and the Classes of Operators in a Normed Space Connected with It.

Definition 6. For a given set of values of the real parameter t let $F(t,x)$ be a mapping of the normed[§] space E into itself. The differential equation

$$\frac{dx}{dt} = F(t,x) \qquad (t_o \leqq t < t_1)$$

for the vector function $x = x(t)$ together with the initial condition

$$x(t_o) = x_o$$

constitutes the *abstract Cauchy problem*.

We shall consider only the linear problem

$$F(t,x) = A(t)x \ ,$$

where $A(t)$ is a function whose values are linear operators in E (an operator-valued function). For simplicity we shall assume that the function $A(t)$ is continuous, and that the solution $x(t)$ belongs to the class of vector-valued functions with a continuous derivative on $[t_o,t_1]$. For definiteness we shall put $t_o = 0$, $t_1 = \infty$. Thus we write

$$\frac{dx}{dt} = A(t)x \qquad (0 \leqq t < \infty) \ ,$$

$$x(0) = x_o \ .$$

[§]The existence of a norm is unnecessary for many problems of this section.

In the subsequent work the term "abstract Cauchy problem" will refer only to this linear problem.

51. For any solution to the abstract Cauchy problem we have the estimate

$$\|x(t)\| \leq \|x_o\| \exp \int_o^t \|A(s)\| \, ds \; .$$

52. If the solution to the abstract Cauchy problem exists, then it is unique.

53. The series

$$I + \int_o^t A(s)ds + \int_o^t A(s)ds \int_o^s A(s_1)ds_1 + \ldots$$

is uniformly convergent on each segment $[0,\alpha]$ $(0 < \alpha < \infty)$.

The sum of this series will be denoted by U_t; for each t it is a linear operator in E.

54. The operator-valued function U_t on $[0,\infty)$ is continuous and has a continuous derivative, and

$$\frac{dU_t}{dt} = A(t)U_t \; .$$

Further, it is clear that $U_o = 1$.

55. The formula $x(t) = U_t x_o$ gives the solution to the abstract Cauchy problem.

And so the abstract Cauchy problem under consideration has for any initial vector x_o a unique solution.

Definition 7. The operator U_t is called *evolutive*.

56. The solutions of the abstract Cauchy problem arising from all possible initial vectors form a linear space relative to the natural operations of addition and scalar multiplication.

57. The space of solutions is isomorphic to the basic space E.

Of special interest is the case where A(t) is a constant operator: A(t) = A.

Under this assumption (which is maintained also later unless the contrary is indicated) we have:

58. $U_{t_1+t_2} = U_{t_1} U_{t_2}$.

Thus the set $\{U_t\}_{t\geq o}$ is a one-parameter semigroup. It can be described by the explicit formula:

59. $U_t = s^{At}$.

Therefore

60. The following formula holds:

$$_o\int^\infty U_t e^{-\lambda t} dt = -R_\lambda(A) \qquad (\mathrm{Re}\lambda > \alpha)$$

where $\alpha = \limsup\limits_{t\to\infty} \dfrac{\log\|e^{At}\|}{t}$.

This formula connects the evolutive operator U_t with the resolvent of the operator A. The integral on the left is the Laplace transform of the function U_t.

One consequence is:

61. $\alpha = \sup\limits_{\lambda\in\sigma(A)} \mathrm{Re}\,\lambda$.

All one-parameter semigroups of operators can be obtained essentially in the same way as (59):

62. If U_t $(t \geq 0)$ is a continuous one-parameter semigroup and $U_o = 1$, then there exists a unique operator A such that $U_t = e^{At}$.

Definition 8. The operator A is called the *infinitesimal operator* of the semigroup.

For the proof of (62) one can use the Laplace transform and II,(133). Actually, we have

63. If U_t $(t \geq 0)$ is a continuous one-parameter semigroup, then its Laplace transform

$$V(\lambda) = {_o\int^\infty} U_t e^{-\lambda t} dt$$

satisfies in some halfplane Re $\lambda > \beta$ the equation

$$\frac{V(\lambda) - V(\mu)}{\lambda - \mu} = -V(\lambda)V(\mu) .$$

Note that from (62) follows

64. Any continuous one-parameter semigroup U_t for which $U_o = I$ has a continuous derivative.

Continuity and the condition $U_o = I$ are understood to be fulfilled for all subsequently considered one-parameter semigroups of operators.

Definition 9. The semigroup U_t is called *compressive* if the operator U_t is a compression for each t.

65. A semigroup U_t in a Euclidean space is compressive iff its infinitesimal operator has the form A = iD where D is a dissipative operator (cf. (38)).

Definition 10. Starting again from (3) and (4), we shall now call an operator in an arbitrary normed space *dissipative* if no points of its spectrum lie in the open lower halfplane and if its resolvent satisfies the inequality

$$\|R_\lambda\| \leq \frac{1}{|\text{Im } \lambda|} \qquad (\text{Im } \lambda < 0) .$$

66. A semigroup U_t in a normed space is compressive iff its infinitesimal operator has the form A = iD where D is a dissipative operator.

Definition 11. The semigroup U_t is called isometric if the operator U_t is an *isometry* for each t.

67. The semigroup U_t in a Euclidean space is isometric iff its infinitesimal operator has the form A = iS, where S is a self-adjoint operator (cf. III,(196),(197).

Definition 12. Starting from IV,(236),(237) we shall call an operator in a normed space *conservative* if its spectrum lies on the real axis and if its resolvent satisfies the inequality

$$\|R_\lambda\| \leq \frac{1}{|\text{Im } \lambda|} \qquad (\text{Im } \lambda \neq 0) .$$

68. A semigroup U_t in an arbirary normed space is isometric iff its infinitesimal operator has the form $A = iS$ where S is a conservative operator.

The dissipative and conservative properties are easily interpreted in terms of the abstract Cauchy problem: In the first case the norms of the solutions do not increase; in the second case the norms of the solutions are conserved.

Definition 13. The *abstract Cauchy problem* is called *stable* if the norm of the evolutive operator is bounded:

$$\sup_t \|U_t\| < \infty .$$

For stability it is sufficient (but of course not necessary) that the operator $-iA$ be dissipative.

69. The abstract Cauchy problem is stable iff the spectrum of the operator A lies in the halfplane $\text{Re } \lambda \leq 0$ and iff the part of the operator corresponding to imaginary points of the spectrum is of scalar type (i.e., iff the dimension of each Jordan cell that belongs to an imaginary eigenvalue equals one; cf. II,D21 and the remarks following (104)).

Definition 14. An *operator* A which satisfies the conditions indicated in (69) is called *stable*.

In Ch. VIII,D22 we encountered the idea of a strictly stable operator. To such an operator corresponds an abstract Cauchy problem which is asymptotically stable, meaning

$$\lim_{t\to\infty} U_t = 0 .$$

70. If the abstract Cauchy problem is stable, then there

exists a norm in E relative to which the operator -iA is dissipative.

In Euclidean space this result can be strengthened:

71. If the abstract Cauchy problem in a Euclidean space is stable, then the operator -iA is similar to a dissipative one.

Theorems (69)-(71) are the continuous analogs of some results in Ch. IV, Sect. 8. We now add some remarks for the case of a strictly stable operator.

72. The operator A in a Euclidean space is strictly stable iff there exists an operator $C > 0$ such that the operator -iCA is strictly dissipative, i.e., the imaginary part (III,(128)) of the operator -iCA is positive (*Lyapunov's Theorem*).

The sufficiency is clear. For the proof of necessity one can first check:

73. If the operator A is strictly stable, then for any $B > 0$ the operator

$$C = 2\int_0^\infty e^{A^* t} B e^{At}\, dt$$

satisfies Lyapunov's Equation

$$A^* C + CA^* = -2B \ .$$

It is interesting that some of the results obtained remain valid for the general case of a variable operator A(t).

74. If the operator -iA(t) is dissipative for each value of t, then U_t is a compression for each t.

75. If the operator -iA(t) is conservative for each value of t, then U_t is isometric for each t.

Theorems (74) and (75) illustrate the so-called "law of permanence of the functional equation" (cf. E. Hille, Analytic Function Theory, vol. II, p. 31). We note, however, that the problem of stability cannot be solved by means of that law. Theorem (69) can be replaced only by weaker assertions. For example:

76. If there exists an $\omega > 0$ such that for all t the resolvent $R_\lambda = R_\lambda(A(t))$ satisfies the inequality

$$\|R_\lambda\| \leq \frac{1}{\lambda+\omega} \qquad (\lambda > 0) ,$$

then the abstract Cauchy problem is asymptotically stable and, in addition, $\|U_t\| \leq e^{-\omega t}$ $(t \geq 0)$.

Let us note that

77. In a Euclidean space

$$\|R_\lambda\| \leq \frac{1}{\lambda+\omega} \qquad (\lambda > -\omega)$$

is equivalent to the condition that the Hausdorff set (III, D54) of the operator A lies in the halfplane Re $\lambda \leq -\omega$.

Theorem (77) suggests a generalization of the notion of the Hausdorff set applicable in an arbitrary normed space. Take any ray $\lambda = \rho e^{i\theta}$ $(0 \leq \theta < 2\pi)$, consider all those real numbers α for which

$$\|R_\lambda(A)\| \leq \frac{1}{\rho-\alpha} \qquad (\lambda = \rho e^{i\theta},\ \rho > \alpha) ,$$

and denote the lower bound of the numbers α by $\alpha(\theta;A)$.

78. In a Euclidean space the Hausdorff set of the operator A coincides with the intersection of the halfplanes

$$\mathrm{Re}(\lambda e^{-i\theta}) \leq \alpha(\theta;A) \qquad (0 \leq \theta < 2\pi) .$$

Definition 15. Let E be a normed space and A a linear operator in E. As the *Hausdorff set* (or the *field of values*) of the operator A we shall now designate the intersection of the halfplanes

$$\mathrm{Re}(\lambda e^{i\theta}) \leq \alpha(\theta;A) \qquad (0 \leq \theta < 2\pi).$$

It is obviously convex and contains the spectrum.

79. For any operator A

$$\limsup_{\substack{|\zeta|\to\infty \\ \arg\zeta=\theta}} \frac{\log\|e^{A\zeta}\|}{|\zeta|} = \alpha(\theta;A),$$

which is the support function of the convex hull of the spectrum of A.

Thus $\alpha(\theta;A)$ is the indicator of growth of the entire function $e^{A\zeta}$, and the Hausdorff set coincides with its indicator diagram.[§]

4. Pseudometrics

Let E once more denote a Euclidean space and let $J \in \mathbb{M}(E)$ be a self-adjoint operator with a two-point spectrum $\lambda = \pm 1$, and let $P_\pm$ be the orthoprojectors onto the respective characteristic subspaces $E_\pm$. We write $\dim E_+ = \kappa$.

Definition 16. The linear functional

$$[x,y] = (Jx,y)$$

is called a *pseudometric* (also an *indefinite metric*). A space E which is endowed with a pseudometric is called *Pontrjagin space* (or an *indefinite space*). The present section is an introduction to Sections 5 and 6, where we shall study some classes of operators in a Pontrjagin space. Here we first introduce some necessary definitions.

Definition 17. In agreement with the terminology of Ch. III, Sect. 2, a subspace $L \subset E$ is called *J-nonnegative*,

[§]For the notions of growth indicator (order) and indicator diagram cf. B. Ja. Levin, Distribution of Zeros of Entire Functions, Am. Math. Soc., Providence, 1964.

(*J-positive, J-nonpositive, J-negative, J-neutral*) if $[x,x] \geqq 0$ (respectively $[x,x] > 0$ for $x \neq 0$, $[x,x] \leqq 0$, $[x,x] < 0$ for $x \neq 0$, $[x,x] = 0$) for $x \in L$. The J-orthogonal complement to the subspace L will be denoted by $L^{(\perp)}$.

Definition 18. A vector $x \in L$ is called an *isotropic vector* of the subspace L, if $[x,y] = 0$ for all $y \in L$. The set N of all isotropic vectors of the subspace L is a subspace; it is called the *isotropic subspace* of the subspace L.

The theorems of this section border on the propositions in Ch. III, Sect. 2, and are partially contained in them, but they are formulated in terms of the geometry of Pontrjagin spaces.

80. For any subspace L we have

$$(L^{(\perp)})^{(\perp)} = L .$$

81. The subspaces L and $L^{(\perp)}$ possess a common isotropic subspace $N = L \cap L^{(\perp)}$.

82. The equality $L + L^{(\perp)} = E$ holds iff $L \cap L^{(\perp)} = 0$.

83. If L is an arbitrary J-nonnegative subspace, then the orthoprojector P realizes a bijective mapping of the subspace L onto P_+L.

84. A J-nonnegative subspace L is maximal iff $P_+L = E_+$.

85. In order that a J-neutral subspace L be maximal, it is necessary and sufficient that at least one of the relations $P_+L = E_+$, $P_-L = E_-$ be satisfied.

86. All maximal J-nonnegative and J-positive subspaces have dimensions κ. All maximal J-nonpositive and J-negative subspaces have dimension $n-\kappa$. All maximal J-neutral subspaces have dimension $\min(\kappa,n-\kappa)$.

87. Any maximal J-neutral subspace L is either a maximal J-

nonnegative or a maximal J-nonpositive subspace, or both.

Definition 19. If L is both a maximal J-nonnegative and a maximal J-nonpositive subspace, then it is called *hypermaximal J-neutral.*

88. The subspace L is hypermaximal J-neutral iff $L = L^{(\perp)}$. This is possible only for $n = 2\kappa$.

89. If L is a maximal J-nonnegative (J-positive) subspace, then $L^{(\perp)}$ is a maximal J-nonpositive (J-negative) subspace.

Definition 20. If for the subspace L there is an operator $K \in \mathbb{M}(E)$ such that

$$KE_+ \subset E_-, \quad \text{Ker } K \supset E_-, \quad L = (I + K)E_+,$$

then K is called an *angular operator of the subspace L relative to* E_+. An angular operator relative to E_+ is defined analogously. An angular operator is not unique.

90. If K is some angular operator of the subspace L relative to E_+, then K^* is an angular operator of the subspace $L^{(\perp)}$ relative to E_-.

91. A subspace L is maximal J-nonnegative iff there exists an angular operator K for the subspace L relative to E_+ which is a compression.

92. Under the conditions of (91) the subspace L will be maximal J-positive iff $\|Kx\| < \|x\|$ $(x \in E,\ x \neq 0)$.

93. Under the conditions of (91) the subspace L will be maximal J-neutral iff $\|Kx\| = \|x\|$ $(x \in E)$.

5. Pseudo-selfadjoint and Pseudo-unitary Operators

We shall now introduce the analog of self-adjoint and unitary operators in a Pontrjagin space.

Definition 21. (a) An operator S is called *pseudo-selfadjoint* (or *J-selfadjoint*) if

$$[Sx,y] = [x,Sy] \quad (x,y \in E),$$

i.e., $S = JS^*J$. (b) An operator U is called *pseudo-unitary* (or *J-unitary*) if

$$[Ux,Uy] = [x,y] \quad (x,y \in E),$$

i.e., $U^*JU = J$.

95. If the subspace L reduces (II,D32) a pseudo-selfadjoint operator S, then the subspace $L^{(\perp)}$ also reduces S.

96. If the subspace L reduces a pseudo-unitary operator U, then the subspace $L^{(\perp)}$ also reduces U.

97. The spectrum of a pseudo-selfadjoint operator is symmetric relative to the real axis.

98. The spectrum of a pseudo-unitary operator is symmetric relative to the unit circumference (i.e., if $\lambda \in \sigma(U)$, then $\bar{\lambda}^{-1} \in \sigma(U)$).

99. If S is a pseudo-selfadjoint operator and $\omega \in \sigma(S)$, then its Cayley transform

$$U = (S - \bar{\omega}I)(S - \omega I)^{-1}$$

is a pseudo-unitary operator and $1 \notin \sigma(U)$.

Conversely,

100. If U is a pseudo-unitary operator and $1 \notin \sigma(U)$, then its Cayley transform

$$S = (\omega U - \bar{\omega}I)(U - I)^{-1}$$

is a pseudo-selfadjoint operator.

101. If S is a pseudo-selfadjoint operator, then the operator $U = e^{iS}$ is pseudo-unitary.

Conversely,

102. If U is a pseudo-unitary operator, then there exists a pseudo-selfadjoint operator S such that $U = e^{iS}$.

103. If the spectrum of a pseudo-selfadjoint operator S is positive, then the operator S_1 with positive spectrum which

satisfies the equation $S_1^2 = S$ is also pseudo-selfadjoint.

The existence and uniqueness of the operator S_1 is guaranteed by theorem II,(151).

104. If the pseudo-selfadjoint operator S is pseudo-nonnegative, i.e., $[Sx,x] \geq 0$ $(x \in E)$, then its spectrum is real.

Definition 22. Two subspaces are said to be *obliquely linked* if no vector $x \neq 0$ of one of them is J-orthogonal to the other subspace.

105. The root subspaces (II,D27) $W_\zeta(U)$ and $W_{1/\bar{\zeta}}(U)$ of the pseudo-unitary operator U are obliquely linked $(\zeta \in \sigma(U))$.

106. The root subspaces $W_\lambda(S)$ and $W_{\bar{\lambda}}(S)$ of the pseudo-selfadjoint operator S are obliquely linked $(\lambda \in \sigma(S))$.

107. If $\lambda \neq \bar{\mu}^{-1}$, then the root subspaces $W_\lambda(U)$ and $W_\mu(U)$ of the pseudo-unitary operator U are J-orthogonal. In particular, for $|\zeta| \neq 1$ the root subspace $W_\zeta(U)$ is J-orthogonal to itself.

108. If $\lambda \neq \bar{\mu}$, then the root subspaces $W_\lambda(S)$ and $W_\mu(S)$ of the pseudo-selfadjoint operator S are J-orthogonal. In particular, for $\lambda \neq \bar{\lambda}$ the root subspace $W_\lambda(S)$ is J-orthogonal to itself.

109. The multiplicity of each eigenvector ζ $(|\zeta| \neq 1)$ of a pseudo-unitary operator does not exceed $\min(\kappa, n-\kappa)$.

110. The multiplicity of each eigenvector $\lambda (\lambda \neq \bar{\lambda})$ of a pseudo-selfadjoint operator does not exceed $\min(\kappa, n-\kappa)$.

111. The number of those eigenvalues of a pseudo-unitary operator that do not lie on the unit circle $|\zeta| = 1$ does not exceed $2 \cdot \min(\kappa, n-\kappa)$.

112. The number of distinct nonreal eigenvalues of a pseudo-selfadjoint operator does not exceed $2 \cdot \min(\kappa, n-\kappa)$.

Propositions (109-(112) can be generalized:

113. If U is a pseudo-unitary operator and if L is the sum of all its root subspaces $W_\zeta(U)$ for $|\zeta| > 1$ (or for $|\zeta| < 1$),

then dim L $\leqq$ min(κ,n-κ).

114. If S is a pseudo-selfadjoint operator and L is the sum of all its root subspaces $W_\lambda(S)$ for Im $\lambda > 0$ (or for Im $\lambda < 0$), then dim L $\leqq$ min(κ,n-κ).

6. Invariant Subspaces of Pseudo-selfadjoint and Pseudo-unitary Operators.

This section is devoted to the existence and the properties of invariant subspaces whose vectors have constant sign (= sign-constant subspaces). We shall investigate initially a special class of operators (the so-called strict pseudo-dilations), whose set of limit points contains the pseudo-unitary operators. In this way the existence of those subspaces for pseudo-unitary operators can be established by a limit process. The result so obtained can be transferred to pseudo-selfadjoint operators by the Cayley transformation (99).

Definition 23. An operator B is a *pseudo-dilation* if

$$[Bx,Bx] \geqq [x,x] \qquad (x \in E).$$

A pseudo-dilation B is *strict* if equality holds only for x = 0. *Pseudo-compression* and *strict pseudo-compression* are defined analogously.

Obviously, if B is an invertible pseudo-compression (strict pseudo-compression), then B^{-1} is a pseudo-dilation (strict pseudo-dilation).

115. If B is a strict pseudo-dilation, then the direct sum of those of its root subspaces that correspond to eigenvalues λ for which $|\lambda| < 1$ ($|\lambda| > 1$) is a J-negative (J-positive) subspace.

The proof can be obtained by iterating the inequality

$$[Bx,Bx] > [x,x] .$$

116. Any strict pseudo-dilation possesses a maximal J-positive invariant subspace M_+ ($\dim M_+ = \kappa$) and a maximal J-negative invariant subspace M_- ($\dim M_- = n-\kappa$). Therefore $M_+ \oplus M_- = E$, and the spectrum $\sigma(B|M_+)$ lies in the domain $|\lambda| > 1$, while the spectrum $\sigma(B|M_-)$ is in the domain $|\lambda| < 1$.

117. Any pseudo-dilation B possesses a maximal J-nonnegative invariant subspace in which lie all its eigenvectors corresponding to eigenvalues of modulus no less than one.

The proof can be obtained by a passage to the limit.

118. Any invertible pseudo-dilation B possesses a maximal J-nonnegative invariant subspace M_+ ($\dim M_+ = \kappa$) and a maximal J-nonpositive invariant subspace M_- ($\dim M_- = n-\kappa$). It follows at the same time that $\sigma(B|M_+)$ is in the closed domain $|\lambda| \geqq 1$, and $\sigma(B|M_-)$ in the closed domain $|\lambda| \leqq 1$.

The formulation of the corresponding propositions for pseudo-compressions presents no difficulties and will be omitted.

Since any pseudo-unitary operator is an invertible pseudo-dilation, it is true that

119. Any pseudo-unitary operator possesses maximal J-nonnegative and J-nonpositive invariant subspaces.

But since the operator U is simultaneously an invertible pseudo-compression, then for $\kappa \leqq n-\kappa$

120. Any pseudo-unitary operator U possesses κ-dimensional J-nonnegative subspaces M_e and M_i such that $\sigma(U|M_e) \subset \{\lambda \,|\, |\lambda| \geqq 1\}$ and $\sigma(U|M_i) \subset \{\lambda \,|\, |\lambda| \leqq 1\}$.

121. The subspace M_e contains all root subspaces $W_\zeta(U)$ with $|\zeta| > 1$, and the subspace M_i contains all root subspaces $W_\zeta(U)$ with $|\zeta| < 1$.

Definition 24. The direct sum M of all root subspaces of the pseudo-unitary operator U corresponding to eigenvalues ζ with $|\zeta| \neq 1$, is called the *hyperbolic invariant subspace* of the operator U.

122. The subspaces $M \cap M_e$ and $M \cap M_i$ are obliquely linked and

$$M = (M \cap M_e) \oplus (M \cap M_i).$$

123. If M is a hyperbolic invariant subspace of a pseudo-unitary operator, then $M \oplus M^{(\perp)} = E$.

124. The spectrum $\sigma(U|M^{(\perp)})$ lies on the unit circle.

125. The equality

$$\dim (M \cap M_e) = \dim(M \cap M_i)$$

holds, and there are bases $\Delta_e = \{g_j\}_1^r$ and $\Delta_i = \{h_k\}_1^r$ in $M \cap M_e$ and $M \cap M_i$ such that

$$[g_j, h_k] = \delta_{jk} \qquad (j,k = 1,\dots,r).$$

In the basis $\{g_1,\dots,g_r, h_1,\dots,h_r\}$ the block-matrix of the pseudo-unitary operator U has the form

$$\begin{vmatrix} V & 0 \\ 0 & (V^*)^{-1} \end{vmatrix} .$$

126. The pseudo-unitary operator U has identical Jordan structures in the root subspaces $W_\zeta(U)$ and $W_{1/\bar\zeta}(U)$ for $|\zeta| \neq 1$.

127. Any pseudo-selfadjoint operator S possesses maximal J-nonnegative and J-nonpositive invariant subspaces.

128. Any pseudo-selfadjoint operator S possesses (for $\kappa \leq n-\kappa$) κ-dimensional J-nonnegative subspaces L_i and L_e such that $\sigma(S|L_i)$ lies in the closed domain $\operatorname{Im} \lambda \geq 0$, and $\sigma(S|L_e)$ lies in the closed domain $\operatorname{Im} \lambda \leq 0$.

129. The subspace L_i (L_e) contains all root subspaces $W_\lambda(S)$

with $\operatorname{Im}\lambda > 0$ ($\operatorname{Im}\lambda < 0$).

Definition 25. The direct sum L of all root subspaces of a pseudo-selfadjoint operator S corresponding to non-real eigenvalues is called the *hyperbolic invariant subspace* of the operator S.

130. The subspaces $L \cap L_i$, $L \cap L_e$ are obliquely linked, and

$$L = (L \cap L_i) \oplus (L \cap L_e).$$

131. If L is the hyperbolic invariant subspace of the pseudo-selfadjoint operator S, then $L \oplus L^{(\perp)} = E$.

132. The spectrum $\sigma(S|L^{(\perp)})$ is real.

133. The equality

$$\dim (L \cap L_i) = \dim(L \cap L_e)$$

holds, and there exist bases $\Delta_i = \{g_j\}_1^r$ and $\Delta_e = \{h_k\}_1^r$ in $L \cap L_i$ and $L \cap L_e$ such that

$$[g_j, h_k] = \delta_{jk} \qquad (j,k = 1,\dots,r).$$

In the basis $\{g_1,\dots,g_r, h_1,\dots,h_r\}$ the block-matrix of the pseudo-selfadjoint operator has the form

$$\begin{vmatrix} T & 0 \\ 0 & T^* \end{vmatrix}.$$

134. The pseudo-selfadjoint operator S has the same Jordan structure in the root subspaces $W_\lambda(S)$ and $W_{\bar\lambda}(S)$ for $\operatorname{Im}\lambda \neq 0$.

In concluding this section let us note that the theorems on the existence of sign-constant invariant subspaces hold also under weaker assumptions than those made above. Thus, for example in (115) it can be assumed without changing the proof that

$$[Bx,Bx] > [x,x]$$

is satisfied only for $[x,x] \geqq 0$ $(x \neq 0)$. Also the following theorem (135) bears a more general character: Let A be an arbitrary operator and Δ its Jordan basis. Pick from Δ the J-positive eigenvectors and the vectors associated to them (II,(91)), denote the linear hull of the chosen vectors by M_+, and construct the subspace M_- similarly from the J-negative eigenvectors.

135. If the operator A satisfies the condition

$$[Ax,Ax] > 0 \qquad ([x,x] \geq 0,\ x \neq 0),$$

then the subspace M_+ is maximal J-positive, M_- is maximal J-negative, and $M_+ \oplus M_- = E$.

For the proof one can use lemmas (136) and (137).

136. If the moduli of the eigenvalues of some operator $T \in \mathbb{M}(E)$ are distinct, then for any vector $x \in E$ ($x \neq 0$) one can find a numerical sequence $\{\alpha_k\}_1^\infty$ such that the limit $\lim_{k\to\infty} \alpha_k T^k x$ exists and is nonzero.

This limit is an eigenvector of the operator T.

137. The set of operators A satisfying the condition $[Ax,Ax] > 0$ $([x,x] \geq 0,\ x \neq 0)$ is open in $\mathbb{M}(E)$.

By passing to the limit in (135) one obtains

138. If the operator A satisfies the condition $[Ax,Ax] \geq 0$ $([x,x] \geq 0)$, then it possesses a maximal J-nonnegative invariant subspace.

At last we want to point out the connection between the questions considered and the "operational" linear fractional transformations.

139. Let the operator A satisfy the conditions of (136) and let it be invertible. If L is any maximal J-nonnegative subspace, then AL is also a maximal J-nonnegative subspace.

140. If under the conditions of (136) K is an angular operator (cf. D20) of the subspace L relative to E_+ then the operator

$$\Phi(K) = (A_{11} + A_{12}K)(A_{21} + A_{22}K)^{-1}.$$

where $A_{11} = P_-AP$, $A_{12} = P_-AP_-$, $A_{21} = P_+AP_+$, $A_{22} = P_+AP_-$, is an angular operator of the subspace AL relative to E.

For $\|K\| \leq 1$, the inequality $\|\Phi(K)\| \leq 1$ holds by (91), and theorem (140) reduces to the existence of a fixed point in an endomorphism of the "operator disc" $\|K\| \leq 1$.

7. The Quadratic Pencil of Self-adjoint Operators

Definition 26. A quadratic pencil of self-adjoint operators is a *quadratic trinomial*

$$K(\lambda) = \lambda^2 I + \lambda B + C$$

with self-adjoint coefficients B and C. If the equation

$$K(\lambda)x = 0$$

has a nontrivial solution $\lambda = \lambda_o$, then the complex number λ_o is called an eigenvalue of the pencil $K(\lambda)$. The set of eigenvalues of the pencil $K(\lambda)$ is called its spectrum and is denoted by $\sigma(K)$. The number of distinct eigenvalues of a self-adjoint pencil does not exceed 2n. Define an operator A in the space $\tilde{E} = E \times E$ by means of the block-matrix

$$\begin{bmatrix} 0 & \sqrt{C} \\ -\sqrt{C} & -B \end{bmatrix},$$

where $\sqrt{C}$ is any square root of C. The operator A is said to be associated with the pencil $K(\lambda)$.

141. $\sigma(K) = \sigma(A)$.

142. The spectrum $\sigma(K)$ is symmetric with respect to the real axis.

143. If $C \leq 0$, then the spectrum $\sigma(K)$ is real.

144. If $B > 0$ (≥ 0), then the spectrum $\sigma(K)$ lies in the open (closed) left halfplane.

145. If $B^2 < 4\theta^2 C$ $(0 < \theta < 1)$, then the spectrum $\sigma(K)$ lies in the domain $|\pi/2 \pm \arg\lambda| < \alpha$, where $\alpha = \arcsin\theta$.

The quadratic operator equation

$$Z^2 + BZ + C = 0 \qquad (*)$$

can also be associated with the pencil $K(\lambda)$:

146. If the operator Z is a root of the equation (*), then $\sigma(Z) \subset \sigma(K)$.

147. If Z_1 is a root of the equation (*) and $Z_2 = -(B + Z_1^*)$, then

$$K(\lambda) = (\lambda I - Z_2^*)(\lambda I - Z_1).$$

Thus the operator Z_2 is also a root of the equation (*) and we have the "operational Vieta formulas"

$$Z_1 + Z_2^* = -B, \quad Z_2^* Z_1 = C.$$

The roots Z_1, Z_2 are called conjugate.

Definition 27. If $C \geq 0$ and for all $x \neq 0$

$$(Bx,x)^2 > 4(Cx,x)(x,x),$$

then the pencil $K(\lambda)$ is called *overdamped.*

148. The eigenvalues of an overdamped pencil $K(\lambda)$ are negative.

149. If $K(\lambda)$ is an overdamped pencil, then the operator A associated with it is J-selfadjoint; here the operator J is defined in $\tilde{E} = E \times E$ by the block-matrix

$$\begin{bmatrix} I & 0 \\ 0 & -I \end{bmatrix}.$$

In the following problems of this section the pencil $K(\lambda)$ is assumed to be overdamped.

151. The root Z_1 satisfies the inequality $Z_1^* Z_1 \leq C$.

The conjugate root Z_2 satisfies the inequality $Z_2^* Z_2 \geq C$.

152. The roots of the equation (*) which satisfy the inequalities of (151) are unique and conjugate.

153. $Z_2 - Z_1 > 0$.

154. The operators Z_1 and Z_2 are self-adjoint in the Friedrichs space $E[Z_2 - Z_1]$ (VIII,D31).

155. $\sigma(K) = \sigma(Z_1) \cup \sigma(Z_2)$.

We shall number the eigenvalues $\lambda_k^{(1)}$ and $\lambda_k^{(2)}$ ($k = 1,\ldots,n$) of the operators Z_1 and Z_2 in the usual way. The eigenvalues $\lambda_1^{(1)},\ldots,\lambda_n^{(1)}$, $\lambda_1^{(2)},\ldots,\lambda_n^{(2)}$ of the quadratic pencil $K(\lambda)$ satisfy minimax properties, which are analogs of the corresponding properties for a self-adjoint operator S, but now the role of (Sx,x) is played by the pair of functionals

$$p^{(1,2)}(x) = -(Bx,x) \mp \sqrt{(Bx,x)^2 - 4(Cx,x)(x,x)} .$$

156. $$\lambda_1^{(1)} = \max_{\substack{\|x\|=1 \\ x\in E}} p^{(1)}(x), \quad \lambda_{n-k+1}^{(1)} = \max_L \min_{\substack{\|x\|=1 \\ x\in L}} p^{(1)}(x) \quad (k=2,\ldots,n),$$

$$\lambda_1^{(2)} = \max_{\substack{\|x\|=1 \\ x\in E}} p^{(2)}(x), \quad \lambda_{n-k+1}^{(2)} = \max_L \min_{\substack{\|x\|=1 \\ x\in L}} p^{(2)}(x) \quad (k=2,\ldots,n),$$

where L runs over the set of subspaces of codimension k-1. (*Duffin's formulae*).

Here

157. $\lambda_1^{(1)} < \lambda_n^{(2)}$.

8. Linear Fractional Transformations with Operator Coefficients

We have already met with these transformations in (140) and in Ch. IX, Sect. 7. Here we shall consider the same transformations in connection with "operator circles".

Definition 28. An *operator circle* is a set of operators $Z \in \mathbb{M}(E)$ determined by the equation

$$Z^*AZ + Z^*B + B^*Z + C = 0 \qquad (1)$$

or by

$$ZAZ^* + ZB + B^*Z^* + C = 0 \qquad (1^*)$$

where $A > 0$ and $C = C^*$.

The operator circles (1) and (1^*) are dual to each other. In the following we restrict our attention to circles of the form (1).

158. An operator circle is a nonempty set iff $B^*A^{-1}B - C \geq 0$. The last inequality is assumed to hold in all following problems.

159. The general solution of equation (1) has the form

$$Z = Z_o + R_1UR_2 \; ,$$

where

$Z_o = -A^{-1}B$, $R_1 = A^{-\frac{1}{2}}$, $R_2 = (B^*A^{-1}B - C)^{\frac{1}{2}}$, and $U \in \mathbb{U}(E)$.

The operator Z_o is called the center of the operator circle, and R_1, R_2 are its left and right radii.

160. If $Z_o \in \mathbb{M}(E)$ and $R_1 > 0$, $R_2 > 0$, then the point $Z = Z_o + R_1UR_2$ runs over an operator circle with center at the point Z_o and left and right radii R_1 and R_2, when U runs through the set of unitary operators.

161. The linear function $W = W_o + FZG$ with any $W_o \in \mathbb{M}(E)$ and invertible coefficients F and G transforms an operator circle into an operator circle.

162. The inversion $W = Z^{-1}$ transforms an operator circle into an operator circle if $C > 0$.

In the simplest case, when equation (1) has the form $Z^*Z = 1$, the operator circle is transformed into the set $\mathbb{U}(E)$ of unitary operators.

Simplifying the language by

Definition 29. The domain

$$\mathbb{K} = \{Z \mid Z^*Z < I\}$$

is called the *open operator unit disc* and its closure the *closed operator unit disc*,

$$\overline{\mathbb{K}} = \{Z \mid Z^*Z \leq I\},$$

we can say that $\mathbb{K}$ is transformed by the inversion (162) into the set of strict compressions. The operator disc $\mathbb{K}$ and its closure $\overline{\mathbb{K}}$ are nothing else but the open and closed unit spheres of the space $\mathbb{M}(E)$ in the operator norm. Let us emphasize, however, that the circle $Z^*Z = I$ does not coincide with the boundary of the disc $Z^*Z \leq I$, i.e., $\overline{\mathbb{K}}$. That circle contains only a part of the boundary (namely, the set of all extreme points of $\overline{\mathbb{K}}$ considered as a convex body, cf. (11)).

We now wish to study the linear fractional transformations of the disc $\mathbb{K}$ onto itself.

The following theorem is a generalization of theorem (20):

163. For any $B \in \mathbb{K}$, the transformation

$$W = (I - BB^*)^{-\frac{1}{2}}(Z - B)(I - B^*Z)^{-1}(I - B^*B)^{\frac{1}{2}}$$

maps each of the sets $\mathbb{K}$, $\overline{\mathbb{K}}$, and $\mathbb{U}(E)$ bijectively onto itself.

For the proof one can use

164. $I - W^*W = T^*(I - Z^*Z)T$, where $T = (I - B^*Z)^{-1}(I - B^*B)^{\frac{1}{2}}$.

For brevity we shall denote the transformation (163) by $W = \Phi[Z;B]$.

165. If $W = \Phi[Z;B]$ $(Z, B \in \mathbb{K})$, then

$$Z = \Phi[W;-B].$$

166. If $Z \in \overline{\mathbb{K}}$, $B \in \mathbb{K}$, and $U \in \mathbb{U}(E)$, then

$$\Phi[UZ;B] = U\Phi[Z;U^*B],$$

$$\Phi[ZU;B] = \Phi[Z;BU^*]U .$$

167. If $Z \in \mathbb{K}$ and $B_1, B_2 \in \mathbb{K}$, then

$$\Phi[\Phi[Z;B_1];B_2] = V^{-1}\Phi[Z;\Phi[B_2;-B_1]]V ,$$

where

$$V = (I - \Phi[B_2;-B_1]^*\Phi[B_2;-B_1])^{-\frac{1}{2}}C \quad \text{with}$$

$$C = (I - B_1^*B_1)^{\frac{1}{2}}(I + B_2^*B_1)^{-1}(I - B_2^*B_2)^{\frac{1}{2}} .$$

168. The operator V is unitary.

We come to the result:

169. Let $\mathbb{E}$ be the set of transformations of the form

$$U_1\Phi[\cdot;B]U_2,$$

where $B \in \mathbb{K}$ and $U_1, U_2 \in \mathbb{U}(E)$. This set satisfies the following conditions:

1) each transformation from $\mathbb{E}$ maps the disc $\mathbb{K}$ (and also the disc $\bar{\mathbb{K}}$ and the set $\mathbb{U}(E)$) bijectively onto itself;
2) the inverse of each transformation from $\mathbb{E}$ is also in $\mathbb{E}$;
3) the product of two transformations from $\mathbb{E}$ also belongs to $\mathbb{E}$.

Thus $\mathbb{E}$ is a group.

170. If the linear fractional transformation

$$W = (A_{11} + A_{12}Z)(A_{21} + A_{22}Z)^{-1}$$

maps the disc $\mathbb{K}$ bijectively onto itself, then it belongs to $\mathbb{E}$.

As is well known, in the scalar case a stronger assertion holds: if the analytic function w(z) maps the unit disc bijectively onto itself, then it has the form

$$w(z) = \gamma\,\frac{z-\beta}{1-\bar{\beta}z} \qquad (|\gamma| = 1, |\beta| < 1).$$

This proposition can also be extended to the operator case.[§]

[§]cf. C. L. Siegel, Symplectic Geometry, Amer. J. Math. 45 (1943), 1-86. Reprinted in bound form by Academic Press, New York, 1964.

abstract Cauchy problem D6
angular operator of L relative to E D20
asymptotically stable Cauchy problem, following D14

Bohr disc D5

compressive operator, following (10)
compressive semigroup D9
conservative operator D12

dissipative operator in Euclidean space D1
dissipative operator in normed space D10
Duffin's formulae (156)

evolutive operator D7

Foiaş' Theorem (44)

Hausdorff set of operator in normed space D15
hyperbolic invariant subspace of ps.-unit. operator D24
hyperbolic invariant subspace of ps.-s.adj. operator D25
hypermaximal J-neutral subspace D19

indicator diagram of Hausdorff set, following (79)
infinitesimal operator of semigroup D8
isometric semigroup D11
isotropic subspace D18
isotropic vector D18

J-positive, etc. subspace D17

Lyapunov's Theorem (72)

v. Neumann's Theorem for Euclidean space (34)

obliquely linked subspaces D22
operational linear fract. transformation (139)ff.
operator circle D28
operator unit disc (open, closed) D29
operator-valued function, following D6
overdamped pencil of self-adjoint operators D27

permanence of functional equation, following (75)
Pontrjagin space D16
pseudo-compression (strict) D23
pseudo-dilation (strict) D23
pseudometric (indefinite metric) D16
pseudo-selfadjoint (J-selfadjoint) D21a
pseudo-unitary (J-unitary) D21b

quadratic pencil of s.-adj. operators D26

sign-constant subspace, preceding D23
simple dissipative operator D2
simple compressive operator D3
spectral set of operator in Euclidean space D4
stable abstract Cauchy problem D13
stable operator D14
Sz.-Nagy's Theorem (27)

GENERAL LITERATURE

Algebra

1. Birkhoff, G. Lattice Theory. 3rd ed., Amer. Math. Soc. Colloq. Pub. XXV, Providence, R.I., 1967.
2. Bourbaki, N. Éléments de Mathématique. Livre II: Algèbre. Hermann, Paris, 1951-
3. Baer, R. Linear Algebra and Projective Geometry. Academic Press, New York, 1952.
4. van der Waerden, B.L. Algebra I, II. F. Ungar Pub. Co., New York, 1972.
5. Kurosh, A.G. Lectures on General Algebra. Chelsea Pub. Co., New York, 1963.
6. Kurosh, A.G. Theory of Groups. 3rd ed. (Russian). Nauka, Moscow, 1967. 2nd ed. (2 vol.) in Engl. translation , Chelsea Pub. Co., New York, 1959-60.

Analysis

7. Bourbaki, N. Éléments de Mathématique. Livre IV: Fonctions d'une variable réelle. Hermann, Paris, 1951-
8. Dieudonné, J. Foundations of Modern Analysis. Academic Press, New York, 1960.
9. Rudin, W. Principles of Mathematical Analysis. 2nd ed., McGraw-Hill, New York, 1964.

Topology

10. Aleksandrov, P.S. Theory of Sets and Functions. Chelsea Pub. Co., New York, 1974.
11. Bourbaki, N. General Topology (2 vols.). Addison-Wesley, Reading, Mass., 1966.
12. Hausdorff, F. Set Theory. Chelsea Pub. Co., New York, 1960.

Linear Algebra

13. Bourbaki, N. Éléments de Mathématique. Livre II. Algèbre. (Linear and Multilinear Algebra) Hermann, Paris, 1951-
14. Gantmacher, F.R. Matrix Theory. Chelsea Pub. Co., New York, 1960.
15. Gel'fand, I.M. Lectures on Linear Algebra. 4th ed. (Russian), Nauka, Moscow, 1971. 2nd ed. in Engl. translation, J. Wiley-Interscience, New York, 1961.
16. Mal'cev, A.I. Foundations of Linear Algebra. W. H. Freeman, San Francisco, 1963.
17. Halmos, P.R. Finite-Dimensional Vector Spaces. 2nd ed., Van Nostrand, Princeton, 1958.

18. Shilov, G.E. Linear Algebra. Revised English ed., Prentice-Hall, N.J., 1971. (Note: This edition contains an English version of Nr. 3 in the Special Literature to Chapter I.)
19. Sperner, E. and Schreier, D. Introduction to Modern Algebra and Matrix Theory. Chelsea Pub. Co., New York, 1951-61.
20. Schneider, H. and Barker, G. P. Matrices and Linear Algebra. 2nd ed., Holt, Rinehart, and Winston, New York, 1973.
21. Greub, W. Linear Algebra. 3rd ed., Springer, New York, 1967.
22. Greub, W. Multilinear Algebra. Springer, New York, 1967.

Functional Analysis

23. Akhiezer, N.I. and Glazman, I.M. Theory of Linear Operators in Hilbert Space (2 vol.). F. Ungar, New York, 1961. 2nd ed., Nauka, Moscow, 1966 (Russian).
24. Banach, S. Théorie des Opérations Linéaires. Chelsea Pub. Co., New York, 1955.
25. Bourbaki, N. Éléments de Mathématique. Livre V. Espaces vectoriels topologiques. Hermann, Paris, 1951-
26. Vulikh, B.Z. Introduction to Functional Analysis. Addison-Wesley, Reading, Mass., 1963.
27. Dunford, N. and Schwartz, J.T. Linear Operators, vol. I. J. Wiley, New York, 1958.
28. Dunford, N. and Schwartz, J.T. Linear Operators, vol. II. J. Wiley, New York, 1963.
29. Dunford, N. and Schwartz, J.T. Linear Operators, vol. III. J. Wiley, New York, 1971.
30. Day, M.M. Normed Linear Spaces. Springer, New York, 1962.
31. Yosida, K. Functional Analysis. 3rd ed. Springer, New York, 1971.
32. Kantorovich, L.V. and Akilov, G.P. Functional Analysis in Normed Spaces. Macmillan, New York, 1964.
33. Kolmogorov, A.N. and Fomin, S.V. Elements of the Theory of Functions and of Functional Analysis. Graylock Press, Rochester, N.Y., 1957-61.
34. Lyusternik, L.A. and Sobolev, V.J. Elements of Functional Analysis. F. Ungar, New York, 1961.
35. Maurin, K. Methods of Hilbert Space (2nd ed.). Polish Scientific Publishers, Warsaw, 1972.
36. Plesner, A.I. Spectral Theory of Linear Operators (2 vol.) F. Ungar, New York, 1969.
37. Raikov, D.A. Vector Spaces. P. Noordhoff, Groningen, 1965.
38. Riesz, F. and Sz.-Nagy, B. Functional Analysis. F. Ungar, New York, 1955.
39. Smirnov, V.I. Course in Higher Mathematics. Vol. V. Addison-Wesley, Reading, Mass., 1964.

SPECIAL LITERATURE§

To Chapter I

1. Aronszajn, N. Quadratic forms on vector spaces, Proc. Intern. Symposium on Linear Spaces 1960. Pergamon Press, London, 1961, [8,9,10].
1a. Aronszajn, N. and Fixman, U. Algebraic spectral problems. Studia Math. XXX (1968), 273-338, [8,9,10].
2. Gel'fand, I.M. Abstrakte Funktionen und lineare Operatoren. Mat. Sb. 4 (1938), 235-286, [12].
3. Gel'fand, I.M. and Shilov, G.E. Categories of finite-dimensional spaces. Vestnik M.G.U., ser. mat. N. 4 (1963), 27-48 (Russian). (See also Nr. 18 of Gen. Lit.).
4. Gohberg, I.C. and Kreĭn, M.G. Fundamental results about defect numbers, root vectors. and the index of linear operators. U.M.N. XII, Nr. 2 (74) (1957), 43-118 (Russian), [4,7,8].
5. Grothendieck, A. La Théorie de Fredholm, Bull. Soc. Math France. Nr. 84 (1956). 319-384, [4,7].
6. Cartan, A. and Eilenberg, S. Homological Algebra. Princeton University Press, 1956.
7. Riesz, F. Über lineare Funktionalgleichungen. Acta Math. 41 (1918), 71-98, [4,7].
8. Skornjakov, L.A. Complemented Modular Lattices and Regular Rings. Oliver & Boyd, Edinburgh, 1964, [3].
9. Šmul'jan, V.L. Linear topological spaces and their connection with Banach spaces. DAN SSSR 22 (1939), 475-447. [12]. (Russian).
10. Dunford, N. Integration in general analysis. Trans. Amer. Math. Soc., 37 (1935), 441-453, [12].
11. Fredholm, I. Sur une classe d'équations fonctionelles. Acta Math. 27 (1903), 365-390, [4,7].
12. Grothendieck, A. Produits tensoriels topologiques et spaces nucléaires. Memoirs Amer. Math. Soc., 16, (1955), [8].
13. Hilbert, D. Grundzüge einer allgemeinen Theorie der linearen Integralgleichungen. Teubner, Leipzig, 1912. (There is a Chelsea reprint.)
14. von Neumann, J. On complete topological spaces. Trans. Amer. Math. Soc. Nr. 37 (1937), 1-20, [11].
15. Schauder, J. Über lineare vollstetige Funktionaloperationen. Studia Math., 2 (1930), 183-196, [4,7].

§The numbers in [] refer to sections; papers without [] refer to the chapter as a whole.

16. Steinitz, E. Bedingt konvergente Reihen und convexe Systeme I, II, III. J. reine und angew. Math. 143, 144, 145 (1913-1915), [12].

To Chapter II

1. Aronszajn, N. and Smith, K.T., Invariant subspaces of completely continuous operators. Ann. of Math. (2) 60, 345-350 (1954), [2].
2. Brodskiĭ, M.S. On Jordan cells of infinite-dimensional operators (Russian), DAN SSSR 111 (1956), 926-929, [4,5].
3. Brodskiĭ, M.S. On unicellular integral operators and a theorem of Titchmarsh. UMN XX, Nr. 5 (1965) 189-192, [4,5] (Russian).
4. Weyl, H. The Classical Groups. Princeton University Press, 1946.
5. Gohberg, I.C. and Kreĭn, M.G. Introduction to the theory of linear nonselfadjoint operators in Hilbert space. Amer. Math. Society, Providence, R.K., 1969.
6. Gohberg, I.C. and Kreĭn, M.G. Theory and application of Volterra operators in Hilbert space. Amer. Math. Soc., Providence, R.I., 1970.
7. Dunford, N. A survey of the theory of spectral operators. Bull. Amer. Math. Soc. 64 (1958), 217-274, [3,8].
8. Jacobson, N. Theory of Rings. Amer. Math. Soc., Providence, R.I., 1943 [1,6].
9. Jacobson, N. Lie Algebras. J. Wiley-Interscience, New York, 1962, [7].
10. Lappo-Danilevskiĭ, I.A. Mémoires sur la théorie des systèmes des équationes différentielles linéaires. Chelsea Pub. Co., New York, 1953, [5,6,10].
11. Ljubič, Ju. I. and Matsaev, V.I. On operators with separated spectrum. Mat. Sb. 56 (1962), 433-468, [3]. (Russian)
12. Naĭmark, M.A. A continuous analog of Schur's lemma and its application to Plancherel's formula for complex classical groups. Izv. AN SSSR, ser. mat. 20 (1956), 3-16 [1]. (Russian)
13. Riesz, F. Sur les fonctions des transformations hermitiennes dans l'espace de Hilbert. Acta Sci. Math. Szeged, 7 (1935), 147-159, [6].
14. Čebotarev, N.G. Introduction to the Theory of Algebras. Gostekhizdat 141, [1]. (Russian)
15. Friedrichs, K.O. Perturbations of Spectra in Hilbert Space. Lectures in appl. math., Vol. 3, Amer. Math. Soc., 1965, [9].
16. Halmos, P.R. Commutators of operators, I; Amer. J. Math. 74 (1952), 237-240; II, 76 (1954), 195-198, [7].

17. Kato, T. Perturbation Theory for Linear Operators. Springer, New York, 1966, [9].
18. von Koch, H. Sur quelques points de la théorie des determinants infinis. Acta Math. 24 (1900), 89-112, [10].
19. Helson, H. Lectures on Invariant Subspaces. Academic Press, New York, 1964, [2,3].
20. von Neumann, J. Zur Algebra der Funktionaloperatoren und der Theorie der normalen Operatoren. Math. Ann. 102 (1929), 370-427, [1-6].
21. von Neumann, J. Über Funktionen von Funktionaloperatoren. Ann. Math. 32 (1931), 191-226, [5,6].
22. von Neumann, J. Über einen Satz von Herrn M. Stone. Ann. Math. 33 (1932), 567-573, [6].
23. Putman, C. R. On commutators of bounded matrices. Amer. J. Math. 73 (1951), 127-131, [7].
24. Putman, C.R. On the spectra of commutators. Proc. Amer. Math. Soc. 5 (1954), 929-931, [7].
25. Schur, I. Neue Begründung der Theorie der Gruppencharaktere, Sitzungsber. Preuss. Akad. 1905, 406, [1,6].
26. Sikorski, R. On Leżański's determinants of linear equations in Banach spaces. Studia Math. 14 (1958), 24-48, [10].
27. Stampfli, G. Sums of projections. Duke Math. J. 31, 3 (1964), 455-461, [8].
28. Sz.-Nagy, B. Spektraldarstellung linearer Transformationen des Hilbertschen Raums. Ergebnisse der Math. Vol. 5. Springer, Berlin, 1942, [6]. (There is an Edwards reprint.)
29. Taylor, A.E. Spectral theory of unbounded closed operators. Proc. symp. Oklahoma (1951), 261-275, [5].
30. Wermer, J. The existence of invariant subspaces. Duke Math. J. 19, 4 (1952), 615-622, [3].

To Chapter III

1. Aronszajn, N. Quadratic forms etc. = Nr. 1 to Ch. I, [1,2].
2. Brodskiĭ, M.S. On the triangular representation of completely continuous operators with a one-point spectrum. U.M.N. XVI, Nr. 1 (1961), 135-141, [9]. (Russian)
3. Weyl, H. The Classical Groups etc. = Nr. 4 to Ch. II.
4. Gohberg, I.C. and Kreĭn, M.G. Introduction etc. = Nr. 5 to Ch. II.
5. Courant, R. and Hilbert, D. Methods of Mathematical Physics, vol. I. Wiley-Interscience, New York, 1953.
6. von Neumann, J. Mathematische Grundlagen der Quantenmechanik. Springer, Berlin, 1932. English translation, Princeton Univ. Press, 1955.

7. Sahnovič, L.A. Investigation of the triangular model of nonselfadjoint operators. Izv. vuzov, Matematika, Nr. 4 (1959), 141-149, [9]. (Russian)
8. Carleman, T. Sur les équations intégrales singulières à noyau réel et symétrique. Uppsala, 1923, [5].
9. Fischer, E. Über quadratische Formen mit reellen Koeffizienten. Monatsh. Math. Physik 16 (1905), 234-249, [8].
10. Frobenius, G. Über unitäre Matrizen. Sitzungsber. Preuss. Akad. XVI (1911), 373-378, [6].
11. Gram, J. P. Über die Entwickelung reeller Funktionen in Reihen mittelst der Methode der kleinsten Quadrate. Crelles J. XCIV (1883), 41-73, [3].
12. Hausdorff, F. Der Wertevorrat einer Bilinearform, Math. Z. 3 (1919), 314-316, [10].
13. Hildebrandt, S. Über den numerischen Wertebereich eines Operators. Math. Ann. 163, No. 3 (1966), [10].
14. Kaczmarz, S. Angenäherte Auflösung von Systemen linearer Gleichungen. Bull. Internat. Acad. Polon. Sci., Ser. A (1937), 355-357, [5].
15. Fan, Ky. Maximum properties and inequalities for the eigenvalues of a completely continuous operator. Proc. Nat. Acad. Sci. U.S.A. 37 (1951), 760-766,[9].
16. Mirsky, L. The Spread of a Matrix. Mathematica 3, Nr. 6 (1956), 127-130, [9].
17. von Neumann, J. Allgemeine Eigenwerttheorie Hermitescher Funktionaloperatoren. Math. Ann. 102 (1929-1930), 49-131.
18. von Neumann, J. Zur Algebra etc. = Nr. 20 to Ch. II.
19. Schur, I. Über die charakteristischen Wurzeln einer linearen Substitution mit einer Anwendung auf die Theorie der Integralgleichungen. Math. Ann. 66 (1909), 488-510, [9].
20. Weierstrass, K. Zur Theorie der bilinearen und quadratischen Formen, Monatsh. Akad. Wiss., Berlin (1867), 310-338, [1,2,5].
21. Weyl, H. Das asymptotische Verteilungsgesetz der Eigenwerte linearer partieller Differentialgleichungen. Math. Ann. 71 (1912), 441-479, [8].
22. Wielandt, H. An extremum property of sums of eigenvalues. Proc. Amer. Math. Soc. 6, Nr. 1 (1955), 106-110,[8].

To Chapter IV

1. Akhiezer, N.I. Theory of Approximation. F. Ungar Pub. Co., New York, 1956,[6].
2. Belickiǐ, G.R. Chains of matrix norms. D.A.N. SSSR 151, Nr. 1 (1963), 9-10, [10] (Russian). English translation in Sov. Math.-Doklady.

3. Belickiĭ, G.R. Automorphisms of order lattices on the set of matrix norms. Mat. Sbornik 73 (1967), [10] (Russian). English translation in Math. USSR-Sbornik.
4. Bernstein, S.N. Sur le problème inverse de la théorie de la meilleure approximation des fonctions continues. C.R. Acad. Sci. 206 (1938), 1520-1523, [6].
5. Paley, R.E.A.C. and Wiener, N. Fourier Transforms in the Complex Domain. Amer. Math. Society, Providence, R.I., 1934.
6. Gelfand, I., Raikov, D., and Shilov, G. Commutative Normed Rings. Chelsea Pub. Co., New York, 1964, [9,10].
7. Gohberg, I.C. and Markus, A.S. Two theorems on the aperture of subspaces of a Banach space. U.M.N. 14, Nr. 5 (1959), 135-140, [7]. (Russian)
8. Gurariĭ, V.I., Spaces of universal distribution. D.A.N. S.S.S.R. 163 (1965), 1050-1053, [5]. (Russian)
9. Gurariĭ, V.I. On the obliqueness of subspaces and the existence of an orthogonal basis in a Banach space. Uč.zap.Har'kov.mat. ob-va 30 (1964), 34-37, [7]. (Russian)
10. Gurariĭ, V.I. On aperture and obliqueness of subspaces of a Banach space. Teorija. funkcii, funkc. an. i pril. Nr. 1 (1965), 194-204, [6,7]. (Russian)
11. Gurariĭ, V.I., Kadec, M.I., and Macaev, V.I. On the distance between finite-dimensional analogs of the space L_p. Mat. sb. 170 (1966), 481-489, [5]. (Russian)
12. Kadec, M.I. Proof of the topological equivalence of all separable infinite-dimensional Banach spaces. Funkcional'nyĭ analiz i ego priloženija 1, Nr. 1 (1967), 61-70, [5]. (Russian)
13. Kadec, M.I. Homeomorphisms of some Banach spaces. D.A.N. S.S.S.R. 92 (1953), 465-468, [6]. (Russian)
14. Kolmogorov, A.N. Zur Normierbarkeit eines allgemeinen topologischen linearen Raumes. Studia Math. 5 (1934), 29-33, [1,2,3].
15. Kolmogorov, A.N. On inequalities between the upper bounds of sequences of derivatives of an arbitrary function on an infinite interval. Uč. zap. M.G.U. 30 (1939), 3-16, [11]. (Russian)
16. Krasnosel'skiĭ, M.A. and Rutickiĭ, Ya. B. Convex Functions and Orlicz Spaces. P. Noordhoff, Groningen, 1961.
17. Kreĭn, M.G. On Bari bases of Hilbert space. U.M.N. XII, Nr. 3 (1957), 333-341, [7]. (Russian)
18. Kreĭn, M.G., Krasnosel'skiĭ, M.A., and Mil'man, D.P. On defect numbers of linear operators in Banach spaces and some geometric questions. Sb. trudov In-ta mat. A.H. S.S.S.R. 11 (1948), 97-112, [7]. (Russian)

19. Kreĭn, M.G., Mil'man, D.P., and Rutman, M.A. On a property of bases in a Banach spaces. Uč. zap. Har'k. mat. ob-va, ser. 4, 16 (1940), 106-110, [7]. (Russian)
20. Levitan, B.M. Almost periodic functions. Gostakhizdat, Moscow, 1953, [8]. (Russian)
21. Ljubič, Ju.I. Almost periodic functions in the spectral analysis of operators. D.A.N. S.S.S.R. 132 (1960), 518-520, [8]. (Russian)
22. Ljubič, Ju.I. On operator norms of matrices. U.M.N. XVII, Nr. 4 (1963), 161-164, [10]. (Russian)
23. Ljubič, Ju.I. On inequalities between powers of linear operators. Izv. A.N. S.S.S.R., ser. mat. 24 (1960), 825-864, [11]. (Russian)
24. Ljubič, Ju.I. A theorem about operators of class K. Teorija funkciĭ, funkc. an. i pril., Nr. 1 (1965), 212-218, [11]. (Russian)
25. Pietsch, A. Nukleare lokalkonvexe Räume. Akademie-Verlag, Berlin, 1965.
26. Solomjak, M.Z. On orthogonal bases in Banach spaces. Vestnik L.G.U., ser. mat. 1 (1957), 27-36, [6]. (Russian)
27. Uryson, P.S. Sur un espace métrique universel, Bull. de Sci. Math. 51 (1927), 1-38, [5].
28. Halmos, P.R. Lectures on Ergodic Theory. Chelsea Pub. Co., New York, 1956, [8].
29. Hardy, G.H., Littlewood, J.E., and Pólya, G. Inequalities. Cambridge Univ. Press, 1964, [1,5,11].
30. Hausdorff, F. Zur Theorie der linearen metrischen Räume. J. reine u. Angew. Math. 167 (1932), 294-311, [1].
31. Ascoli, G. Sugli spazi lineari metrici e le loro varietà lineari. Ann. Mat. Pura Appl. 10 (1932), 33-81, 203-232, [4].
31a. Bohnenblust, F. Convex regions and projections in Minkowski spaces. Ann. Math. 39 (1938), 301-308, [10].
32. Banach, S. Sur les operations dans les ensembles abstraits et leur application aux équations intégrales, Fund. Math. 3 (1922), 133-181.
33. Dvoretzky, A. and Rogers, C.A. Absolute and unconditional convergence in normed linear spaces. Proc. Nat. Acad. Sci. USA, 36 (1950), 192-197, [1].
34. Hahn, H. Über lineare Gleichungssysteme in linearen Räumen. Crelles J. CLVII (1927), 214-229, [4].
35. James, R.C. Orthogonality and linear functionals in normed linear spaces. Trans. Amer. Math. Soc. 61 (1947), 261-292, [4].
36. Jordan, P. and von Neumann, J. On inner products in linear metric spaces. Ann. Math. 36 (1935), 719-723, [1].

37. Mazur, S. Über konvexe Mengen in linearen normierten Räumen. Studia Math. 4 (1933), 70-84, [4].
38. Mazur, S. and Ulam, S. Sur les transformations isométriques d'espace vectoriel normés. C.R. Acad. Sci. Paris 194 (1932), 946-948, [5].
39. Minkowski, H. Gesammelte Abhandlungen. v. II, 1911.
40. von Neumann, J. On complete topological spaces. Trans. Amer. Math. Soc. 37 (1937), 1-20, [1,2,3].
41. Rutovitz, D. Some parameters associated with finite-dimensional Banach spaces. J. Lond. Math. Soc. 40 (1965), 241-255, [5].
42. Sz.-Nagy, B. Perturbations des transformations auto-adjointes dans l'espace de Hilbert. Comm. Math. Helv. 19 (1947), 347-366, [7].
43. Sz.-Nagy, B. On uniformly bounded linear transformations in Hilbert space. Acta Sci. Math. Szeged 11 (1947) 152-157, [8].
44. Wiener, N. Limit in terms of continuous transformations. Bull. Soc. Math. France 50(1922), 124-134, [1].
45. Schneider, H. and Strang, G. Comparison theorems for supremum norms. Numer. Math. 4 (1962), 15-20, [10].

To Chapter V

1. Weyl, H. The Classical Groups etc. = Nr. 4 of Sp. Lit. to Ch. II.
2. de Rham, G. Variétés Différentiables. Hermann et Cie., Paris, 1960, [3].
3. Jacobson, N. Lie Algebras. J. Wiley-Interscience, New York, 1962, [7].
4. Cartan, E. Leçons sur la géométrie des espaces de Riemann. 2nd ed., Gauthier-Villars, Paris, 1946, [3].
5. Cartier, P. Demonstration algébrique de la formule de Hausdorff. Bull. Soc. Math. France 84 (1956), 241-249, [7].
6. Cartier, P. Remarques sur le théorème de Birkhoff-Witt. Ann. Scuola Norm. Sup. Pisa (3) 12 (1958), 1-4, [7].
7. Whitney, H. Geometric Integration Theory. Princeton Univ. Press, 1957, [3].
8. Favard, J. Cours de géométrie différentielle locale. Gauthier-Villars, Paris, 1957.
9. Čebotarev, N.G. Theory of Lie Groups. O.N.T.I., 1940 [7]. (Russian)
10. Sirokov, P.A. Tensor Calculus. Gostekhizdat, 1934 (Russian).
11. von Neumann, J. Some matrix inequalities and metrization of matrix space. Izv. In-ta mat. i mek. Tomsk. un-ta 1 (1937), 286-300, [6].

12. Schatten, R. A theory of cross-spaces, Princeton Univ. Press, 1950, [6].
13. Weil, A. Sur les théorèmes de Rham, Comm. Math. Helv. 26 (1952), 119-145, [3].
14. Schneider, H. Bound norms. Linear Algebra Appl. 3 (1970), 11-21, [5].
15. Taylor, A.E. A geometric theorem and its applications to biorthogonal systems. Bull. Amer. Math. Soc. 53 (1947), 614-616, [5].

To Chapter VI

1. Daleckiĭ, Ju. L. Integration and differentiation of operators depending upon parameters, U.M.N. XII, Nr. 1 (1957), 182-186, [5]. (Russian)
2. Daleckiĭ, Ju.L. and Kreĭn, S.G. Integration and differentiation of functions of hermitian operators and applications to the theory of perturbations. Tr. Voronezh. un-ta, Nr. 1 (1956), 81-105, [5]. (Russian)
3. Cartan, É. Theory of Spinors. M.I.T. Press, Cambridge, Mass., 1967, [3].
4. Nemyckiĭ, V.V. The method of fixed points in analysis. U.M.N. 1 (1936), 141-175, [4]. (Russian)
5. Perov, A.I. On multidimensional linear equations with constant coefficients, D.A.N. S.S.S.R. 154 (1964), 1266-1269, [4]. (Russian)
6. Smul'yan, Ju.L. On the differentiability of the norm in a Banach space. D.A.N. S.S.S.R. 27 (1940), 643-648, [4]. (Russian)
7. Steinitz, E. Bedingt konvergente Reiken etc. = Nr. 16 of Sp. Lit. to Ch. I.
8. Kato, T. Perturbation Theory etc. = Nr. 17 of Sp. Lit. to Ch. II, [5].
9. Nashed, M.Z. Some remarks on variations and differentials. Amer. Math. Monthly 73 (1966), Supplement, [4].

To Chapter VII

1. Ahiezer, N. and Kreĭn, M. Some questions in the theory of moments. Amer. Math. Soc., Providence, R.I. 1962, [3].
2. Weyl, H. The elementary theory of convex polyhedra. Annals of Math. Studies, Nr. 24, 3-18, Princeton Univ. Press, 1950, [2,3].
3. Gohberg, I.C. and Kreĭn, M.G. Introduction to the Theory etc.= Nr. 5 of Sp. Lit. to Ch. II, [5].
4. Dmitriev, N.A. and Dynkin, E.B. Characteristic roots of stochastic matrices. Izv. A.H. S.S.S.R., ser. mat. 10 (1946), 167-184, [4]. (Russian)

5. Karpelevič, F.I. On characteristic roots of matrices with nonnegative elements, Izv. A.H. S.S.S.R., ser. mat. 15 (1951), 361-383, [4]. (Russian)
6. Kreĭn, M.G. On some questions in the geometry of convex ensembles, which belong to a linear, normed, and complete space. D.A.N. S.S.S.R. XIV (1937), 5-8, [3]. (Russian)
7. Kreĭn, M.G. About positive additive functionals in linear normed spaces. Uč. zap. Har'k. mat. ob-va. XIV (1937), 227-237, [3]. (Ukrainian)
8. Kreĭn, M.G. and Mil'man, D.P. On extreme points of regularly convex sets, Studia Math. 9 (1940), 133-138, [4].
9. Kreĭn, S.G. and Petunin, Ju.I. Scales of Banach spaces. U.M.N. XXI, Nr. 2 (1966), 89-168, [7]. (Russian)
10. Lidskiĭ, V.B. On eigenvalues of sums and products of symmetric matrices. D.A.N. S.S.S.R. 75 (1950), 769-772, [6]. (Russian)
11. Markus, A.S. Eigenvalues and singular numbers of sums and products of linear operators. U.M.N.XIX, Nr. 4 (1964), 93-123, [4,5]. (Russian)
12. Mityagin, B.S. An interpolation theorem for modular spaces. Mat. sb. 66 (1965), 473-482, [7]. (Russian)
13. Nudel'man, A.A. and Švarcman, P.A. On the spectrum of the product of unitary matrices. U.M.N. XIII, Nr. 6 (1958), 111-117, [6]. (Russian)
14. Parodi, M. La localisation des valeurs caractéristiques des matrices et ses applications. Gauthier-Villars, Paris, 1959, [6].
15. Romanovsky, V.I. Discrete Markov Chains. Wolters-Noordhoff Pub. Co., Groningen, 1970, [4].
16. Fan, Ky. On Systems of Linear Inequalities, in Annals of Math. Studies 38, 99-156, Princeton Univ. Press, 1956, [3].
17. Hadwiger, H. and Delbrunner, H. Combinatorial Geometry in the Plane (translated with additions by V. Klee). Holt, Rinehart, and Winston, New York, 1964, [2].
18. Helly, E. Über Mengen konvexen Körper mit gemeinschaftlichen Punkten. Jahrb. Deut. Math. Verein. 32 (1923), 175-176, [2].
19. Amir-Moez, A. Extreme properties of eigenvalues of a Hermitian transformation and singular values of the sum and product of linear transformations. Duke Math. J. 23 (1956), 463-476, [5,6].
20. Birkhoff, G. Tres observaciones sobre el algebra lineal, Rev. Univ. Nac. Tukuman 5 (1946), 147-151, [4].
21. Bonnesen, T. and Fenchel, W. Theorie der konvexen Körper. Springer, Berlin, 1934. There is a Chelsea reprint.

22. Eidelheit, M. Zur Theorie der konvexen Mengen in linearen normierten Räumen. Studia Math. 6 (1936), 104-111, [3].
23. Farkas, J. Theorie der einfachen Ungleichungen. Crelles J. 124 (1901), 1-27, [3].
24. Horn, A. On the eigenvalues of sums of Hermitian matrices. Pacific J. Math. 12 (1962), 225-241, [5,6].
25. Horn, A. On the eigenvalues of a matrix with prescibed singular values. Proc. Amer. Math. Soc. 5 (1954), 4-7, [5].
26. Horn, A. On the singular values of products of completely continuous operators. Proc. Nat. Acad. U.S.A. 36 (1950), 374-375, [5,6].
27. John F. On symmetric matrices whose eigenvalues satisfy linear inequalities. Proc. Amer. Math. Soc. 17, 5 (1966), 1140-1145, [6].
28. Ky Fan. On a theorem of Weyl concerning eigenvalues of linear transformations. Proc. Nat. Acad. Sci. U.S.A. 35 (1949), 652-655, [5].
29. Ky Fan. Maximum properties and inequalities for eigenvalues of completely continuous operators. Proc. Nat. Acad. Sci. U.S.A. 37 (1951), 760-766, [5,6].
30. Minkowski, H. Abhandlungen etc. = Nr. 39 of Sp. Lit. to Ch. IV.
31. Mirsky, L. Symmetric gauge functions and unitarily invariant norms. Quart. J. Math. 11 (1960), 50-59, [5,6,7].
32. Ostrowski, A. Sur quelques applications des fonctions convexes et concaves au sens de I. Schur. J. math. pur. et appl. 31 (1952), 253-292, [5].
33. Rado, R. An inequality. J. London Math. Soc. 27 (1952), 1-6, [4].
34. Riesz, M. Sur les maxima des formes bilineaires et sur les fonctionelles linéaires. Acta Math. 49 (1926), 465-497, [7].
35. Schatten, R. Norm ideals of completely continuous operators. Ergebnisse der Mathematik. N.F. Nr. 27, Springer, Berlin, 1961, [7].
36. Weyl, H. Inequalities between the two kinds of eigenvalues of a linear transformation. Proc. Nat. Acad. Sci. U.S.A. 35 (1949), 408-411, [5].
37. Wielandt, H. An extremum property of sums of eigenvalues. Proc. Amer. Math. Soc. 6 (1955), 106-110, [5,6].

To Chapter VIII

1. Beckenbach, E.F. and Bellman, R. Inequalities. Springer, New York, 1965, [5].
2. Gantmacher, F.R. and Kreĭn, M.G. Oszillationsmatrizen, Oszillationskerne und kleine Schwingungen mechanischer

Systeme. Akademie-Verlag, Berlin, 1960, [5].

3. Gale, D. The Theory of Linear Economic Models. McGraw-Hill, New York, 1960, [2,3].
4. Gohberg, I.C. and Krein, M.G. Introduction etc. = Nr. 5 of Sp. Lit. to Ch. II, [4].
5. Dubovickiĭ, A.Ja. and Milyutin, A.A. Extreme problems in the presence of restrictions. Zhurn. vyč. mat. u mat. fiz. 5 (1965), 395-493, [3]. (Russian)
6. Kantorovič, L.V., On partially ordered linear spaces and their applications in the theory of linear operators. D.A.N. S.S.S.R. IV (1935), 11-14, [1]. (Russian)
7. Kantorovič, L.V., Vulih, B.Z. and Pinsker, A.G. Functional Analysis in Partially Ordered Spaces. Gostekhizdat, 1950. (Russian)
8. Karlin, S. Mathematical Methods in the Theory of Games, Programming, and Economics. Addison-Wesley, Reading, Mass., 1959, [2,3].
9. Kreĭn, M.G. On Bari bases in Hilbert spaces. U.M.N. XII, Nr. 3 (1957), 333-341, [4]. (Russian)
10. Kreĭn, M.G. and Rutman, M.A. Linear operators leaving invariant a cone in a Banach space, U.M.N. III, Nr. 1 (1948), 3-95. (Russian). English trans.: Amer. Math. Soc. Translations, Ser. I, vol. 10, 199-325, [5].
11. Ky Fan, On Systems etc. = Nr. 16 in Sp. Lit. to Ch. VII, [2].
12. Šmul'yan, Ju.L. Hellinger's operator integral. Mat. sb. 49 (1959), 381-430, [7]. (Russian).
13. Judin, A. Solution of two problems in the theory of partially ordered spaces. D.A.N. S.S.S.R. XXIII (1939), 418-422, [1]. (Russian)
14. Davis, C. Notions generalizing convexity for functions defined on spaces of matrices, in Proc. Symp. Pure Math. VII: Convexity, 187-201, Amer. Math. Soc., Providence, R.I., 1963, [9].
15. Frobenius, G. Über Matrizen aus nicht negativen Elementen. Sitzungsber. Preuss. Akad. Wiss. 1912, 456-477, [5].
16. Haar, A. Die Minkowskische Geometrie und die Annäherung an stetige Funktionen. Math. Ann. 78 (1918), [3].
17. Hoffman, A.J. and Wielandt, H.W. The variation of the spectrum of a normal matrix. Duke Math. J. 20 (1953), 37-39, [4].
18. Kraus, F. Über konvexe Matrixfunktionen, Math. Z. 41 (1936), 18-42, [9].
19. Ky Fan and Hoffman, A.J. Some metric inequalities in the space of matrices. Proc. Amer. Math. Soc. 6 (1955), 111-116, [4].
20. Löwner, K. Über monotone Matrixfunktionen, Math. Z. 41 (1936), 18-42, [9].

21. Motzkin, T.S. Beiträge zur Theorie der linearen Ungleichungen. Jerusalem, 1936, [2].
22. von Neumann, J. and Morgenstern, O. Theory of Games and Economic Behavior, Princeton Univ. Press, 1953, [2,3].
23. Perron, O. Jacobischer Kettenbruchalgorithmus. Math. Ann. 64 (1907), 1-76, [5].
24. Sz.-Nagy, B. Sur les lattis linéaires de dimension finie, Comm. Math. Helv. 17 (1944-45), 209-213, [1].

To Chapter IX

1. Ahiezer, N.I. Some questions in the theory of moments. Amer. Math. Soc., Providence, R.I., 1962, [9].
2. Ahiezer, N.I. and Kreĭn, M.G. On some questions in the theory of moments. D.N.T.V.U., 1938, [9]. (Russian)
3. Birman, M.S. On the spectrum of singular boundary problems. Mat.sb. 55 (1961), 125-170, [6]. (Russian)
4. Brodskiĭ, M.S. and Livsic, M.S. Spectral analysis of nonselfadjoint operators and piecewise defined systems. U.M.N. XIII, Nr. 1 (1958), 3-86, [8].
5. Glazman, I.M. Direct Methods for Qualitative Spectral Analysis of Singular Differential Operators, Fizmatgiz, 1963, [6]. (Russian)
6. Glazman, I.M. On a class of solutions to the classical moment problem. Uč. zap. Har'k. Mat. ob-va, XX (1959), 95-98, [9]. (Russian)
7. Glazman, I.M. and Naĭman, P.B. On the convex hull of orthogonal systems of functions. D.A.N. S.S.S.R. 102 (1955), 445-448, [9]. (Russian)
8. Krasnosel'skiĭ, M.A. On selfadjoint extensions of Hermitian operators. U.M.Zh. Nr. 1 (1949), 21-38, [4]. (Russian)
9. Kreĭn, M.G. Basic aspects of the representation theory of Hermitian operators with defect index (m,m). U.M.Zh. Nr. 2 (1949), 3-66, [4]. (Russian)
10. Kreĭn, M.G. Theory of self-adjoint extensions of semibounded Hermitian operators and its application, I. Mat. sb. 20 (62) (1947), 431-495; II. Mat. sb. 21 (63) (1947), 365-404, [5]. (Russian)
11. Kreĭn, M.G. and Krasnosel'skiĭ, M.A. Basic theorems on the extensions of Hermitian operators and their applications to the theory of orthogonal polynomials and the moment problem. U.M.N. II, Nr. 3 (1947), 60-106, [5,9]. (Russian)
12. Livšic, M.S. On a class of linear operators in Hilbert space. Mat. sb. 19 (1946), 239-262, [7]. (Russian)
13. Livšic, M.S. On the theory of isometric operators with equal defect numbers. D.A.N. S.S.S.R. 58 (1947), 13-15, [7]. (Russian)

14. Livšic, M.S. About the spectral decomposition of linear nonselfadjoint operators. Mat. sb. 34 (1954), 145-199, [8]. (Russian)
15. Livšic, M.S. Operators, Oscillations, and Waves. Nauka, Moscow, 1966, [8]. (Russian)
16. Naĭmark, M.A. Spectral functions of a symmetric operator. Izv. A.N. S.S.S.R. 4 (1940), 277-318, [7]. (Russian)
17. Naĭmark, M.A. Extremal spectral functions of a symmetric operator. Izv. A.N. S.S.S.R. 11 (1947), 327-344, [7]. (Russian)
18. Straus, A.B. Characteristic functions of linear operators. Izv. A.N. S.S.S.R. 24 (1960), 43-74, [7]. (Russian)
19. Friedrichs, K.O. Spektraltheorie halbbeschränkter Operatoren. Math. Ann. 109 (1934), 465-487, 685-713; 110 (1935), 777-779, [5].
20. Kilpi, Y. Über selbstadjungierte Fortsetzungen symmetrischer Transformationen im Hilbertschen Raum, Ann. Acad. Fenn. 23 (1959), [5].
21. von Neumann, J. Allgemeine Eigenwerttheorie etc. = Nr. 17 of Sp.Lit. to Ch. III.
22. Penrose, R. On best approximate solutions of linear matrix equations. Proc. Cambr. Phil. Soc. 52 (1956), 17-19, [3].

To Chapter X

1. Bellman, R. Stability Theory of Differential Equations. McGraw-Hill, New York, 1953, [3]. There is a Dover reprint.
2. Glazman, I.M. On the decomposability in terms of eigenelements of dissipative operators. U.M.N. XIII, Nr. 3 (1958), 179-181, [1]. (Russian)
3. Ginzburg, Ju.I. and Iohvidov, I.S. Studies in the geometry of infinite-dimensional spaces with a bilinear metric. U.M.N. XVIII, Nr. 4 (1962), 3-56, [4]. (Russian)
4. Iohvidov, I.S. and Kreĭn, M.G. Spectral theory of operators in a space with an indefinite metric. Tr. Mosk. mat. ob-va 5 (1956), 367-432; 8 (1959), 413-496, [4,5,6]. (Russian)
5. Kacnel'son, V.E. On conditions for a basis of root vectors in some classes of nonselfadjoint operators. Funkc. an. i pril. 1, Nr. 2 (1967), 39-59, [1]. (Russian)
6. Kacnel'son, F.E. and Macaev, V.I. On spectral sets of operators in a Banach space and the estimation of functions of finite-dimensional operators. Teorija funkciĭ, funkc. an. i pril. Nr. 3 (1966), 3-10, [2].
7. Kreĭn, M.G. Lectures on stability theory for solutions of differential equations in a Banach space. Izd-vo A.H. U.S.S.R., 1964, [3]. (Russian)

8. Kreĭn, M.G. and Langer, G.K. On some mathematical principles in the linear theory of damped vibrating con-
   tinuua. Tr. mezhd. simp. po prim. teoriĭ funkciĭ v
   meh. spl. sredy, (1965), 283-322, [7]. (Russian)
9. Lumer, G. and Phillips, R.S. Dissipative operators in
   a Banach space. Pacific J. Math. 11 (1961), 679-698,[3].
10. Ljubič, Ju.I. Classical and local Laplace transform in
    the abstract Cauchy problem. U.M.N. XXI, Nr. 3 (1966),
    3-51, [3]. (Russian)
11. Ljubič, Ju.I. Conservative operators. U.M.N. XX, Nr. 5
    (1965), 221-225, [3]. (Russian)
12. Markus, A.S. Some tests for complete systems of eigen-
    vectors of a linear operator in a Banach space. Mat. sb.
    70 (1966), 526-561, [1]. (Russian)
13. von Neumann, J. Eine Spektraltheorie fűr allgemeine
    Operatoren eines unitáren Raumes. Math. Nachr. 4 (1951),
    258-281, [2].
14. Pontryagin, L.S. Hermitian operators in a space with in-
    definite metric. Izv. A.N. S.S.S.R. 8 (1944), 243-280,
    [4,5,6]. (Russian)
15. Potapov, V.P. Multiplicative structure of J-nonexpanding
    matrix functions. Tr. Mosk. mat. ob-va 4 (1955),
    125-236, [8]. (Russian)
16. Pjateckiĭ-Šapiro, I.I. Automorphic functions and the geo-
    metry of classical domains. Gordon and Breach, New
    York, 1963, [8].
17. Sz.-Nagy, B. von. Sur les contractions de l'espace de
    Hilbert I. Acta Sci. Math. Szeged 15 (1953), 87-92;
    II. ibidem, 18 (1957), 1-14, [1].
18. Phillips, R.S. Dissipative operators and hyperbolic
    systems of partial differential equations. Trans. Amer.
    Math. Soc. 90 (1959), 193-254, [1].
19. Foiaş, C. Sur certains théorèmès de J. von Neumann
    concernant les ensembles spectraux Acta Sci. Math.
    Szeged 18 (1957), 15-20, [2].
20. Hua, L.K. Harmonic Analysis of Functions of Several Com-
    plex Variables in the Classical Domains. Translations
    of Mathematical Monographs, vol. 6. Am. Math. Soc.,
    Providence, R.I., 1963, [8].
21. Šmul'yan, Ju.L. Nonexpanding operators in a finite-
    dimensional space with an indefinite metric, U.M.N.
    XVIII, Nr. 6 (1963), 225-230, [5,6].
22. Duffin, R.J. A minimax theory for overdamped networks,
    J. Rat. Mech. and Anal. 4 (1955), 221-233, [7].
23. Hille, E. Le problème abstrait de Cauchy. Rend. Sem. Mat.
    Univ. e Politech. Torino, 12 (1953), 95-103, [3].

Dictionary of General Terms

Group: a set G for whose elements $g \in G$ a binary relation, called (group) multiplication, is defined. This means that each ordered pair $g_1, g_2 \in G$ determines a product $g = g_1 g_2 \in G$, which satisfies the following group axioms:

(1) $(g_1 g_2)g_3 = g_1(g_2 g_3)$:group multiplication is associative ;

(2) an identity element $e \in G$ exists such that $ge = eg = g$ for any $g \in G$;

(3) to every $g \in G$ there is an inverse element such that $gg^{-1} = g^{-1}g = e$.

It follows easily that G contains exactly one identity element, and every g has exactly one inverse.
The group axioms imply that the equations

$$gx = h, \quad yg = h \tag{*}$$

have the unique solutions $x = g^{-1}h$, $y = hg^{-1}$ obtained from (*) by division from the left, respectively from the right.
A group is called abelian if the group multiplication is commutative, i.e., if $g_1 g_2 = g_2 g_1$ for every pair $g_1, g_2 \in G$.

A group homomorphism of G into G' is a mapping $\varphi: G \to G'$ which "respects" group multiplication: $\varphi(g_1 g_2) = \varphi(g_1)\varphi(g_2)$. This clearly implies $\varphi(e) = e'$ and $\varphi(g^{-1}) = [\varphi(g)]^{-1}$.

Category (in the narrow sense of category of sets): a class of sets $\{M_\gamma\}_{\gamma\in\Gamma}$ and a family of mappings (morphisms) $H_{\beta\alpha} = \{\varphi^\nu_{\beta\alpha}\}$ satisfying the following conditions:

(1) for every $\nu \in N$, $\varphi^\nu_{\beta\alpha}$ maps $M_\alpha \to M_\beta$. (The index set N is unrestricted; if $\alpha \neq \beta$, N may be empty).

(2) the composition $\varphi^\mu_{\gamma\beta}\circ\varphi^\nu_{\beta\alpha}$ belongs to the family of mappings $H_{\gamma\alpha}$ for any triplet α,β,γ, for which the composition is defined.

(3) $H_{\alpha\alpha}$ is never empty; it must at least contain the identity mapping $M_\alpha \to M_\alpha$.

Quasiorder: a binary relation defined on a set M for pairs $x,y \in M$ and denoted by $x \prec y$. It must satisfy

(1) $x \prec x$ for all $x \in M$ (reflexivity);

(2) $x \prec y$ & $y \prec z \Rightarrow x \prec z$ (transitivity).

A set endowed with a quasiorder relation is called quasi-ordered. Elements x and y of such a set for which either $x \prec y$ or $y \prec x$ are called comparable. The quasiorder becomes an order if $x \prec y$ & $y \prec x \Rightarrow x = y$.

Ring: a set R on which are defined two binary operations, addition and multiplication, satisfying:

(1) R is an abelian group for the addition;

(2) a left and a right distributive law holds for the composition of the two operations, i.e.,

$$(r_1 + r_2)r = r_1r + r_2r, \quad r(r_1 + r_2) = rr_1 + rr_2.$$

The abelian group in (1) is also known as the additive group of the ring. Its identity element is called the zero element and is usually denoted by 0. The element inverse to r in the additive group of R is here called the additive inverse to r (also the opposite of r) and denoted by -r. Hence

$$r + 0 = 0 + r = r \quad \text{and} \quad r + (-r) = (-r) + r = 0.$$

In the additive group one speaks of "subtraction" instead of "division" and writes r-s for r + (-s).
A ring is commutative (associative), if the ring multiplication is commutative (associative).
If R is a ring and the set R-{0} is an abelian group also in multiplication, R is called a field.

Metric space: a linear space E endowed with a metric, i.e., a function $\rho(x,y)$, $x,y \in E$, whose range is the set of reals. $\rho(x,y)$ satisfies:

(1) $\rho(x,y) > 0$ if $x \neq y$, and $\rho(x,x) = 0$;

(2) $\rho(y,x) = \rho(x,y)$;

(3) $\rho(x,z) \leqq \rho(x,y) + \rho(y,z)$ (triangle inequality).

The quantity $\rho(x,y)$ is known as the distance between the points x and y.
Let $x_o \in M \subset E$ and $\varepsilon > 0$. The sets

$$\{x|\rho(x,x_o) < \varepsilon\}, \quad \{x|\rho(x,x_o) \leqq \varepsilon\}, \quad \{x|\rho(x,y) = \varepsilon\}$$

are called the open ball, the closed ball, and the sphere respectively. The point x_o is the center and ε is the radius. A sequence $\{x_k\}_1^\infty$ of points in a metric space $E^{(m)}$ is called convergent (to $x \in E^{(m)}$) if $\lim_{k\to\infty} \rho(x_k,x) = 0$. A sequence $\{x_k\}_{k=1}^\infty$, $x_k \in E^{(m)}$, which satisfies $\lim_{j,k\to\infty} \rho(x_k,x_j) = 0$, is called fundamental, and every convergent sequence in $E^{(m)}$ is indeed fundamental, but the converse need not be true. If it is true in some $E^{(m)}$, then $E^{(m)}$ is called a complete metric space.

Equivalence relation: a binary relation defined on some set M and written $x \sim y$, which satisfies

(1) $x \sim x$ (reflexivity);

(2) $x \sim y \Rightarrow y \sim x$ (symmetry);

(3) $x \sim y \;\&\; y \sim z \Rightarrow x \sim z$ (transitivity).

Every equivalence relation on M produces a decomposition of M into equivalence classes $(K_\alpha)_{\alpha \in A}$:

$$M = \bigcup_{\alpha \in A} K_\alpha \quad \text{and} \quad K_\alpha \cap K_\beta = \phi, \quad \text{if } \alpha \neq \beta.$$

No class must be empty, and it is the characteristic property of decomposition into classes that $x \sim y$ is true iff x and y belong to the same class K_α.

Semigroup: a set endowed with an associative binary relation (multiplication).

Topological space: a set X together with a topology τ, often written (X,τ). This means that X contains a family of subsets $\{G_\alpha\}_{\alpha \in A}$, called the open sets in X, which obey the following postulates:

(1) the union of any family of open sets $\{G_\alpha\}_{\alpha \in A' \subset A}$ is open;

(2) the intersection of any finite family of open sets $\{G_n\}_{n=1}^{N}$ is open;

(3) X and ϕ are open.

A set F in a topological space is called closed if its complement X-F is open. The properties of closed sets are dual to (1)-(3):

(1') the intersection of any family of closed sets is closed;

(2') the union of any finite family of closed sets is closed;

(3') ϕ and X are closed.

A topological space (X,τ) is called connected, if it contains no sets that are at the same time open and closed, except X and ϕ.

A closed set $F \subset (X,\tau)$ is called connected if it cannot be decomposed into two nonempty closed parts without common points.

Let B be a subset of the family $\{G_\alpha\}_{\alpha \in A}$, disregarding $G_\alpha = \phi$. B is called a base of the topology τ, and its elements G_α are called neighborhoods, if each nonempty open set of (X,τ) is the union of elements from B.

An element $G_\alpha \in B$ is called a neighborhood of the point x, if $x \in G_\alpha$.

Let X be some set, and B a family of nonempty subsets of X, with the properties

(1) if V_1 and $V_2 \in B$, then there is a set $V_3 \in B$ such that

$$V_3 \subset V_1 \cap V_2 ;$$

(2) for every $x \in X$ there is a $V \in B$ such that $x \in V$.

Then a topology τ can be defined on X such that B is the base of τ, and there is only one such τ. This construction is known as "giving a topology by means of neighborhoods". In every metric space $E^{(m)}$ a natural neighborhood base consists of the sets

$$V(x_o,\varepsilon) = \{x \mid \rho(x,x_o) < \varepsilon\}$$

i.e., the family of all open balls.

If a set X is endowed with two topologies τ_1 and τ_2, and if every set M open in τ_1 is also open in τ_2, then τ_1 is called weaker than τ_2 (or τ_2 stronger than τ_1). If τ_1 is weaker than τ_2, and τ_2 is weaker than τ_1, then $\tau_1 = \tau_2$.

Let M be some subset of (X,τ). The point $x \in X$ is a limit point of M, if every open set containing x also contains some

$y \in M$, where $y \neq x$. The set of all limit points of M is often denoted by M' and called the derived set of M. The set $\overline{M} \underset{def}{=} M \cup M'$ is called the closure of M. It is also the intersection of all closed sets that enclose M. The set $\partial M \underset{def}{=} \overline{M} - M$ is called the boundary of M, and its elements are the boundary points of M. The set $\text{Int } M \underset{def}{=} M - \partial M$ is called the interior of M, and its elements are the interior points of M.

The sets $H_\alpha \subset X$, $\alpha \in A$, are said to form a covering of M, if $M \subset \bigcup_{\alpha \in A} H_\alpha$. A covering is open if all H_α are open.

A topological space is separated (or a Hausdorff space, or a T_2-space), if for any two points $x_1, x_2 \in X$, $x_1 \neq x_2$, open sets G_1 and G_2 exist such that

$$x_1 \in G_1, \; x_2 \in G_2, \; G_1 \cap G_2 = \phi.$$

A set M in a separated topological space is called compact (or a compactum), if every open covering $\{H_\alpha\}_{\alpha \in A}$ of M contains some finite covering $\{H_{\alpha_n}\}_{n=1}^{N}$.

A compact set (in a Hausdorff space) is always closed. A set M is relative-compact[§] if its closure $\overline{M}$ is compact.

The topological product of two topological spaces (X, τ_1) and (Y, τ_2) is defined as the Cartesian product $X \times Y$; the topology of the product space consists of the sets $G \times H$, where $G \subset X$ and $H \subset Y$ are open sets from τ_1 and τ_2, respectively.

A mapping $\varphi : X \to Y$ of topological spaces is called continuous if the full inverse image $\varphi^{-1} H$ is open in X for every H open in Y.

[§] precompact in the Russian text.

$\subset$	non-strict inclusion
Int F	interior of set F
∂F	boundary of F
E	linear space
n	dimension of E
C^n	complex arithmetical n-tuple space
R^n	real arithmetical n-tuple space
sp Γ	linear hull of vector system Γ
Co	convex hull
E/L	factor space of E modulo subspace L
$\mathrm{Hom}(E,E_1)$	space of homomorphisms from E into E_1
$\mathbb{B}(E_1,E_2)$	space of bilinear functionals on $E_1 \times E_2$
$\mathbb{M}(E)$	space of operators in E
$\sigma(A)$	spectrum of operator A
tr(A)	trace of A
$R_\lambda(A)$, R_λ	resolvent of A
$A_{\vert L}$	restriction of A to L
$\delta(A)$	defect number of A
$\tilde{A}$	extension of A
P	projector
$\bar{P}$	projector complementary to P
$P^\perp$	projector orthogonal to P
S	self-adjoint operator; also Hermitian operator
U	unitary operator
A,B	commutator of A and B
$\oplus, \Sigma^{\oplus}$	direct sum
$\boxplus, \boxplus_{k=1}^{n}$	orthogonal sum
(x,y)	scalar product of vectors x and y
$\Vert\cdot\Vert$	norm
$\mathbb{S}(E)$	space of self-adjoint operators in E
$\mathbb{U}(E)$	space of unitary operators in E
dim L	dimension of L
codim L	codimension of L
Ker h	kernel of homomorphism h
Im h	image of h
rk h	rank of h
def h	defect of h
ind h	index of h
E'	conjugate space to E
h'	conjugate homomorphism to h
K'	conjugate wedge to wedge K
A'	transpose of matrix A
$L^\perp$	subspace orthogonal to L
E^*	Hermitian-conjugate space to E

A^*	Hermitian-conjugate operator to A
h^*	Hermitian-conjugate homomorphism to h
$\rho(A)$	spectral radius of A
$\times$	Cartesian product
$\otimes, \bigotimes_{k=1}^{n}$	tensor product
$\wedge$	external product
$A\nabla B$	A and B commute
$A\nabla\nabla B$	A commutes with every C, which commutes with B
$\dot{>}$	quasiorder (binary relation); special cases $\dot{>}$ and $\dot{\gg}$ on p. 333
$\geq, \succ$	special case of $\dot{>}$ in E: $x \geq 0$ admits $x = 0$; $x \succ 0$ excludes $x = 0$
$>$	special case of $\dot{>}$, if order in E is determined by solid positive wedge K; then $x > 0$ means $x \in \text{Int } K$
$\simeq$	isomorphism between linear spaces
$\approx$	similarity between operators